Inquiring into Inquiry Learning and Teaching in Science

Edited by

Jim Minstrell

Emily H. van Zee

American Association for the Advancement of Science
Washington, DC

This publication was supported by the American Association
for the Advancement of Science. Any interpretations and conclusions
are those of the authors and do not necessarily represent the views
of the American Association for the Advancement of Science.

ISBN 0-87168-641-4
AAAS Publication 00-1S

Contents

iv   Contents

Contributors

Sandra K. Abell
  Professor of Science Education
  Department of Curriculum
    and Instruction
  Purdue University
  West Lafayette, IN

Bruce Alberts
  President
  National Academy of Sciences
  Washington, DC

Karen Amati
  Science Teacher
  Lessenger Middle School
  Detroit Public Schools
  Detroit, MI

Gail Anderson
  Principal
  Pine Valley Elementary School
  Metropolitan School District
    of Warren County
  Pine Village, IN

Doris Ash
  Science Educator
  Institute for Inquiry
  The Exploratorium
  San Francisco, CA

Phyllis Blumenfeld
  Professor
  School of Education
  University of Michigan
  Ann Arbor, MI

Rodger W. Bybee
  Executive Director
  Biological Sciences Curriculum Study
  Colorado Springs, CO
  (formerly Executive Director, Center
    for Science, Mathematics, and
    Engineering Education
    National Research Council)

Susan Carpenter
  Researcher
  Wisconsin Center for Education
    Research
  University of Wisconsin
  Madison, WI

Audrey B. Champagne
  Professor
  School of Education and Department
    of Chemistry
  University at Albany
  State University of New York

Janice Chezem
  Third-Grade Teacher
  Earhart Elementary School
  Lafayette School Corporation
  Lafayette, IN

Lezlie S. DeWater
  Visiting Lecturer
  Department of Physics
  University of Washington
  Seattle, WA
  (on leave from Seattle Public Schools)

Fred N. Finley
  Professor of Science Education
  College of Education and Human
    Development
  University of Minnesota
  Minneapolis, MN

Kathleen M. Fisher
  Professor of Biology
  San Diego State University
    And Center for Research in
    Mathematics and Science
    Education
  San Diego, CA

John R. Frederiksen
  Principal Scientist
  Educational Testing Service
  Oakland, CA

David Hammer
  Professor
  Departments of Physics and
    Curriculum and Instruction
  University of Maryland
  College Park, MD

Christine Klein
  Program Manager
  Investigative Learning Center
  St. Louis Science Center
  St. Louis, MO

Vicky L. Kouba
  Professor
  School of Education
  University at Albany
  State University of New York

Joseph Krajcik
  Professor

School of Education
  University of Michigan
  Ann Arbor, MI

Rhonda Hawkins
  Sixth-Grade Teacher
  Brandywine Elementary School
  Prince Georges County Public
    Schools
  Brandywine, MD

Marlene Hurley
  Professor
  School of Education
  University of Wisconsin - Superior
  Superior, WI

Marletta Iwasyk
  Kindergarten/First-Grade Teacher
  Orca at Columbia Alternative School
  Seattle Public Schools
  Seattle, WA

Akiko Kurose
  First-Grade Teacher
  Laurelhurst Elementary School
  Seattle Public Schools
  Seattle, WA
  (deceased)

Rebecca Kwan
  First-Grade Teacher
  East Silver Spring Elementary School
  Montgomery County Public Schools
  Silver Spring, MD

Diantha Lay
  Fourth-Grade Teacher
  Judith A. Resnik Elementary School
  Montgomery County Public Schools
  Gaithersburg, MD

Richard Lehrer
  Professor
  School of Education
  University of Wisconsin
  Madison, WI

Ron Marx
  Professor
  School of Education
  University of Michigan
  Ann Arbor, MI

Lillian C. McDermott
  Professor
  Department of Physics
  University of Washington
  Seattle, WA

J. Randy McGinnis
  Professor
  School of Education
  University of Maryland
  College Park, MD

Kathleen E. Metz
  Professor
  School of Education
  University of California - Riverside
  Riverside, CA

Jim Minstrell
  Research Scientist
  Talaria Inc.
  Seattle, WA
  (retired from Mercer Island School
    District)

Constance Nissley
  Science Teacher
  Green Acres School
  Rockville, MD

M. Cecilia Pocoví
  Lecturer
  Facultad de Ciencias Exactas
  Universidad Nacional de Salta
  Salta, Argentina

Angie Putz
  First-Grade Teacher
  County View Elementary School
  Verona, WI

Deborah L. Roberts
  Mathematics Teacher
  Silver Spring International
    Middle School
  Silver Spring, MD
  (formerly First-Grade Teacher,
    Rolling Terrace Elementary)

Leona Schauble
  Professor
  School of Education
  University of Wisconsin
  Madison, WI

Dorothy Simpson
  Physics Teacher
  Mercer Island High School
  Mercer Island, WA

Elliot Soloway
  Professor
  College of Engineering
  University of Michigan
  Ann Arbor, MI

Emily H. van Zee
  Professor
  School of Education
  University of Maryland
  College Park, MD

Gerald F. Wheeler
  Executive Director
  National Science Teachers Association
  Arlington, VA

Barbara Y. White
  Professor
  Graduate School of Education
  University of California
  Berkeley, CA

Judy Wild
  Fourth-Grade Teacher
  Sacred Heart School
  Bellevue, WA
  (recently retired)

Brian E. Woolnough
  University Lecturer in Science
    Education
  Department of Educational Studies
  Oxford University
  United Kingdom

Anat Zohar
  Senior Lecturer
  School of Education
  Hebrew University of Jerusalem
  Jerusalem, Israel

Introduction

Jim Minstrell and Emily H. van Zee

"Teachers of science plan an inquiry-based science program for their students," according to Teaching Standard A recommended by the National Research Council (NRC) in the *National Science Education Standards* (1996) (p. 30). In planning this inquiry about inquiry learning and teaching, we chose to focus on three questions: Why inquiry?, What does inquiry look like?, and What are some of the issues associated with shifting toward inquiry-based practices? In inviting authors to contribute, we tried to include many perspectives—from scientists, teachers, researchers, professional development specialists, and administrators. We particularly tried to include authors who are working with students from diverse cultural backgrounds and socioeconomic circumstances. Settings ranged from science lessons with primary students to informal meetings with experienced teachers. We asked authors to define what they mean by inquiry teaching and learning and then to address a relevant question or issue in the context of their own practices. We invite the reader to join us in pondering these responses.

WHY INQUIRY?

In the first section of the book, leaders of the science and science education communities reflect upon what they mean by inquiry and why they think inquiry should be emphasized in school science. These chapters also provide historical and philosophical perspectives on the current reform efforts.

How do scientists think about inquiry? Bruce Alberts, president of the National Academy of Sciences, reflects upon meaningful aspects of his early education and notes that these were associated with "struggling to meet a challenge in which my own initiative was needed to acquire an understanding." He provides some examples from inquiry curricula and also some counterexamples, including college science labs that he found "utterly boring." Alberts includes a delightful passage from Richard Feynman's account of a conversation Feynman had as a child with his father. Alberts uses this to illustrate ways to develop a student's inquisitiveness; he then challenges college faculty to develop courses that nurture such habits of mind. He suggests that scientists have a responsibility to volunteer in schools, provide professional development for teachers, and form a political force advocating reform. He also encourages young scientists to consider teaching at the K-12 level as a way of reinvigorating the schools.

How do teachers think about inquiry? Gerald F. Wheeler, Executive Director of the National Science Teachers Association, comments on three faces of inquiry. Some teachers seem to view inquiry simply as a teaching strategy for motivating students by engaging them in hands-on activities. This is not enough. Students need to learn how to question the phenomena, that is, to engage in a dialogue with the material world. Wheeler's teaching goal is to place students in situations that enable them to practice having such dialogues. Also important is to see the structure of inquiry itself as a content to be learned. Students need to become aware of the nature of scientific ways of knowing. As they design and conduct investigations, they should recognize the need to identify assumptions, to use critical and logical thinking, to base inference on evidence, and to consider alternative explanations.

What is the history of inquiry approaches to science instruction? An overview is provided by Rodger W. Bybee, former Executive Director of the Center for Science, Mathematics, and Engineering Education at the National Research Council, who now directs the Biological Sciences Curriculum Study. Bybee begins by presenting three versions of inquiry in action and presenting the reader with a quiz to assess interpretations. Then he traces the history of inquiry teaching from late in the nineteenth century to the present, including views expressed by John Dewey, Joseph Schwab, F. James Rutherford, and agencies such as the American Association for the Advancement of Science, as presented in Project 2061's publications, and the National Academy of Sciences in the *National Science Education Standards* (NRC, 1996). Bybee distinguishes between two ways in which the *Standards* use the term "inquiry": to refer to content and to teaching strategies. The content standards include understanding fundamental abilities and concepts associated with science as inquiry. Bybee recommends starting with a standards-based perspective, What is it we want students to learn? and then asking Which teaching strategies provide the best opportunities to accomplish that outcome? and What assessment strategies are appropriate and provide the best evidence of students' attaining the outcomes?

What philosophical bases underlie a conception of science as inquiry? Fred N. Finley, a professor of science education, and M. Cecilia Pocoví, a scientist from Argentina, review how the scientific method is typically presented in science textbooks and why teachers might choose this traditional view of scientific inquiry. They reflect upon successes associated with the development of the scientific method and its relation to the development of intellectual freedom, new forms of government, and technological advances. Then these authors reconsider each element of the traditional scientific method taught in schools in the context of issues raised by recent philosophical debates about the nature of scientific inquiry. They recommend, for example, that students learn about the effect that preconceptions and theories of the world have upon questions formulated, observations made, and interpretations developed. Students should learn that scientific inquiry does not always involve experimentation, that there are many contexts in which other approaches are more appropriate. Students also should become aware of the importance of the discussions, arguments, and modifications typical of the presentation of new ideas in a scientific community.

WHAT DOES INQUIRY LOOK LIKE?

This section presents examples of inquiry teaching and learning in several contexts. These include elementary and secondary classrooms, professional development programs in a variety of settings, and college science courses for teachers. Each chapter contributes specific instances and insights to our general inquiry about inquiry learning and teaching.

How might elementary school teachers shift toward more inquiry-based practices? Teachers who have been primarily using textbooks might engage students in more hands-on activities. The *National Science Education Standards* (NRC, 1996) notes, however, that providing more experiences with natural phenomena is not enough: students also need opportunities to talk together about what they think. Especially important are opportunities to formulate theories and to consider evidence that confirms or disconfirms these ideas. A university researcher, Sandra K. Abell, and two third-grade teachers, Gail Anderson and Janice Chezem, provide examples of shifts in practice toward greater emphasis on science as argument and explanation. They reflect upon what they learned about inquiry teaching and learning as they engaged students in thinking together about whether sounds are produced by vibrating objects.

How can teachers design classrooms to support inquiry? A team of university researchers, Richard Lehrer, Susan Carpenter, and Leona Schauble, and a first-grade teacher, Angie Putz, present a vision of inquiry teaching and learning that demonstrates ways to encourage and shape student questioning. They trace the chain of inquiry during a year-long investigation initiated by the children's curiosity about changes in the color of apples. The children designed strategies for testing their ideas about ripening, invented ways to record their observations, extended their investigation to decomposing, compared rates of change for several kinds of fruits and vegetables, and constructed models of phenomena they had decided to track. The teacher's design tools included asking questions that pushed students' questions farther, establishing norms of argumentation based on evidence, focusing upon displays and inscriptions invented by students, and engaging students in evolving chains of inquiry.

How can college faculty foster teachers' inquiries about inquiry learning and teaching? Emily H.van Zee reviews some of the literature generated by teachers reflecting upon their own practices, particularly those who are documenting and articulating ways in which they teach science through inquiry. Then she describes ways that she engages prospective teachers in learning how to do research as they learn how to teach in courses on methods of teaching science in elementary schools. She also discusses the formation and structure of the Science Inquiry Group, teachers who are developing case studies of their own teaching practices. Deborah L. Roberts, a graduate of the course and founding member of the Science Inquiry Group, reflects upon the road she traveled as a teacher who first learned, and now teaches, science through the process of inquiry.

What do teachers inquire about teaching and learning science as inquiry? In the set of case studies included here, teachers formulated issues to examine, collected data such as videotapes of instruction and copies of their students' work, and developed interpretations of their own teaching practices. Many of these teachers are working with students from diverse cultures. Marletta Iwasyk reflects upon ways in which she helped her primary students learn how to ask productive questions of one another. Akiko Kurose presents questions that her first graders asked in a context in which they had had extensive observational experience. Rebecca Kwan comments upon ways in which she modified her curriculum in order to follow up on a first grader's unexpected question. Constance Nissley describes a regularly scheduled Choice Time in which elementary students could follow their own curiosities. Judy Wild reflects upon the development of her fourth graders' conceptual understanding of electric circuits. Diantha Lay reports upon an inquiry conference that she organized for her fourth graders to share their science projects with students from other schools. Rhonda Hawkins recounts ways in which three sixth graders were competent science inquirers even though they were not able to communicate their understandings through writing. Dorothy Simpson identifies strategies to foster collaborative conversations among high school physics students.

How can teachers use the results of research on inquiry teaching and learning? Educational research traditionally has provided the basis for design of new instructional methods and materials that teachers then implement in their classrooms. David Hammer, a professor of physics education, describes a different use of research that evolved in a series of meetings with high school physics teachers. He and the teachers discussed snippets that the teachers had selected from tapes of their instruction, samples of students' work, and so forth. They also read reports of research on learning and tried to use insights from these in interpreting the data under discussion. In this way, perspectives from educational research enriched the perceptions and judgments of the teachers as they developed their interpretations. Conversations about the snippets and summaries of teacher perceptions provide models for both teachers and researchers of ways to engage in insightful discussions of theory and practice.

What principles guide the practice of inquiry in informal learning environments? Doris Ash and Christine Klein, museum science educators, describe and compare two museum settings. One is an Institute for Inquiry in which teachers learn science through long-term inquiry activities based upon their own questions. The other is a "museum" school where middle school students do research in the authentic context of enriching the exhibits. The authors compare learning in informal and formal environments, present vignettes from their two settings, define common principles, suggest ways of implementing these principles in other contexts, and include resources for putting these principles into practice. They emphasize two elements in building a community of inquiry: an ethos of questioning and scaffolding. According to these authors, learning is a social process driven by the learners' curiosity. In facilitating inquiry, knowing when and how to intervene is critical.

How can college science faculty prepare teachers to develop an inquiry-based science program? College faculty provide implicit models of science teaching by the ways that they structure their courses. If teachers are to teach science by inquiry, they need to have experiences learning science by inquiry in the college courses required for their majors. A university professor and an elementary school teacher provide two perspectives on the need for special science courses for teachers. Lillian C. McDermott, a professor of physics, discusses why traditional college science courses are inadequate for preparing teachers to teach science at any level—elementary, middle, or senior high school. She then describes the intellectual objectives and instructional approach of special physics courses for teachers. These courses served as the setting for development of a curriculum for college courses for teachers, *Physics by Inquiry* (McDermott, 1996). Taught entirely in the laboratory, these courses develop not only knowledge of subject matter but also knowledge of difficulties that students may encounter in learning these topics. Lezlie S. DeWater reflects upon what she experienced initially as a participant and then as a staff member in these courses. In particular, she discusses how she questions and listens to her students as she guides them in making sense of the world around them.

What strategies can college professors use to implement inquiry-based instruction? Kathleen M. Fisher, a professor of biology, reviews reasons for modeling such teaching, comments on when to avoid inquiry approaches, describes several inquiry-based strategies, and summarizes ways that she has adapted a lecture course for active learning. She also discusses six features of inquiry-based learning: eliciting prior knowledge, prediction, engagement with a phenomenon, group work, higher order thinking, and student-centered classes. Then she describes

SemNet®, a computer program that students can use to create a map of ideas having many complex interconnections. She closes the chapter by reviewing some of the evidence for the need for change in the ways we teach and learn.

WHAT ISSUES ARISE WITH INQUIRY LEARNING AND TEACHING?

This section examines some of the issues that teachers may consider in shifting toward inquiry-based instruction. These include using technology to support inquiry, incorporating metacognitive strategies, attempting inquiry with young children, addressing students' reasoning difficulties, teaching students with disabilities, clarifying instructional goals, and assessing learning.

In what ways can technology support students' inquiries? University researchers, Joseph Krajcik, Phyllis Blumenfeld, Ron Marx, and Elliot Soloway, describe instructional, curricular, and technological supports for inquiry in science classrooms. They provide examples of ways that learning technologies can enhance the formulation of questions, design of investigations, collection and display of data, development of analyses, and presentation of findings. The Investigators' Workshop, for example, includes computational tools such as Model-It that help students to build, test, and evaluate models of dynamic systems. These authors emphasize the roles of metacognition and collaboration in inquiry. Karen Amati is a science and technology resource teacher who provides a detailed account of using Model-It with urban middle school students. She describes how Model-It prompts students to develop explanations rather than memorize definitions or bits of information. She also comments upon the role of the teacher as a facilitator of learning.

Can students learn to assess their own reasoning as they construct and revise theories? Researchers, Barbara Y. White and John R. Frederiksen, collaborated with teachers in developing and testing a computer-enhanced science curriculum in urban middle schools. The ThinkerTools Inquiry Curriculum enables students to learn about the processes of scientific inquiry and modeling as they construct and revise theories about force and motion. Students evaluate their own and one another's research in a reflective process that includes assessing whether they are reasoning carefully and collaborating well. This process is called "metacognitive facilitation." The ThinkerTools curriculum was effective in reducing the performance gap between low and high achieving students.

Is inquiry-based instruction appropriate for young children? Kathleen E. Metz, a professor of education, challenges the traditional assumption that young children are not developmentally ready to engage in abstract thinking.

She suggests that the ability to reason competently depends upon the depth of children's knowledge. Such knowledge includes not only conceptual understanding of the domain but also knowledge of the enterprise of empirical inquiry, of methodologies specific to a domain, of ways to represent and analyze data, and of the use of tools such as binoculars, thermometers, and computers. This author then describes a project to help young children build knowledge that will empower their independent inquiry in biology. She provides examples of a curriculum module in animal behavior, children's reflections upon their inquiries, and teachers' perspectives on the value and challenges of this approach.

How can teachers address students' reasoning difficulties? Anat Zohar, a professor of science education, considers various challenges that students encounter such as matching research problems to appropriate experimental designs, controlling variables, applying the logic of hypothesis testing, and differentiating between experimental results and conclusions. She advocates teaching such reasoning skills systematically and provides an example from the Thinking in Science project. This curriculum explicitly teaches scientific reasoning in subjects that are part of the regular science syllabus. Activities include investigation of microworlds, learning activities promoting argumentation skills about bio-ethical dilemmas in genetics, and open-ended inquiries. The curriculum builds upon examples with which children are familiar from everyday life, provides opportunities to practice reasoning skills in several contexts, and engages students in metacognitive activities that lead to generalizations about reasoning formulated by the students themselves.

Can students with disabilities learn science as inquiry? Professor of science education, J. Randy McGinnis, reviews the literature in four areas: portrayals of inquiry learning by instructors teaching science to students with disabilities, reasons for using inquiry-based instruction for students with disabilities, evidence that such instruction is appropriate for these students, and implications for teachers. The latter include developing inquiry-based instruction while establishing differing expectations for student assessment based upon the objectives in the students' Individualized Education Plan. Close collaboration with special educators is advisable. Also recommended were providing structure through use of a student notebook with a format, introduction of key vocabulary and material by the teacher, student generation of predictions or hypothesis on what will be learned from an experiment, participation in experimental activities, oral presentations by the learning groups on the data they collected, elicitation of summary statements, and group construction of conclusions.

What is the purpose of "practical work" in school science? Brian E. Woolnough, a science educator from Great Britain, asserts that much practical work is "ineffective, unscientific...boring...time wasting...and unstimulating" because students do the experiments by following step-by-step procedures to verify known principles with little intellectual curiosity, purpose, or motivation. Woolnough distinguishes between acquiring scientific knowledge through prescribed laboratories and learning to do science. He advocates engaging students in authentic science activities of a problem-solving investigative nature that develops their expertise in working like scientists. The CREST program (CREativity in Science and Technology) provides an example of a program that has stimulated many students to become involved in genuine scientific and technological activities. The outcomes of such student projects include motivation, challenge, ownership, success, and self-confidence as well as acquisition of scientific knowledge and skills.

How can inquiry learning be assessed? University researchers, Audrey B. Champagne, Vicky L. Kouba, and Marlene Hurley, reflect upon the complexity of assessment at all levels. They distinguish between scientific inquiry as practiced by scientists and science-related inquiries as practiced by science literate adults and K-12 students. Science-related inquiries include information-based investigations to assist in decision making and to evaluate claims as well as experimentation to test theories and laboratory-based investigations. Champagne, Kouba, and Hurley delineate projects, abilities, and information assessed during four phases of laboratory investigations: when questions are generated, an investigation is planned, data are collected and interpreted, and conclusions argued and reported. In addition, they discuss decisions, assessment strategies, and individuals responsible for assessments that inform classroom practices and report student progress. The authors provide a similar matrix for planning and evaluation of K-12 programs and courses.

WHAT HAVE WE LEARNED ABOUT INQUIRY?

In the epilogue, Jim Minstrell reflects upon what we have learned about inquiry through the process of reading and talking and thinking with the authors and each other. He identifies some common themes embedded in the chapters of this book but points out that inquiry is complex. It likely involves integrating several of these themes into a coherent view of teaching and learning that closely approximates the activities of scientists as they attempt to make sense of their experiences. To summarize and make these themes more real, Minstrell uses a vignette to discuss them in the context of his own teaching practices.

What is inquiry? We knew when we started this project that we were unlikely to come to a definitive answer. What we have gained, however, is a much deeper appreciation of its complexity. We invite you and your colleagues to join us in this inquiry about inquiry learning and teaching in science.

REFERENCES

McDermott, L. 1996. *Physics by inquiry*. New York: Wiley.

National Research Council. 1996. *National science education standards*. Washington, DC: National Academy Press.

Part 1

Why Inquiry?

Some Thoughts of a Scientist On Inquiry

Bruce Alberts

What do we mean when we emphasize that much of science should be taught as inquiry?

It is certainly easy to recognize another, much more familiar type of science teaching, in which the teacher provides the student with a large set of science facts along with the many special science words that are needed to describe them. In the worst case, a teacher of this type of science is assuming that education consists of filling a student's head with a huge set of word associations —such as mitochondria with "powerhouse of the cell," DNA with "genetic material," or motion with "kinetic energy." This would seem to make preparation for life nearly indistinguishable from the preparation for a quiz show, or the game of trivial pursuit.

If education is simply the imparting of information, science, history, and literature become nearly indistinguishable forms of human endeavor, each with a set of information to be stored in one's head. But most students are not interested in being quiz show participants. Failing to see how this type of knowledge will be useful to them, they often lack motivation for this type of "school learning." Even more important to me is the tremendous opportunity that is being missed to use the teaching of science to provide students with the skills of problem solving, communication, and general thinking that they will need to be effective workers and citizens in the 21st century.

SOME EXAMPLES OF INQUIRY

If I think back to those aspects of my early education that have meant the most to me, I associate all of them with struggling to achieve an understanding that required my own initiative: writing a long report on "The Farm Problem" in seventh grade in which I was forced to explain why our government was paying farmers for not growing a crop; being assigned to explain to my eighth-grade class how a television set works; or in ninth grade grappling with the books on spectroscopy in the Chicago public library in order to prepare a report on its uses in chemistry.

What I mean by teaching science as inquiry is, at a minimum allowing students to conceptualize a problem that was solved by a scientific discovery, and then forcing them to wrestle with possible answers to the problem before they are told the answer. To take an example from my field of cell biology: the membrane that surrounds each cell must have the property of selective permeability—letting foodstuffs like sugars pass inward and wastes like carbon dioxide pass out, while keeping the many large molecules that form the cell tightly inside. What kind of material could this membrane be made of, so that it would have these properties and yet be readily able to expand without leaking as the cell grows? Only after contending with this puzzle for a while will most students be able to experience the pleasure that should result when the mechanism that nature derived for enclosing a cell is illustrated and explained. Classroom research with long-term followup shows that students are more likely to retain the information that they obtain in this way—incorporating it permanently into their view of the world (see, for example, G. Nuthall & A. Alton-Lee, 1995).

But there is much more. Along with science knowledge, we want students to acquire some of the reasoning and procedural skills of scientists, as well as a clear understanding of the nature of science as a distinct type of human endeavor. For some aspects of science knowledge that are more accessible to direct study than is the nature of the cell membrane, we therefore want students not only to struggle with possible answers to problems, but also to suggest and carry out simple experiments that test some of their ideas. One of the skills we would like all students to acquire through their science education is the ability to explore the natural world effectively by changing one variable at a time, keeping everything else constant. This is not only the way that scientists discover which properties in our surroundings depend on other properties; it also represents a powerful general strategy for solving many of the problems that are encountered in the workplace, as well as in everyday life in our society.

As an example, a set of fifth-grade science lessons developed by the Lawrence Hall of Science concentrates on giving students extensive experience in manipulating systems with variables. In this case, eight weeks of lessons come in a box along with a teacher's guide with instructions on how to teach with these materials (1993). The class starts by working in groups of four to construct a pendulum from string, tape, and washers. After each group counts the number of swings of its pendulum in 15 seconds with results that vary among pendulums, the class is led to suggest further trials that eventually trace the source of this variability to differences in the length of the string. Hanging the pendulums with different swing counts on a board in the front of the room makes clear the regular relation between pendulum length and swing rate, allowing each group to construct a pendulum with a predictable number of swing counts. This then leads to graphing as a means of storing the data for reuse in future pendulum constructions. A teacher could also exploit this particular two-week science lesson to acquaint students with the history of time keeping, emphasizing the many changes in society that ensued once it became possible to divide the day and night into reliable time intervals through the invention of pendulum clocks (Boorstin, 1985).

Contrast this science lesson with more traditional instruction about pendulums, in which the teacher does all of the talking and demonstrating, the students displaying their knowledge about which variables—length, weight, starting swing height—affect swing rate by filling in a series of blanks on a ditto sheet. A year later, the students are unlikely to remember anything at all about pendulums; nor have they gained the general skills that are the most important goal of the hands-on experience: recognition of the power of changing one variable at a time; the ability to produce graphs to store and recall information; the realization that everyone can carry out interesting experiments with everyday materials.

THE IMPORTANCE OF MOTIVATION

Why are we so often fascinated to watch a live sporting event, sitting on the edge of our seats as the tension builds in a close contest? And why, in comparison, do we have so little interest in watching the same event replayed on television, where the final outcome is already known? I conclude that human beings like to confront the unknown. Other types of games demonstrate that we also like challenging puzzles. Solving puzzles calls for playing out the consequences of a gamble—following particular pathways selected by our free will. Properly constructed, inquiry in education motivates students for the same reasons—it confronts them with an unknown puzzle, which can be solved only by a process that involves risk taking.

I use this conjecture to explain why essentially every scientist whom I know remembers being utterly bored by the cookbook laboratories common to college biology, chemistry, and physics courses. My own experience is typical. After two years as a premedical student, I could stand these required labs no longer. I therefore petitioned out of the laboratory attached to the physical chemistry course at Harvard, seizing on an opportunity to spend afternoons in my tutor's research laboratory. This experience was so completely different that it soon caused me to forget about applying to medical school. Within a year I had decided to go to graduate school in biophysics and biochemistry, in preparation for a career in science.

Extensive studies have been carried out that examine the motivation and value systems of the students in American schools. One of these extended over a period of 10 years and involved 20,000 middle-class Americans in grades 6 to 10. The results have been published in the academic literature, but they were also presented for public consumption in a book (Steinberg et al., 1997). The results are extremely distressing to those, like myself, who believe that the future of this nation will depend primarily on the quality of the education that our young people are receiving. Fully 40% of the students studied were categorized as "disengaged" from learning. These students attended school regularly, but did not think that any studying that they did there was relevant or important. And only 15% of all students said that their friends would think better of them if they did well in their academic studies.

Who is to blame for this state of affairs? Some of the onus must be on parents who pay too little attention to what their adolescents are doing in school. But having been a parent who was once frightened by the overwhelming influence of peer attitudes on my own children's values, I have to see this as a much more complex issue. What are our children being taught in grades 6 to 10? Would we ourselves find the curriculum interesting and motivating? Speaking as a scientist who has examined what is taught as science in grades 6 to 10, for most schools I must answer with a resounding no! In general, the curriculum is built around dull, vocabulary-laden textbooks, which are impossible to understand in any real sense of the word. Most of these textbooks have clearly been written by people who either lack any deep understanding of the material being taught, or are constrained by their publishers from making their book interesting to study or to read. In such a situation, is it any wonder that school becomes an institution in which peer values discourage academic performance?

A MAJOR CHALLENGE FOR OUR SCHOOLS

Inquiry is in part a state of mind, and in part a skill that must be learned from experience. The state of mind is inquisitiveness—having the curiosity to ask "Why" and "How." The good news is that young children are naturally curious. But if their incessant "Why" is dismissed by adults as silly and uninteresting, given only a perfunctory "just because" or "I don't know," children can lose the gift of curiosity that they began with, and develop into passive, unquestioning adults. Visit any second-grade classroom and you will generally find a room full of energy and excitement, with kids eager to make new observations and to try to figure things out. What a contrast with our eighth graders, who so often seem bored and disengaged from learning and from school.

The challenge is to create an educational system that exploits the tremendous curiosity that children initially bring to school, so as to maintain their motivation for learning—not only during their school years, but also throughout their lifetimes. Above all, we need to convince both teachers and parents of the importance of giving encouraging and supportive answers to the many "Why" questions, thereby showing that we value inquisitiveness. I am reminded of the profound effect that Richard Feynman's father had on his development as a scientist. As Feynman (1998) tells it:

> One kid says to me, "See that bird? What kind of bird is that?"
> I said, "I haven't the slightest idea what kind of a bird it is."
>
> He says, "It's a brown-throated thrush. Your father doesn't teach you anything!"
> But it was the opposite. He had already taught me: "See that bird?" he says. "It's a Spencer's warbler." (I knew he didn't know the real name.) "Well, in Italian, it's a *Chutto Lapittida*. In Chinese, it's a *Chung-long-tah,* and in Japanese, it's a *Katano Tekeda.* You can know the name of that bird in all the languages of the world, but when you're finished, you'll know absolutely nothing whatever about the bird. You'll only know about humans in different places, and what they call the bird. So let's look at the bird and see what it's *doing*—that's what counts." (I learned very early the difference between knowing the name of something and knowing something.)
>
> He said, "For example, look: the bird pecks at its feathers all the time. See it walking around, pecking at its feathers?"
> "Yeah."

He says, "Why do you think birds peck at their feathers?"

I said, "Well, maybe they mess up their feathers when they fly, so they're pecking them in order to straighten them out."

"All right," he says. "If that were the case, then they would peck a lot just after they've been flying. Then, after they've been on the ground for a while, they wouldn't peck so much any more—you know what I mean?"

"Yeah."

He says, "Let's look and see if they peck more just after they land."

It wasn't hard to tell: there was not much difference between the birds that had been walking around a bit and those that had just landed. So I said, "I give up. Why does a bird peck at its feathers?"

"Because there are lice bothering it," he says. "The lice eat flakes of protein that come off its feathers."

He continued, "Each louse has some waxy stuff on its legs, and little mites eat that. The mites don't digest it perfectly, so they emit from their rear ends a sugar-like material, in which bacteria grow."

Finally he says, "So you see, everywhere there's a source of food, there's some form of life that finds it."

Now, I knew that it may not have been exactly a louse, that it might not be exactly true that the louse's legs have mites. That story was probably incorrect in *detail*, but what he was telling me was right in *principle*. (pp. 13-15)

Very few children are fortunate enough to have a parent like Feynman's. Much of the responsibility for nurturing the state of mind needed to be an inquiring adult therefore falls to our schools. Maintaining children's initial curiosity about the world requires making them confident that they can use the methods of inquiry to find answers for their questions. This self-confidence can be developed in only one way: from a string of actual successes. It is not enough to encourage students to inquire. They must also have many opportunities to obtain the diverse set of skills needed for repeated success in such experiences.

For our schools, we should seek a curriculum that begins in kindergarten and increases in difficulty so as to provide, at each grade level, challenges appropriate to the students' age. This curriculum should focus on student and class inquiry, rather than on the memorization and regurgitation of facts. At each grade level, the inquiries need to be carefully designed to present students with challenges that are difficult enough to seem almost inaccessible at first, but which allow at least partial success for most students. We want students to see clearly that, as they acquire the tools and habits of inquiry, they are becoming more and more proficient in dealing with the world around them. School then becomes a highly relevant place for students: a place where they recognize that they are learning important skills for their life *outside* of school.

A MAJOR CHALLENGE FOR SCIENTISTS

Instead of merely blaming others for the current state of science education, we scientists need to confront our own failings. Why do the same scientists who remember with distaste their own college laboratory experiences continue to run their own college students through the same type of completely predictable, recipe-driven laboratory exercises that once bored them? I remain mystified, with no good answer. But I am trying to encourage my former university colleagues to think deeply about this question and act accordingly. Perhaps they can think of no alternative. If so, they should spend a few hours examining one of the outstanding science modules based on inquiry that have been developed for elementary schools (see, for example, *Science and Technology for Children*, a joint project of the National Academy of Sciences (NAS) and the Smithsonian Institution at *http://www.si.edu/nsrc*). I see no reason why inexpensive, commercially available college laboratory modules could not be produced that are modeled after such outstanding elementary school examples. A project with this aim could stimulate a badly needed rethinking of what our introductory college science laboratories should be like, and what purposes they are supposed to serve.

We scientists also have a great deal of work to do in addressing the nature of our introductory college science courses. Where in a typical Biology 1 college course is the science as inquiry that is recommended for K-12 science classes in the *National Science Education Standards* (National Research Council [NRC], 1996)? These courses generally attempt to cover all of biology in a single year, a task that becomes ever more impossible with every passing year, as the amount of new knowledge explodes. Yet old habits die hard, and most Biology 1 courses are still given as a fact-laden rush of lectures. These lectures leave no time for inquiry: they even fail to provide students with any sense of what science is, or why science as a way of knowing has been so successful in improving our understanding of the natural world and our ability to manipulate it for human benefit. (For attempts to change this situation, see *Science Teaching Reconsidered*, [NRC, 1997] and *Teaching Evolution and the Nature of Science* [NAS, 1998]).

ON BECOMING A SCIENTIST

Very few students of science will go on to become professional scientists. That is not the primary purpose of current science education reforms. But I am convinced, both by my personal experience and from my extensive interactions with students, that the desired changes in our nation's K-16 science education will also contribute to the production of better scientists. If we stress understanding in addition to knowledge, and if we use inquiry methods that generate scientific habits of mind, students will not need to work in a research laboratory to appreciate the excitement of a life in science. And students with superb memorization skills, who often do well in our current science classes, will not be misled into believing that excelling in science requires the same skills as doing well on an exam.

If young people with outstanding scientific potential are never exposed to scientific inquiry and never given any illustration of what doing science is like, how can they think meaningfully about the possibility of a scientific career? But here we face another conundrum. Because of the way that we teach science in our colleges and universities, most science teachers in our schools—including former science majors—have never participated in scientific inquiry themselves. Is it any wonder that so many teachers are unable to teach their children according to the recommendations of the *Benchmarks for Science Literacy* (American Association for the Advancement of Science, 1993) or the *National Science Education Standards* (NRC, 1996), even when supplied with outstanding hands-on science curricula?

Faced with this dilemma, some suggest that we retreat from our ambitious education goals and settle for what all teachers can teach—science as memorization, evaluated by multiple-choice examinations that stress the recall of word associations. But I am convinced that we need not settle for a second-class education for our children, and that indeed we cannot do so without giving up our hope of remaining the world's leading nation.

As president of the National Academy of Sciences for the past six years, I have been trying to convince my many scientific colleagues across the United States that they must stop being part of the problem and instead become part of the answer. Our nation is blessed with the world's strongest scientific and engineering community, and very few places in our nation lack experts in scientific inquiry. These working scientists and engineers need to connect intimately to our local K-12 education systems—as volunteers to help teachers and school districts, as providers of professional development, and as a stable local political force advocating for a new type of science education (see *http://www.nas.edu/rise*).

But we need something more. The necessity of hiring two million of our nation's 3.5 million teachers in the next decade (National Commission on Teaching and America's Future, 1996), coupled with the imminent retirement of the bolus of science teachers and leaders who were produced in the era immediately after Sputnik, requires the entry of a new generation of talented scientists into our nation's K-12 teaching corps. Ideally, they would become teachers with a deep understanding of both science and inquiry—and form a natural bridge between the culture of science and that of the schools.

In the abstract, there would seem to be little chance of finding large numbers of such talented people and moving them into our K-12 school systems. But these are not normal times for scientists. Over the course of the past 40 years, the flourishing scientific enterprise in the United States has developed a dependence on an ever-increasing influx of young trainees who, serving as graduate students and postdoctoral fellows, perform most of the research that is carried out in our universities and publicly funded research institutes. As these people have aged and formed their own laboratories, they too have wanted young trainees to staff their laboratories. Because most professors will produce many potential new professors over the course of their careers, this system cannot be sustained over the long run unless either the number of science faculty at universities keeps increasing, or many other types of positions are developed in our society for Ph.D. scientists. Such concerns, triggered by an increasing frustration expressed by the young scientists looking for traditional employment, caused the National Research Council to carry out a major study to track the current career paths of life scientists (1998). The findings reveal that, over the past decade, the number of Ph.D.s awarded in the life sciences has been increasing at a rate of about 4% a year, whereas the number of research positions for them in universities, research institutes, and industry has been increasing at only about 3% a year. The result is a widening ever-increasing pool of poorly paid postdoctoral researchers who are spending longer and longer times in temporary positions.

From the twenty thousand or so present postdoctoral researchers in the life sciences and an expected growth in their numbers, could we generate a new generation of outstanding science teachers at the K-12 level who really understand inquiry? My own contacts with these young scientists have convinced me that many of them are willing to try. But they will do so only if efficient training programs become available to provide them with the additional skills that they need to teach well, if we in the scientific community demonstrate our support for their career change and continue to treat them as colleagues, and if school systems are willing to hire and support them once they have been trained.

I view the current situation as a terrific, one-time opportunity for scientists who want to help reinvigorate our school systems. With the proper preparation and support, these scientists can immediately introduce inquiry into the curriculum, and they can help generate new types of professional development experiences for other teachers in their schools. The National Academy of Sciences has begun to focus on this critical issue, which I believe to be of utmost importance for the future of science, as well as for the future of our schools.

REFERENCES

American Association for the Advancement of Science. 1993. *Benchmarks for science literacy.* New York: Oxford University Press.

Boorstin, D.J. 1985. *The discoverers.* New York: Random House.

Feynman, R.P. 1998. *The making of a scientist, What do you care what other people think?* New York: Bantam Books.

Lawrence Hall of Science. 1993. Variables. Module in *Full option science system.* Chicago: Encyclopedia Britannica Educational Corp.

National Commission on Teaching and America's Future. 1996. *What matters most: Teaching for America's future.* New York: Author.

National Academy of Sciences. 1998. *Teaching evolution and the nature of science.* Washington, DC: National Academy Press.

National Research Council. 1996. *National science education standards.* Washington, DC: National Academy Press.

National Research Council. 1997. *Science teaching reconsidered.* Washington, DC: National Academy Press.

National Research Council. 1998. *Trends in the early careers of life scientists.* Washington, DC: National Academy Press.

Nuthall, G., and A. Alton-Lee. 1995. Assessing classroom learning: How students use their knowledge and experience to answer classroom achievement test questions in science and social studies. *American Educational Research Journal* 31: 185-223.

Steinberg, L., B. Brown, and S. Dornbusch. 1997. *Beyond the classroom.* Cambridge, MA: Touchstone Books.

The Three Faces of Inquiry

Gerald F. Wheeler

The word "inquiry" comes up often in conversations about reform in science education. But "inquiry" is an elastic word, stretched and twisted to fit people's differing worldviews. Inquiry itself a core tenet of the standards, and the ambiguity surrounding it is a threat to reform efforts.

THE FIRST FACE ENGAGES STUDENTS

One image of inquiry has a classroom with children engaged in a hands-on activity. The noise level and class demographics vary with teacher style but there's nearly always a perceivable high level of energy in the classroom. When asked, these teachers talk of inquiry from the point of view of a teaching strategy for motivating students. These activities, the experts suggest, are the best teaching strategies for engaging children in the joys of learning science. The research concerning how little is learned when there is no engagement is robust. The argument for engagement is best expressed by that common chant of the 60s:

I see . . . I forget
I hear . . . I remember
I do . . . I understand

There is a danger in this view of inquiry. Unless we know more about the "doing"—*what* is being done, and *how* is it being done—we can't assess the truthfulness of the third assertion. This view of inquiry—commonly heard as a call for more hands-on activities in students' school science—falls short of reform goals because not all hands-on activities are inquiry-based activities.

I've seen children with materials in their hands but little evidence of understanding in their minds. If you give a child a battery and a small motor with two wires, the child will connect the wires to the battery and the motor will turn. If the motor has a propeller on it, the child might make the wind blow in his or her face. Or, more often, turn it on a neighbor.

That's a hands-on experience but a relatively weak inquiry situation. *Benchmarks for Science Literacy*, the American Association for the Advancement of Science publication agrees:

> Hands-on experience is important but does not guarantee meaningfulness. (1993, p. 319)

And in the long run, I don't believe that by itself it even engages the young mind in science.

THE SECOND FACE GETS THE INQUIRY ACTIVITY RIGHT

At first glance, the classroom looks the same: children working with materials. But there's a difference. The children are interacting with the materials—they're doing an experiment.

In this image the child develops questions and devises ways to the answers *via the materials*. While the activities may have started with that important component of "messing around," they eventually take on the purpose of the searching for answers.

In the example of the battery and motor, simply encouraging the child to ask a question—such as what will happen if the wires are reversed—and then giving the opportunity to seek answers moves the inquiry to this next level. Eventually, the questions get more complex, asking, for instance, what changes when more batteries are used. Hands-on activities when enhanced with questioning initiates a richer inquiry opportunity.

There are plenty of questions that are within the grasp of the child. The key is whether the activity allows students to engage in a dialogue with the material world. The inquiry activity involves observing, asking questions, making predictions, and thinking about the results—reflecting on the predictions—and crafting the next move.

As Wendy Saul states in her delightful introductory chapter in *Beyond the Science Kit*:

> Inquiry is realized in the coming together of materials and learner.
> (Saul & Reardon, 1996. p. 7)

Good teachers recognize the ways materials and curiosity relate, and help students as they tentatively or bullishly try to connect their own questions to ways of "finding out." Good teachers also work with children to "make sense" of what they find and construct arguments that seem convincing to others in their scientific community.

My simple definition of the process of science and thus scientific inquiry activities points to what we're trying to get our students to do: Science is the process of talking to the material world. Scientists understand their world by figuring out how to pose questions to the phenomena at hand. In the same way, we want our students to understand their world by learning how to ask the right questions—to the phenomena, not the teacher. It's fortuitous that research on student learning shows that this ability is also a powerful tool for learning science. My teaching goal is to place my students in situations where they can practice having the dialogue.

One of my best inquiry workshops happened by accident.

I was invited to do something on inquiry at a summer institute for middle school teachers. I decided to do a workshop on "Batteries and Bulbs." I hadn't done a workshop on electricity for ages, and I thought it would be fun to show teachers the richness of that topic.

I was told I would be conducting four workshops in succession for the institute's 120 teachers. I didn't want to carry workshop materials for 120 people on the airplane, so I put the tiny flashlight bulbs and some wire into my carry-on and shipped the heavier batteries directly to the hotel.

When I arrived at the hotel at midnight, I informed the registration clerk that there was a package waiting for me. He returned from the back room with a confused look. My workshop materials were nowhere to be found.

Between midnight and 8 AM, I became a scavenger connoisseur. By the time the first workshop began, I had found 8 batteries, 2 packets of paper towels (luckily, different brands), a roll of masking tape, a box of rubber bands, 4 bottles of different dish soaps, some string and washers, and paper coffee cups. My well-planned workshop on inquiry had just been replaced with "Today, we're going to look at a variety of phenomena."

The teachers divided into small teams. I suggested they play a little, come up with some initial questions, and then decide how to ask a question *directly* to the phenomena of interest.

Some teams played with the paper towels: one team looked at the strength of the two brands of towels, another team checked out absorbency, and another found a magnifying glass and examined the towel textures. A couple of teams looked at the stretching properties of rubber bands. Then they measured stretch for different weights using the washers. (They didn't have rulers, so they made their own scales by using string and making knots of units. One team named these units after one of the team members.)

One team got fascinated with the periodic motion of pendula and with a little creative mounting on a wall was able to achieve precise, repeatable trials of amplitude versus period of swing.

This was inquiry in its purest form.

THE THIRD FACE HAS A CONTENT DIMENSION

Engaging in the process of inquiry doesn't require a specific, organized experiment. For the teachers' workshop I wanted to create an authentic experience, having them look at some piece of the material world in a systematic fashion. It didn't matter what phenomena they investigated. Having multiple options, in fact, provided a convincing demonstration of this. Any experiment can lead to interesting investigations about the material world.

With the teachers in the workshop, I enhanced the image of inquiry by talking about the structure of inquiry. Inquiry was a content to be learned. This enhanced image is a crucial part of the two major standards efforts and, frequently, overlooked.

The two major standards efforts, one conducted by the National Research Council (NRC) and the other by the American Association for the Advancement of Science (AAAS) enhanced the image of inquiry further by declaring it to be a content to be learned.

The AAAS Project 2061 publication, *Benchmarks for Scientific Literacy* states in the first chapter:

> If students themselves participate in scientific investigations that progressively approximate good science, then the picture they come away with will likely be reasonably accurate.

> . . . the laboratory can be designed to help students learn about the nature of scientific inquiry. (1993, p. 9)

The *National Science Education Standards* document (NRC, 1996) lists "inquiry" in its content section and defines it as follows:

> Inquiry requires identification of assumptions, use of critical and logical thinking, and consideration of alternative explanations. Students will engage in selected aspects of inquiry as they learn the scientific way of knowing the natural world, but they also should develop the capacity to conduct complete inquiries. (p. 23)

When inquiry becomes a content to be inquired into and learned, the role of the teacher changes from just engaging students to that of stewardship over the development of knowledgeable thinkers about inquiry. But no set of experiments, if done in a vacuum, will lead to a scientifically literate graduating student. Inquiry is a content different from other content. It's not something to be studied for a short time and then left behind. Inquiry has a meta-content character that demands its presence while all the other content is being learned.

Both the traditional basic content and the meta-content, inquiry, must be infused into the children's experiences. We know we cannot plod from curriculum lesson to curriculum lesson (or chapter to chapter in a textbook) without an awareness of what the child is thinking or doing. It is equally important that we do not become driven by the seductive energy of kids' curiosity and jump from interesting to interesting phenomena without any concern for the growth of shared knowledge and a coherence of learning experiences. If left to their own devices, students will merely deepen their own misconceptions.

Expanding on the simple definition of science given above, we can add that the goal of science is to find rules of Nature. It is important that as students work within an inquiry activity they keep to the point of the inquiry—to search out Nature's mysteries. Students need to be challenged for the evidence of an opinion or inference. And they need to learn the basic core of science as outlined in the standards documents.

REFERENCES

American Association for the Advancement of Science. 1993. *Benchmarks for science literacy.* New York: Oxford University Press.

National Research Council. 1996. *National science education standards.* Washington, DC: National Academy Press.

Saul, W., and J. Reardon. (Eds.) 1996. *Beyond the science kit.* Portsmouth, NH: Heinemann Press.

Teaching Science as Inquiry

Rodger W. Bybee

INTRODUCTION

The idea of teaching science as inquiry has a rather long history in science education. There is an equally long history of confusion about what teaching science as inquiry means and, regardless of the definition, its implementation in the classroom. In short, we espouse the idea and do not carry out the practice. Publication of the *National Science Education Standards* (National Research Council, 1996) once again brought science as inquiry to the top of educational goals. The *Standards* answer definitional questions. Teaching science as inquiry, the *Standards* explain, requires imparting not only scientific information but the skills of inquiry and, more deeply, an understanding of what scientific inquiry is about.

INQUIRY IN ACTION

A science teacher wanted to see inquiry in action so she visited three classrooms. Her considerations included the content of lessons, the teaching strategies, the student activities, and what students learned. During five days in each classroom, she made these observations.

Classroom 1

The students engaged in an investigation initiated by significant student interest. A student asked what happened to the water in a watering can. The can was almost full on Friday and almost empty on Monday. One student proposed that Willie the pet hamster had left his cage at night and drunk the water. Encouraged by the teacher to find a way to test this idea, the students covered the water so Willie could not drink it. Over several days they observed that the water level did not drop. The teacher then challenged the students to think about other explanations. The students' questions resulted in a series of full investigations about the disappearance of water from the container. The teacher employed strategies such as asking students to consider alternative explanations, using evidence to form their explanations, and designing simple investigations to test an explanation. The science teacher never did explain evaporation and related concepts.

Classroom 2

In a class studying evolution, the teacher distributed two similar but slightly different molds with dozens of fossil brachiopods. The students measured the lengths and widths of the two populations of brachiopods. The teacher asked whether the differences in length and width might represent evolutionary change. As the students responded, the teacher asked—How do you know? How could you support your answer? What evidence would you need? What if the fossils were in the same rock formation? Are the variations in length and width just normal variations in the species? How would difference in length or width help a brachiopod adapt better? The fossil activity provided the context for students to learn about the relationships between the potential for a species to increase its numbers, the genetic variability of offspring due to mutation and recombination of genes, the finite supply of resources required for life, and the ensuing selection by the environment for those offspring better able to survive and leave offspring. In the end, students learned about changes in the variations of characteristics in a population—biological evolution.

Classroom 3

In this science classroom, students selected from among several books that provided extended discussions of scientific work. Readings included *The Double Helix*, *The Beak of the Finch*, *An Imagined World*, and *A Feeling for the Organism*. Over a three-week period, each student read one of the books as homework. Then, in groups of four, all students discussed and answered the same questions—What led the scientist to the investigation? What conceptual ideas and knowledge guided the inquiry? What reasons did the scientist cite for conducting the investigation? How did technology enhance the gathering and manipulation of data? What role did mathematics have in the inquiry? Was the scientific explanation logically consistent? based in evidence? open to skeptical review? and built on knowledge from other experiments? After reading the books and completing the discussion questions, the groups prepared oral reports on the topic, "The Role of Inquiry in Science."

After completing the classroom visits, the science teacher summarized her observations (see Table 1).

TABLE 1. SUMMARY OF OBSERVATIONS

	Classroom 1	Classroom 2	Classroom 3
CONTENT OF LESSONS	Changing water level in an open container	Investigation of variations in fossils	Stories of scientists and their work
TEACHING STRATEGIES	Challenge students to think about proposed explanations and use evidence to support conclusions	Provide molds of fossils and ask questions about student measurements and observations	Provide questions to focus discussions of readings
STUDENT ACTIVITIES	Design simple, but full investigations	Measure fossils and use data to answer questions	Read and discuss a book about scientific investigations
STUDENT OUTCOMES	Develop the ability to reason using logic and evidence to form an explanation	Understand some of the basic concepts of biological evolution	Understand scientific inquiry as it is demonstrated in the work of scientists

This introduction should have engaged your thinking about teaching science as inquiry. In order to further clarify your thinking, take a few minutes and respond to the questions here. Refer to the passages or summary table as often as necessary. Select the best answers and provide brief explanations for your choices.

1. Which classroom would you cite as furnishing the best example of teaching science as inquiry?
   A. 1
   B. 2
   C. 3
   D. None of the classrooms
   E. All of the classrooms
   EXPLANATION:

2. If teaching science as inquiry is primarily interpreted to mean using laboratory experiences to learn science concepts, which classroom was the best example?
   A. 1
   B. 2
   C. 3
   D. None of the classrooms
   E. All of the classrooms
   EXPLANATION:

3. If students had numerous experiences with the same teaching strategies and the same activities devised by the students as in Classroom 1, but pursued different questions, what would you predict as the results for students?
   A. Their thinking abilities, understanding of the subject, and understanding of inquiry will be <u>higher</u> than students who were in the other two classes.
   B. Their thinking abilities, understanding of the subject, and understanding of inquiry will be <u>lower.</u>
   C. Their thinking abilities would be higher and their understanding of the subject and of inquiry will be lower.
   D. Their understanding of the subject matter will be higher and their thinking abilities along with their understanding of inquiry lower.
   E. All learning outcomes would be the same as for students in the other two classes.
   EXPLANATION:

4. Which of these generalizations about teaching science as inquiry would the observations of the three classrooms suggest to you?

   A. Overuse of one teaching strategy may constrain opportunities to learn the subject.
   B. Differing teaching strategies and student activities may bring differing benefits and trade-offs.
   C. The potential learning outcomes for any one sequence of lessons may be greater than for the sum of the individual lessons.
   D. Teaching strategies may need to differ in accordance with the result sought.
   E. All of the above.

   EXPLANATION:

5. If the teacher continues observing the three classrooms for another week, what would you recommend she look for in order to formulate an answer to the question, What is teaching science as inquiry?

   A. What the students learned about scientific inquiry.
   B. What teaching strategies the teacher used.
   C. What science information, concepts, and principles the students learned.
   D. What inquiry abilities the students developed.
   E. What teachers should know and do to achieve the different learning goals of scientific inquiry.

   EXPLANATION:

6. Drawing on these observations, the science teacher proposes that teaching science as inquiry may have multiple meanings. Which of these would you recommend as a next step in her investigation?

   A. Explore how others have answered the question—What is teaching science as inquiry?
   B. See how the *National Science Education Standards* explain Science as Inquiry.
   C. Elaborate the implications of teaching science as inquiry.
   D. Try teaching science as inquiry in order to evaluate the approach in school science programs.
   E. All of the above.

   EXPLANATION:

As you engaged in the review of the observations in three classrooms, what idea of inquiry did you originally apply? Assuming you are now engaged in questions about teaching science as inquiry, we can proceed to a review of several historical discussions of inquiry.

PERSPECTIVES FROM HISTORY

In the United States, science itself was not valued prior to the mid-nineteenth century. "...faith," Charles Stedman writes, "was at least as important as empirical data and in many instances it dominated the practices of science. This faith was often a complex mixture of Christian theology, idealism, and entrenched traditions" (1987).

In the late nineteenth century, several people brought science into discussions of school and college curricula. Charles W. Eliot, president of Harvard University from 1869 to 1895, articulated the need for science and laboratory approaches in the curriculum. Louis Agassiz, also at Harvard, provided an early example of teaching science as inquiry when he had students come to his lab and study specimens. He directed field trips to the countryside and seashore, encouraged students to make their own collections, and conducted instruction by correspondence with specimen collectors around the country (Stedman, 1987).

John Dewey

In 1909, when the presence of science in the school curriculum was bringing disagreements about what science is and thus how it should be taught, John Dewey addressed the education section of the American Association for the Advancement of Science on the topic "Science as Subject-Matter and as Method" (Dewey, 1910). Dewey's general theme was that science teaching gave too much emphasis to the accumulation of information and not enough to science as a method of thinking and an attitude of mind: "Science teaching has suffered because science has been so frequently presented just as so much ready-made knowledge, so much subject-matter of fact and law, rather than as the effective method of inquiry into any subject-matter." (p. 124)

Notice that in these passages, Dewey refers to aims that include the abilities of inquiry, the nature of science, and an understanding of a subject.

> Surely if there is any knowledge which is of most worth it is knowledge of the ways by which anything is entitled to be called knowledge instead of being mere opinion or guess work or dogma.

Such knowledge never can be learned by itself; it is not information, but a mode of intelligent practice, an habitual disposition of mind. Only by taking a hand in the making of knowledge, by transferring guess and opinion into belief authorized by inquiry, does one ever get a knowledge of the method of knowing. (p. 125)

But that the great majority of those who leave school have some idea of the kind of evidence required to substantiate given types of belief does not seem unreasonable. Nor is it absurd to expect that they should go forth with a lively interest in the ways in which knowledge is improved by a marked distaste for all conclusions reached in disharmony with the methods of scientific inquiry. (p. 127)

Near the conclusion, Dewey makes this powerful statement.

One of the only two articles that remain in my creed of life is that the future of our civilization depends upon the widening spread and deepening hold of the scientific habit of mind; and that the problem of problems in our education is therefore to discover how to mature and make effective this scientific habit. (p. 127)

Some ninety years ago, then, John Dewey articulated as objectives of teaching science as inquiry: developing thinking and reasoning, formulating habits of mind, learning science subjects, and understanding the processes of science. Dewey's *Logic: The Theory of Inquiry*, published in 1938, presents his stages in the scientific method: induction, deduction, mathematical logic, and empiricism. This book no doubt influenced the many science textbooks that treat the scientific method as a fixed sequence as opposed to a variety of strategies whose use depends on the question being investigated and the researchers. Discussions about the role of scientific method in science classrooms and textbooks continue in the community of science educators (Klapper, 1995; Storey & Carter, 1992).

Joseph J. Schwab

In the late 1950s and the 1960s, Joseph Schwab published articles on inquiry (or enquiry, his preferred spelling). Schwab laid the foundation for the emergence of inquiry as a prominent theme in the curriculum reform of that era (Schwab, 1958; 1960; 1966). In 1958 he grounded in science itself his argument for teaching science as inquiry: "The formal reason for a change in present methods of teaching the sciences lies in the fact that science itself has changed. A new view concerning the nature of scientific inquiry now controls research." According to Schwab, scientists no longer conceived science as stable truths to be verified; they were viewing it as principles for inquiry, conceptual structures revisable in response to new evidence. Schwab distinguished between "stable" and "fluid" inquiry. These terms suggest the distinction between normal and revolutionary science as made popular by Thomas Kuhn in his classic of 1970, *The Structures of Scientific Revolutions*. Stable inquiry uses current principles to "fill a...blank space in a growing body of knowledge" (1966), while fluid inquiry is the invention of conceptual structures that will revolutionize science.

Schwab observed that teachers and textbooks were presenting science in a way that was inconsistent with modern science. Schwab in 1966 found that science was being taught "...as a nearly unmitigated rhetoric of conclusions in which the current and temporary constructions of scientific knowledge are conveyed as empirical, literal, and irrevocable truths." A "rhetoric of conclusions, then, is a structure of discourse which persuades men to accept the tentative as certain, the doubtful as the undoubted, by making no mention of reasons or evidence for what it asserts, as if to say, 'this, everyone of importance knows to be true.'" The implications of Schwab's ideas were, for their time, profound. He suggested both that science should be presented as inquiry, and that students should undertake inquiries.

In order to achieve these changes, Schwab argued in1960, science teachers should first look to the laboratory and use these experiences to lead rather than lag behind the classroom phase of science teaching. He urged science teachers to consider three levels of openness in their laboratories. At the primary level, the materials can pose questions and describe methods of investigation that allow students to discover relationships they do not already know. Next, the laboratory manual or textbook can pose questions, but the methods and answers are left open. And on the most sophisticated level, students confront phenomena without questions based in textbooks or laboratories. They are left to ask questions, gather evidence, and propose explanations based on their evidence.

Schwab also proposed an "enquiry into enquiry." Here teachers provide students with readings, reports, or books about research. They engage in discussions about the problems, data, role of technology, interpretation of data, and conclusions reached by scientists. Where possible, students should read about alternative explanations, experiments, debates about assumptions, use of evidence, and other issues of scientific inquiry.

Joseph Schwab had a tremendous influence on the original design of instructional materials—the laboratories and invitations to inquiry—for the Biological Sciences Curriculum Study (BSCS). Schwab's recommendation paid off in the late 1970s and early 1980s when educational researchers asked questions about the effectiveness of these programs. In 1984 Shymansky reported evidence supporting his conclusion that "BSCS biology is the most successful of the new high school science curricula."

F. James Rutherford

In 1964 F. James Rutherford observed that while in the teaching of science we are unalterably opposed to rote memorization and all for the teaching of scientific processes, critical thinking, and the inquiry method, in practice science teaching does not represent science as inquiry. Nor is it clear what teaching science as inquiry means. At times the concept is used in a way that makes inquiry part of the science content itself. At others, authors refer to a particular technique or strategy for bringing about learning of some particular science content.

Rutherford (1964) presented the following conclusions:

1. It is possible to gain a worthwhile understanding of science as inquiry once we recognize the necessity of considering inquiry as content and operate on the premise that the concepts of science are properly understood only in the context of how they were arrived at and of what further inquiry they initiated.

2. As a corollary, it is possible to learn something of science as inquiry without having the learning process itself to follow precisely any one of the methods of inquiry used in science.

3. The laboratory can be used to provide the student experience with some aspects or components of the investigative techniques employed in a given science, but only after the content of the experiments has been carefully analyzed for its usefulness in this regard. (pp. 80-84)

In the end, Rutherford connected to teaching science as inquiry a knowledge base for doing so. Until science teachers acquire "a rather thorough grounding in the history and philosophy of the sciences they teach, this kind of understanding will elude them, in which event not much progress toward the teaching of science as inquiry can be expected." (p. 84)

Project Synthesis

In the late 1970s and early 1980s, the National Science Foundation supported a project that synthesized a number of national surveys, assessments, and case studies about the status of science education in the United States (Harms & Kohl, 1980; Harms & Yager, 1981). One major portion of this review centered on the role of inquiry in science teaching and was completed by Wayne Welch, Leo Klopfer, Glen Aikenhead, and James Robinson in 1981. Their analysis revealed that the science education community was using the term "inquiry" in a variety of ways, including the general categories of inquiry as content and inquiry as instructional technique, and was unclear about the term's meaning. The evidence indicated that "although teachers made positive statements about the value of inquiry, they often felt more responsible for teaching facts, 'things which show up on tests,' 'basics' and structure and the work ethic." Among the teachers surveyed, the main consideration was of inquiry as an instructional technique. For not teaching science as inquiry, not employing it for introducing the content, or not using experiences oriented to inquiry, teachers gave a number of reasons. Among them were problems managing the classroom, difficulty meeting state requirements, trouble obtaining supplies and equipment, dangers that some experiments might pose for students, and concerns about whether inquiry really worked. In conclusion, the authors reported:

> The widespread espoused support of inquiry is more simulated than real in practice. The greatest set of barriers to the teacher support of inquiry seems to be its perceived difficulty. There is legitimate confusion over the meaning of inquiry in the classroom. There is concern over discipline. There is worry about adequately preparing children for the next level of education. There are problems associated with the teachers' allegiance to teaching facts and to following the role models of the college professors. (p. 40)

The portion of Project Synthesis relating to biology concludes: "In short, little evidence exists that inquiry is being used" (Hurd et al., 1980). Costenson and Lawson in1986 presented the results of their survey of a group of biology teachers. Inquiry, some teachers had claimed, takes too much time and energy. It is too slow. The reading is too difficult, and the students are insufficiently mature. Experiments may put students at risk. Inquiry makes it hard to track the progress of students, and to place material in proper sequence. It violates the habits that teachers have developed. And it is too expensive. The objections are similar to what Welch and his colleagues had reported. Similar results would probably be obtained for other disciplines, particularly at the secondary level. They form the substantial barriers between policies, such as that set by the *Standards* in 1996, that recommend science as inquiry and the programs exemplified in BSCS materials that incorporate into teaching science as inquiry the actual practices in science classrooms. "In our opinion," the report on biology declares,

> ...the previous reasons for not using inquiry are not sufficient to prevent its use. However, to implement inquiry in the classroom we see three crucial ingredients: (1) teachers must understand precisely what scientific inquiry is; (2) they must have sufficient understanding of the structure of biology itself; and (3) they must become skilled in inquiry teaching techniques. (p. 158)

The passage makes the important distinction between inquiry as <u>content</u> to be understood first by teachers and then by students and inquiry as <u>technique</u> that teachers are to use to help students learn biology.

Project 2061

In 1985, F. James Rutherford inaugurated Project 2061, a long-term initiative of the American Association for the Advancement of Science (AAAS) to reform K-12 education. It set the stage for the *National Science Education Standards* published in 1996 by the National Research Council (NRC). In the initial years, the project outlined what all students should know and be able to do by the time they complete the twelfth grade. Project 2061 materials such as *Science for All Americans* issued in 1989, and *Benchmarks for Science Literacy* which AAAS published in 1993, have made significant statements about teaching science as inquiry. Rutherford's observations and recommendations presaged in 1964 the place Project 2061 assigns to the nature and history of science and that which it sets for habits of mind.

The lead chapter of *Science for All Americans* outlines recommendations for the nature of science and another provides recommendations for "Historical Perspectives." A chapter on "Habits of Mind" includes categories of values and attitudes, manipulation and observation, communication, and very importantly, skills of critical response.

In a separate chapter on "Effective Learning and Teaching," *Science for All Americans* makes the general recommendation, "Teaching Should Be Consistent With the Nature of Scientific Inquiry," followed by specific advice:

▶ Start with Questions About Nature
▶ Engage Students Actively
▶ Concentrate on the Collection and Use of Evidence
▶ Provide Historical Perspectives
▶ Insist on Clear Expression
▶ Use a Team Approach
▶ Do Not Separate Knowing From Finding Out
▶ Deemphasize the Memorization of Technical Vocabulary (pp. 147-149)

Benchmarks for Science Literacy show specific results of learning about the nature of science, gaining historical perspectives, and acquiring good habits of mind. In addition, there is an excellent research base that indicates what students know and are able to do relative to various benchmarks.

Project 2061 also set in place goals and specific benchmarks for teaching scientific inquiry as content. Included as well are recommendations for using teaching techniques associated with inquiry.

NATIONAL SCIENCE EDUCATION STANDARDS: INQUIRY AS CONTENT

The *National Science Education Standards* (NRC, 1996) present a present-day statement on teaching science as inquiry. Defining what all students should know and be able to do by grade twelve, and what kinds of learning experiences they need to achieve scientific literacy, the document reaffirms the conviction that inquiry is central to the achievement of scientific literacy.

In 1991, the National Research Council was asked by the President of the National Science Teachers Association to coordinate efforts to develop national standards for science education. Between 1991 and 1995, the NRC produced several drafts of standards, submitted those to extensive review, and set in motion a process for developing a national consensus for the standards. In December 1995, the NRC released the *National Science Education Standards*, which presents a vision of a scientifically literate populace by describing what students should know and be able to do after thirteen years of school science. In addition to Content Standards, the document contains standards for Teaching, Professional Development, Assessment, School Science Programs, and the Educational System. Angelo Collins provided in 1995 a detailed history of the science education standards, and elsewhere I have discussed the *Standards* (and *Benchmarks*) and the aim of achieving scientific literacy (Bybee, 1997).

Release of the *Standards* again brought to the forefront in the educational community the issue of teaching science as inquiry. In the *Standards*, scientific inquiry refers to several related, but different things: the ways scientists study the natural world, activities of students, strategies of teaching, and outcomes that students should learn. The *Standards* provide this summary of inquiry:

> ...inquiry is a multifaceted activity that involves making observations; posing questions; examining books and other sources of information to see what is already known; planning investigations; reviewing what is already known in light of experimental evidence; using tools to gather, analyze, and interpret data; proposing the results. Inquiry requires identification of assumptions, use of critical and logical thinking, and considerations of alternative explanations. (p. 23)

The *Standards* use the term "inquiry" in two ways. Inquiry is content, which means both what students should <u>understand</u> about scientific inquiry and the <u>abilities</u> they should develop from their experiences with scientific inquiry. The term also refers to teaching strategies and the processes of learning associated with activities oriented to inquiry.

Here in summary are the standards on content in science as inquiry for grades nine through twelve:

TABLE 2. CONTENT STANDARD FOR SCIENCE AS INQUIRY

As a result of activities in grades 9-12, all students should develop

▶ Abilities necessary to do scientific inquiry.
▶ Understandings about scientific inquiry.

SCIENCE AS INQUIRY: THE ABILITIES

Table 3 presents the abilities students should attain. Note the emphasis on cognitive abilities and critical thinking. Without eliminating activities such as observing, inferring, and hypothesizing, this emphasis differentiates the *Standards* from traditional material that concentrates on processes.

TABLE 3. SCIENCE AS INQUIRY:
FUNDAMENTAL ABILITIES FOR GRADES 9-12

- Identify questions and concepts that guide scientific investigations.
- Design and conduct scientific investigations.
- Use technology and mathematics.
- Formulate and revise scientific explanations and models using logic and evidence.
- Recognize and analyze alternative explanations and models.
- Communicate and defend a scientific argument

Identify questions and concepts that guide scientific investigations. Students should formulate a testable hypothesis and demonstrate the logical connections between the scientific concepts guiding a hypothesis and the design of an experiment. They should demonstrate appropriate procedures, a knowledge base, and conceptual understanding of scientific investigations.

Design and conduct scientific investigations. Designing and conducting a scientific investigation requires introduction to the major concepts in the area being investigated, proper equipment, safety precautions, assistance with methodological problems, recommendations for use of technologies, clarification of ideas that guide the inquiry, and scientific knowledge obtained from sources other than the actual investigation. The investigation may also require students clarification of the question, method, controls, and variables; student organization and display of data; student revision of methods and explanations; and a public presentation of the results with a critical response from peers. Regardless of the scientific investigation performed, students must use evidence, apply logic, and construct an argument for their proposed explanations.

Use technology and mathematics to improve investigations and communications. A variety of technologies, such as hand tools, measuring instruments, and calculators, should be an integral component of scientific investigations. The use of computers for the collection, analysis, and display of data is also a part of this standard. Mathematics plays an essential role in all aspects of an inquiry. For example, measurement is used for posing questions, formulas are used for developing explanations, and charts and graphs are used for communicating results.

Formulate and revise scientific explanations and models using logic and evidence. Student inquiries should culminate in formulating an explanation or model. Models should be physical, conceptual, and mathematical. In the process of answering the questions, the students should engage in discussions and arguments that result in the revision of their explanations. These discussions should be based on scientific knowledge, the use of logic, and evidence from their investigation.

Recognize and analyze alternative explanations and models. The aspect of standard emphasizes the critical abilities of analyzing an argument by reviewing current scientific understanding, weighing the evidence, and examining the logic so as to decide which explanations and models are best.

Communicate and defend a scientific argument. Students in school science programs should develop the abilities associated with accurate and effective communication. These include writing and following procedures, expressing concepts, reviewing information, summarizing data, using language appropriately, developing diagrams and charts, explaining statistical analysis, speaking clearly and logically, constructing a reasoned argument, and responding appropriately to critical comments. (NRC, 1996, pp. 175-76)

SCIENCE AS INQUIRY: THE UNDERSTANDINGS

Table 4 summarizes the fundamental understandings that students should develop as a result of their science education.

TABLE 4. SCIENCE AS INQUIRY:
FUNDAMENTAL CONCEPTS FOR GRADES 9-12

- Conceptual principles and knowledge guide scientific inquiries.
- Scientists conduct investigations for a variety of reasons including discovering new aspects of the natural world, explaining recently observed phenomena, testing conclusions of prior investigations, and making predictions of current theories.

- Scientists rely on technology to enhance the gathering and manipulation of data.
- Mathematics is essential in scientific inquiry.
- Scientific explanations must adhere to criteria, such as logical consistency, rules of evidence open to questioning and based on historical and current knowledge.
- Results of scientific inquiry—new knowledge and methods—emerge from different types of investigations and public communications among scientists.

Conceptual principles and knowledge guide scientific inquiries. Scientists usually inquire about how physical, living, or designed systems function. Historical and current scientific knowledge influence the design and interpretation of investigations and the evaluation of proposed explanations made by other scientists.

Scientists conduct investigations for a wide variety of reasons. For example, they may wish to discover new aspects of the natural world, explain recently observed phenomena, or test the conclusions of prior investigations or the predictions of current theories.

Scientists rely on technology to enhance the gathering and manipulation of data. New techniques and tools provide new evidence to guide inquiry and new methods to gather data thereby contributing to the advance of science. The accuracy and precision of the data, and therefore the quality of the exploration, depends on the technology used.

Mathematics is essential in scientific inquiry. Mathematical tools and models guide and improve the posing of questions, gathering of data, constructing explanations, and communicating results.

Scientific explanations must adhere to criteria. A proposed explanation, for instance, must be logically consistent; it must abide by the rules of evidence; it must be open to questions and possible modification; and it must be based on historical and current scientific knowledge.

Results of scientific inquiry—new knowledge and methods—emerge from different types of investigations and public communication among scientists. In communicating and defending the results of scientific inquiry, arguments must be logical and demonstrate connections between natural phenomena, investigations, and the historical body of scientific knowledge. In addition, the methods and procedures that scientists have used to obtain evidence must be clearly reported to enhance opportunities for further investigation. (NRC, 1996, p. 176)

NATIONAL SCIENCE EDUCATION STANDARDS: INQUIRY AS TEACHING STRATEGIES

I turn to questions that emerge from the discussion of inquiry as content. How do science teachers help students attain the abilities and understanding described in the Science as Inquiry Standards? And what do the *Standards* say about teaching?

Science Teaching Standards

The Science Teaching Standards (see Table 5) provide a comprehensive perspective for science teachers who wish to provide students with the opportunities to experience science as inquiry. The *Standards* advocate the use of diverse teaching techniques:

Although the *Standards* emphasize inquiry, this should not be interpreted as recommending a single approach to science teaching. Teachers should use different strategies to develop the knowledge, understandings, and abilities described in the content standards. Conducting hands-on science activities does not guarantee inquiry, nor is reading about science incompatible with inquiry. Attaining the understanding and abilities described in [the prior section] cannot be achieved by any single teaching strategy or learning experience. (NRC, 1996, pp. 23-24)

TABLE 5. SCIENCE TEACHING STANDARDS

A. Teachers of science plan an inquiry-based science program for their students.

B. Teachers of science guide and facilitate learning.

C. Teachers of science engage in ongoing assessment of their teaching and of student learning.

D. Teachers of science design and manage learning environments that provide students with the time, space, and resources needed for learning science.

E. Teachers of science develop communities of science learners that reflect the intellectual rigor of scientific inquiry and the attitudes and social values conducive to science learning.

F. Teachers of science actively participate in the ongoing planning and development of the school science program.

What Should Science Teachers Know, Value, and Do?

Science teachers should know the differences between inquiry as a description of methods and processes that scientists use; inquiry as a set of cognitive abilities that students might develop; and inquiry as a constellation of teaching strategies that can facilitate learning about scientific inquiry, developing the abilities of inquiry, and understanding scientific concepts and principles.

In placing this discussion of teaching after the discussion of content, I have wanted to make the point that the desired outcomes—learning science as subject and science as inquiry—present the primary answer to the question, "What is teaching science as inquiry?" The very character of science as inquiry lodges in strategies for teaching inquiry.

A PRESENT-DAY PERSPECTIVE ON TEACHING SCIENCE AS INQUIRY

There is, in my view, a rich and thorough intellectual foundation for teaching science as inquiry. That foundation includes work by Bakker and Clark in 1988, Moore in 1993, Duschl in 1994, and in 1997 Hatton and Plouffe, and Mayr.

Constructing a New View of Inquiry

My use of the initial observations of classrooms and questions set the context for this essay. The questions based on those observations allowed you to think deeply about the observations and explore several issues associated with the theme of teaching science as inquiry. Returning to the observations and questions now provides an opportunity to separate inquiry as content and inquiry as teaching strategies and establish a perspective on teaching science as inquiry.

Question one probes the dominant perception of teaching science as inquiry. If your view was that inquiry is primarily activity directed by students, you probably answered A. If it was using laboratory experiences to teach the subject, you probably answered B. Few teachers answer C, for most do not view understanding scientific inquiry as a primary aim of school science. Those who responded D probably explained that some elements of all three classrooms contained inquiry.

Question two emphasizes the conception that most secondary teachers hold of inquiry: inquiry as technique or laboratory experiences for learning science concepts. The best answer is B. In classroom one, students had many opportunities to develop the abilities of inquiry; and students in classroom three developed an understanding of scientific inquiry. But neither of the two classes concentrated on the subjects of science: concepts of life, earth, and physical phenomena.

Question three was designed to probe the idea of inquiry as teaching strategy and engage your thinking about this as a singular approach to teaching science and the implied learning outcomes for students. If you used this approach all the time, what would students learn and what would they not learn? I suggest that the best answer is C. The primary assumption here is that classroom experiences of inquiry alone do not guarantee understanding subjects. Teachers should make explicit connections between the experiences and the content of inquiry and subject.

Question four asks for a generalization about the connection between teaching strategies and learning outcomes. I suggest that the best response is E because each of the others has some basis in practical truth.

In question five, the teacher could look at any of the responses or could look at all. Response E best anticipates a theme of this essay: that science teachers must have some understanding of scientific inquiry and a variety of teaching strategies and abilities to help students learn science subjects and the content of inquiry.

Question six organizes the reader's thinking to other sections of this chapter. The evaluation of my success and yours lies in E and especially D.

TOWARD A STANDARDS-BASED APPROACH
TO TEACHING SCIENCE AS INQUIRY

Most discussions of teaching science as inquiry begin with the assumption that inquiry is a teaching strategy. Science teachers ask, "Should I use full or partial inquiries? Should the approach be guided by the teacher or left to the student?" A standards-based perspective views the situation differently. Such a perspective begins with the educational outcomes—What is it we want students to learn?— and then identifies the best strategies to achieve the outcome. Table 6 provides examples of this perspective. Reading from left to right, the table asks these questions: What content do I wish students to learn? Which teaching techniques provide the best opportunities to accomplish that? What assessment strategies most align with the students' opportunities to learn and provide the best evidence of the degree to which they have done so?

TABLE 6. EXAMPLES OF TEACHING AND ASSESSMENT THAT SUPPORT INQUIRY-ORIENTED OUTCOMES

Standards-Based Educational Outcomes What should students learn?	Teaching Strategies What are the techniques that will provide opportunities for students to learn?	Assessment Strategies What assessments align with the educational outcomes and teaching strategies?
Understanding Subject Matter (e.g., Motions and Forces; Matter, Energy, and Organization in Living Systems; Energy in the Earth System)	Students engage in a series of guided or structured laboratory activities that include developing some abilities to do scientific inquiry but emphasize subject matter (e.g., laws of motion, F=ma, etc.)	Students are given measures that assess their understanding of subject matter. These may include performance assessment in the form of a laboratory investigation, open-response questions, interviews, and traditional multiple choice.
Developing Abilities Necessary to Do Scientific Inquiry (e.g., students formulate and revise scientific explanations and models using logic and evidence)	Students engage in guided or structured laboratory activities and form an explanation based on data. They present and defend their explanations using (1) scientific knowledge and (2) logic and evidence. The teacher emphasizes some inquiry abilities in the laboratory activities used for subject-matter outcomes.	Students perform a task in which they gather data and use that data as the basis for an explanation.

Developing Abilities Necessary to Do Scientific Inquiry (e.g., students have opportunities to develop all the fundamental abilities of the standard)	Students complete a full inquiry that originates with their questions about the natural world and culminates with a scientific explanation based on evidence. The teacher assists, guides, and coaches students.	Students do an inquiry without direction or coaching. The assessment rubric includes the complete list of fundamental abilities.
Developing Understandings about Scientific Inquiry (e.g., scientific explanations must adhere to criteria such as: a proposed explanation must be logically consistent; it must abide the rules of evidence; it must be open to question and possible modification; and it must be based on historical and current scientific knowledge)	The teacher could direct students to reflect on activities from several laboratory activities. Students also could read historical case studies of scientific inquiry (e.g., Darwin, Copernicus, Galileo, Lavoisier, Einstein). Discussion groups pursue questions about logic, evidence, skepticism, modification, and communication.	Students are given a brief account of a scientific discovery and asked to describe the place of logic, evidence, criticism, and modification.

In Table 6, I provide examples that answer questions about teaching science as inquiry. In developing the examples in this table, I tried to hold to a clear understanding of the realities of standards, schools, science teachers, and students. Science teachers must teach the basics of subjects. The content standards for physical, life, and earth and space sciences provide teachers with an excellent set of fundamental understandings that could form their educational outcomes. After identifying the educational results, teachers must consider the effective teaching strategies and recognize that we have a considerable research base for the concepts that students hold about many basic concepts of science. We also have some comprehension of the processes and strategies required to bring about conceptual change (Berkheimer & Anderson, 1989; Hewson, 1984; Hewson & Hewson, 1988; Gazzetti et al., 1993; King, 1994; Lott, 1983). The teaching strategies include a series of laboratory experiences that help students to confront current concepts and reconstruct them so they align with basic scientific concepts and principles such as those in the *Standards*. For teaching science as inquiry, a variety of educators have described methods compatible with standards-based approaches to teaching science as inquiry (American Chemical Society, 1997; Bingman, 1969; Connelly et al., 1977; Layman, Ochoa, & Heikkinen, 1996; Novak, 1963; Hofstein & Walberg, 1995).

Using investigations to learn subjects provides the first opportunities for students to develop the abilities necessary to do scientific inquiry. For teaching science concepts, a series of laboratories might encourage the use of technology and mathematics to improve investigations and communications; the formulation and revision of scientific explanations and models by use of logic and evidence; and the communication and defense of a scientific argument. But science teachers must decide for themselves the appropriate abilities and make them explicit in the course of the laboratory work.

A second educational outcome, very closely aligned with learning subjects, is developing abilities necessary to do scientific inquiry. Laboratories provide many opportunities to strengthen them. These outcomes were in the background of the discussion of subject matter; here they are in the foreground. Science teachers could indeed base the activity on content, such as motions and forces, energy in the earth's system, or the molecular basis of heredity, but they could make several of the fundamental abilities the explicit outcomes of instruction. Over time, students would have ample opportunities to develop all of them. This approach to teaching science as inquiry overlaps and complements the science teacher's effort to cultivate an understanding of science concepts. The teacher structures the series of laboratory activities and provides varying levels of direct guidance.

A further result also sharpens abilities necessary for scientific inquiry. But now students have opportunities to conduct a full inquiry, which they think of, design, complete, and report. They experience all of the fundamental abilities in a scientific inquiry appropriate to their stage of sophistication and their current understanding of science. The science teacher's role is to guide and coach. The classic example of this is the science fair project.

Finally, we come to the aspect of teaching science as inquiry that is most frequently overlooked. I refer to developing understandings about scientific inquiry. On the face of it, this seems like an educational outcome that would be easy to accomplish once the science teacher has decided to instruct by means of an activity or laboratory and has gained an understanding of inquiry. Numerous ways are available of having students identify, compare, synthesize, and reflect upon their various experiences founded in inquiry. Case studies from the history of science provide insights about the processes of scientific inquiry. Developing students' understanding of scientific inquiry is a long-term process that can be implemented with educational activities such as are mentioned here.

Questions of time, energy, reading difficulties, risks, expenses, and the burden of the subject need not be rationalizations for not teaching science as inquiry. Nurturing the abilities of inquiry is consistent with other stated goals for science teaching, for example, critical thinking; and it complements other school subjects, among them, problem solving in mathematics and design in technology. And understanding science as inquiry is a basic component of the history and nature of science itself.

CONCLUSION

Most evidence indicates that science teaching is not now, and never has been, in any significant way, centered in inquiry whether as content or as technique. Probably the closest the science education community came to teaching science as inquiry was during the 1960s and 1970s as we implemented the curriculum programs spurred by Sputnik and provided massive professional development experiences for teachers. The evidence does indicate that these programs were effective for the objectives related to inquiry that were emphasized in that era. Although science educators continue to chant the inquiry mantra, our science classrooms have not been transformed by the incantations.

The *Standards* have restated and provided details of what we mean by teaching science as inquiry. Appropriately viewed, inquiry as science content and inquiry as teaching strategies are two sides of a single coin. Teaching science as inquiry means providing students with diverse opportunities to develop the abilities and understandings of scientific inquiry while also learning the fundamental subjects of science. The teaching strategies that provide students those opportunities are found in varied activities, laboratory investigations, and inquiries initiated by students. Science teachers know this simple educational insight. It is now time to use the *Standards* and begin a new chapter where we act on what we know and teach science as inquiry.

REFERENCES

American Association for the Advancement of Science. 1993. *Benchmarks for science literacy*. New York: Oxford University Press.

American Chemical Society. 1997. *Chemistry in the national science education standards: A reader and resource manual for high school teachers*. Washington, DC: American Chemical Society.

Bakker, G., and C. Clark. 1988. *Explanation: An introduction to the philosophy of science*. Mountain View, CA: Mayfield.

Berkheimer, G., and C.W. Anderson. 1989. The matter and molecules project: Curriculum development based on conceptual change research. Paper presented at the Annual Meeting of the National Association for Research in Science Teaching, San Francisco.

Bingman, R.M. (Ed.) 1969. *Inquiry objectives in the teaching of biology,* position paper, Vol. 1, No. 1. Kansas City, MO: Mid-continent Regional Educational Laboratory and Biological Sciences Curriculum Study.

Bybee, R.W. 1997. *Achieving scientific literacy: From purposes to practice.* Portsmouth, NH: Heinemann Press.

Collins, A. 1995. National science education standards in the United States: A process and a product. *Studies in Science Education* 26:7-37.

Connelly, F. M., M. Finegold, J. Clipsham, and M.W. Wahlstrom. 1977. *Scientific enquiry and the teaching of science.* Toronto, Ontario: The Ontario Institute for Studies in Education.

Costenson, K., and A. Lawson. 1986. Why isn't inquiry used in more classrooms? *The American Biology Teacher* 48 (3):150-158.

Dewey, J. 1910. Science as subject matter and as method. *Science* 121-127.

Dewey, J. 1938. *Logic: The theory of inquiry.* New York: MacMillan.

Duschl, 1994. Research on the history and philosophy of science. In *Handbook of research on science teaching and learning* edited by D. H. Gabel, 443-465. New York: Macmillan.

Gazzetti, B. J., T.E. Snyder, G.V. Glass, and W.S. Gamas. 1993. Promoting conceptual change in science: A comparative meta analysis of instructional interventions from reading education and science education. *Reading Research Quarterly* 28(2):117-158.

Harms, N., and S. Kohl. 1980. *Project synthesis*. Final report submitted to the National Science Foundation. Boulder, CO: University of Colorado.

Harms, N. C., and R.E. Yager. (Eds.) 1981. *What research says to the science teacher*. Vol. 3. Washington, DC: National Science Teachers Association.

Hatton, J., and P. B. Plouffe. 1997. *Science and its ways of knowing*. Upper Saddle River, NJ: Prentice-Hall.

Hewson, M. G. A. 1984. The role of conceptual conflict in conceptual change and the design of science instruction. *Instructional Science* 13:1-13.

Hewson, P., and M.G.A. Hewson. 1988. An appropriate conception of teaching science: A view from studies of learning. *Science Education* 72(5):597-614.

Hofstein, A., and H.J. Walberg. 1995. Instructional strategies. In *Improving science education,* edited by B.J. Fraser and H.J. Walberg. Chicago: University of Chicago Press.

Hurd, P. D., R.W. Bybee, J.B. Kahle, and R. Yager 1980. Biology education in secondary schools of the United States. *The American Biology Teacher* 42(7): 388-410.

King, A. 1994. Guiding knowledge construction in the classroom: Effects of teaching children how to question and how to explain. *American Educational Research Journal* 31(2): 338-368.

Klapper, M. H. 1995. Beyond the scientific method. *The Science Teacher* 36-40.

Kuhn, T. S. 1970. *The structure of scientific revolutions,* rev. ed. Chicago: University of Chicago Press.

Layman, J. W., G. Ochoa, and H. Heikkinen. 1996. *Inquiry and learning: Realizing science standards in the classroom*. New York: College Entrance Examination Board.

Lott, G. W. 1983. The effect of inquiry teaching and advanced organizers upon student outcomes in science education. *Journal of Research in Science Teaching* 20(5): 434-438.

Mayr, E. 1997. *This is biology: The science of the living world*. Cambridge, MA: The Belknap Press of Harvard University Press.

Moore, J. A. 1993. *Science as a way of knowing: The foundations of modern biology*. Cambridge, MA: Harvard University Press.

National Research Council. 1996. *National science education standards*. Washington, DC: National Academy Press.

Novak, A. 1963. Scientific inquiry in the laboratory. *The American Biology Teacher* 342-346.

Rutherford, F. J. 1964. The role of inquiry in science teaching. *Journal of Research in Science Teaching* 2:80-84.

Rutherford, F.J., and A. Ahlgren. 1989. *Science for all Americans*. New York: Oxford University Press.

Schwab, J. J. 1958. The teaching of science as inquiry. *Bulletin of the Atomic Scientists* 14:374-379.

Schwab, J. J. 1960. Enquiry, the science teacher, and the educator. *The Science Teacher* 6-11.

Schwab, J. 1966. *The teaching of science*. Cambridge, MA: Harvard University Press.

Shymansky, J. A. 1984. BSCS programs: Just how effective were they? *The American Biology Teacher* 46(1):54-57.

Stedman, C. H. 1987. Fortuitous strategies on inquiry in the good ole days. *Science Education* 71(5): 657-665.

Storey, R. D., and J. Carter. 1992. Why the scientific method? *The Science Teacher* 18-21.

Welch, W. W., L.E. Klopfer, G.S. Aikenhead, and J.T. Robinson. 1981. The role of inquiry in science education: Analysis and recommendations. *Science Education* 65(1): 33-50.

Considering the Scientific Method of Inquiry

Fred N. Finley and M. Cecilia Pocoví

OVERVIEW

The purpose of this book is to encourage the teaching of scientific inquiry. In order to do so properly, we need to understand the nature of scientific inquiry and to reconsider some of the most common conceptions associated with what the phrase "scientific inquiry" means. To do otherwise is to run the risk of teaching ideas that are incorrect or misleading in the light of what is currently known. This chapter provides a number of considerations toward teaching a rich, interesting, and reasonably current view of scientific inquiry.

In many instances in science education, scientific inquiry is equated or nearly equated with the traditional notion of the scientific method. Yet much of the work in the history and philosophy of science has provided serious, well-grounded criticisms of the traditional scientific method. Many, if not most, philosophers and historians of science would argue that there is no singular scientific method that properly describes either how science does work or how it should. Others would go so far as to argue that there is virtually no meaning to the phrase "scientific method."

Because the traditional version of the method is still taught and modeled in science and science education courses while remaining dominant within science curricula at all levels, and, perhaps most importantly in the thinking of the general public, simply attacking the use of the traditional scientific method would almost certainly fail to provide much assistance to present teachers of science. The traditional idea of the scientific method is just too deeply entrenched in our culture to be replaced quickly or easily. For that reason, the approach that we have taken is to examine the conventional method in the light of developments in the history and philosophy of science and to suggest ideas for improving its use. This scheme cannot take full account of recent criticisms that challenge the very core of the traditional idea of a scientific method, but it offers some improvements in the conception of the method that we present to our students.

WHAT IS TAUGHT?

All science teachers try to present an accurate and coherent view of the science discipline they are teaching. In each case, we have our own particular idea of scientific inquiry, usually that which we were taught early through texts, lectures, textbook problems, and laboratory or field exercises. The most common view we learned is the traditional scientific method, a generalized, single sequence of several steps for solving problems.

The method is often introduced with paragraphs such as these:

> Much of the work done in biology is to solve problems. Problems are not solved by flipping a coin or taking a guess as to the outcome. Scientists use a series of steps called the scientific method to solve problems.
>
> Have you ever tried to turn on a light and found that it didn't work? Maybe you have turned the key in a car's ignition and the car didn't start. If you have had problems like those, you probably have used the scientific method.

The method itself goes something like this, taken nearly verbatim, as were the preceding passages, from high school biology textbooks.

1. Recognize and research the problem.
2. Form a hypothesis—a statement that can be tested.
3. Conduct an experiment in which you control variables to test the hypothesis.
4. Collect, organize, and analyze all relevant data.
5. Form your conclusions—which may lead to another hypothesis.

6. Present the theory…a hypothesis that has been tested again and again by many scientists with similar results each time.

The above method includes the ideas that science is objective and that conclusions are justified by formal logic and unbiased observations. The claim then emerges that the knowledge generated by the method is the truth about the natural world.

The method is usually presented in expository text and lectures explaining the steps. The presentation is occasionally reinforced by historical vignettes, such as a discussion of Torricelli's experimentation with the mercury barometer and Pasteur's experiments with generation. The method, either in whole or in parts, is also taught through the use of laboratory activities and formalized laboratory reports addressing mealworm behavior, perhaps, or the conditions necessary for plant growth, the conservation of matter, the behavior of gases or classification of various organisms. Typical activities would be collecting plants, insects, and rocks, measuring temperature, pressure, pH, current, mineral hardness, and barometric pressure, or classification of rocks and minerals, plants and animals, and elements. Finally, there is occasionally a disclaimer indicating that the steps might not always be used in this order or that some other way might be involved. The most commonly cited other way is probably "by accident" such as in the case of Roentgen's discovery of x-rays or in reverie such as when Kekule purportedly discovered the structure of benzene while gazing into a fire and seeing images of snakes forming rings by chasing their tails.

WHY TEACH THE TRADITIONAL VIEW OF SCIENTIFIC INQUIRY?

The Successes of the Scientific Method

The most obvious reason for continuing this view of the scientific method is that the method has a long history of achievement within Western culture. It was a successful competitor to the belief that all knowledge was revealed by God through chosen messengers. As Galileo and others found out, the idea that an individual could come to know the world through the use of the senses and reasoning was not popular with the church and governments of the age. Yet this way of thinking about ourselves led to intellectual, spiritual, political, and economic freedom. Not only new ways of coming to understand the natural world but new forms of viewing human nature, new systems of government, and new economic systems emerged. The development of the scientific method was one of the greatest intellectual revolutions in the history of the world.

The scientific method was also perceived as resulting in scientific laws and theories that made order out of chaos, rendered nature predictable, and promised individuals that they could utilize, manage, and control the environment for human benefit. The results of the use of the scientific method were seen to be the essential precursor to technological advances of the industrial revolution and with it the leisure time, the amenities, the economic security, and the military defenses of modern democracies. We greatly altered and improved our lives in many ways by using our belief in the scientific method.

The presentation of the scientific method provided a framework, some would say paradigm, for extraordinarily productive research programs. This was true for the sciences, and the philosophy of science as well. Within the sciences, innumerable inquiries relied on the belief that the observations and the logic of the method provide true results.

In the philosophy of science, the scientific method became the object of research itself. Investigators studied how this method improves the confidence of scientists in their claims about the natural world. More specifically studied was the logic of the scientific method—how it was that the method could be said to provide knowledge that was proven to be true or false. Along the way, the ways in which the method is fallible were also defined: when it would not resolve the question of whether a knowledge claim was logically true.

Complementary studies of the relationships between conceptual knowledge and the method, especially in the last half of the twentieth century, elaborated the understanding of how much of modern belief about the world depended on the traditional scientific method. Researchers studied whether observations are unbiased reflections of sensory impressions, whether the method could generate new and trustworthy laws and concepts, how the public and private lives of scientists influence their research, how the culture of a particular time does so, how intricately technology and the sciences are intertwined, and how many other factors influence modern beliefs about the natural world. All of this research was possible and necessary because the traditional scientific method was highly regarded in the sciences and many other areas of western culture, and because its relentless practice of skeptical examination and reexamination could even turn to questioning the method itself.

Because the development of the scientific method was directly and positively related to developing intellectual freedom, new forms of government, technological development, and economic and social history, the method became deeply engrained in modern culture. The scientific method therefore became a part of what we taught to our students.

The Scientific Method and "Best Thinking"

The public and educators came to perceive the scientific method as a highly valuable way of thinking in many (if not all) of the circumstances of daily life. John Dewey in 1916 and a host of others argued that the scientific method was the pinnacle of human thought. At the very least, educators have argued that the critical thinking skills that make up the scientific method are important in themselves.

There is evidence that faith in the traditional scientific method goes back at least as far as the late 1600s in children's didactic literature, in the writings of Locke and Rousseau as they became popularized at the end of the 1700s, and in the rationales late in the nineteenth century for developing mental capacities by the exercise of disciplined scientific thought. The nature study movement, the Progressive Education movement, curriculum development and teacher education programs of the National Science Foundation, and the more recent *Benchmarks for Science Literacy* of 1993 and *National Science Education Standards* of 1996 have insisted on the necessity for mastering the method.

Our profession has always sought something we could teach that would make problem solving easier and critical thinking more precise. We thus have been seeking a magical intellectual prescription—a way of thinking that is learnable in a finite time with limited resources, and applicable to multiple areas of our life. We have good social, political, and economic reasons to believe that the scientific method can be that prescription and there has been extensive additional support from various philosophical and psychological theories. The method appears to be simple to understand and simple to teach. There are a few steps to learn, they can be used over and over again, and each step in its own right seems to be achievable by students at various ages and levels of knowledge and skill.

The Scientific Method, Psychology, and Science Education

The status and promise of the scientific method guaranteed that it would be part of what is taught in science classes. But there is another reason. The traditional method was also closely related to the psychologies that have declared what our students should learn and how they should be taught.

The behaviorism that J. B. Watson proposed in 1924, E. L. Thorndyke in 1932, and B. F. Skinner in 1971 dominated the profession's view of human learning from the early 1900s up through the early 1970s. Behaviorism was predicated on a number of ideas that were consistent with the traditional view of the scientific method. It embodied deep commitment to empiricism. In behaviorism, this translates into the idea that people need only to depend upon direct observations of overt behavior in association with direct observations of environmental stimuli: that to understand learning, we only need to examine stimuli, responses, and patterns of explicit rewards. The behaviorists were deeply committed as well to formal, logical experimentation for testing hypotheses. To establish laws of learning, they used their own experimentation (often using animals as substitutes for human beings). Thorndyke's law of effect, which locates motivation in the arousal and satisfaction of primitive needs and desires, is one of the most prominent examples. Behaviorists, in sum, strove to be objective and scientific by emulating their colleagues in the natural sciences.

Since both the psychology of learning and the philosophy of science during the first sixty or seventy years of the century shared the most central beliefs about how people have come to learn about the natural and human world, it was inevitable that science education would become committed to the scientific method and behaviorism alike. After the Second World War, many educators utilized behaviorist theories to help define what students should learn—by stating behavioral objectives; how the objectives should be sequenced—in learning hierarchies; the conditions under which the students would learn—according to the laws of rewards in behavior modification; and how they should be tested—objectively and strictly by reference to the behaviors stated in the objectives. Even in the face of significant challenges from cognitive psychology beginning in the late 1960s, behaviorism has remained prominent (yet sometimes camouflaged) in many of our ways of thinking about science curriculum, instruction, and assessment.

NEW CONSIDERATIONS OF THE SCIENTIFIC METHOD

The most recent analyses of the scientific method, however, as well as various aspects of the studies that have occurred since the publication of Bacon's *Novum Organum* in 1620, have provided significant challenges to the traditional formulation. Understanding what has been learned in recent years about the scientific method provides a much richer and deeper understanding of the traditional scientific method. This more complete understanding is what our students need to learn.

Objectivity of the Scientific Method

The scientific method was originally and continues to be put forward as unbiased by prejudices and preconceptions, dependent only on the observations from nature. These observations were thought to be directly related to unambiguous sensory impressions—what people see, hear, taste, smell, and feel. The method was considered to be free of any particular theory, and in fact was expected to generate correct theory. The idea that theory follows from the method and is uninfluenced by anything other than the observations (and logic) is no longer defensible from either a philosophical or a psychological perspective. It's now known that at any level beyond the fundamental physiological responses, people always have preconceptions that filter what they sense: that is, their observations.

This is the case at even the basic level. Different people viewing the same sheet of colored paper may report quite different observations. One sees blue, another green, and another aquamarine. While this difference in observation may seem trivial, in the classification of various natural objects it can become quite important. In the identification of rock types in thin section under a microscope, one person sees beautifully colored crystalline patterns; another sees a specific type of igneous rock. A geologist traveling across the landscape is likely to see layered sedimentary beds that she interprets as a simple onlap-offlap sequence resulting from successive transgressions and regressions of continental seas. Her traveling companion may see nothing but rocks in layers. But being a biologist, he may catch dramatic differences between flora and fauna in ravines and those on the plateau and interpret those as the result of differences in local climatic and soil conditions between the two locations. Both people literally see the world by means of what they already know and believe about the natural world.

Preconceptions or a theory of what is in the world, how it works, how it looks, feels, smells, tastes, and sounds, is a major determinant of what people observe and how they interpret those observations. These theories are in large part socially constructed from formal schooling—plate tectonics, Darwinian evolution, kinetic molecular theory; from the cultural and linguistic notions of a particular society—"Close the door: You're letting the heat out and the cold in"; from what features of a phenomenon the interpreter has or has not encountered directly— you see moving objects slowing down, not traveling on and on forever in a straight line until some force acts on them as physicists say they do, and you do not regularly observe and notice the conservation of mass.

In each step of the traditional scientific method, prior knowledge and belief condition observations, and this directly confronts the very foundation of that traditional view. As F. Suppe summarized in 1977 the research of Russell Norwood Hanson, "one's scientific view of the world is theory laden, viewed through a conceptual pattern. Part of this view is a function of the meaning one attaches to terms within a context; part of it is a function of the law-like generalizations, hypotheses, and methodological presuppositions that one holds in context." (p.163)

Recommendation. Students should be taught that whatever aspect of the scientific method is being considered, it is influenced by their initial conceptions or theories of the natural world—what is there, how it works, how it can be observed, and how results can be interpreted. In addition, they should learn that their observations are dependent on the ideas they use while making them, that observations alone are not infallible determinants of the validity of a scientific law, and that the ideas we ask them to learn are essential to their being able to understand the natural world on their own.

Recognize and Research the Problem

Recognizing a problem must begin with an understanding of what defines something as problematic. L. Lauden in 1977 identified two major types of problems, empirical and conceptual. When problems are proposed in science classrooms, they are almost always empirical. Conceptual problems are rarely if ever considered.

What is commonly referred to as an empirical problem exists when someone expects a certain observation such as particular yield from a chemical experiment and obtains something else. The anomalous difference between what initial theory predicts and what is observed constitutes the problem. People expect to see that a deck of cards contains red diamonds and red hearts and black spades and black clubs. Only in relation to this standard expectation do they see a red spade or a black diamond as a problem. If they expected that the colors would be red spades and black diamonds, they would see no problem at all. The history of science is replete with examples that fit this form. Roentgen's discovery of x-rays occurred because he recognized that a barium platinocyanide screen glowed while he was working with cathode rays when he thought it should not have done so. Lavoisier's discovery of oxygen as a distinct species was possible only because he expected that the phlogiston theory was flawed and was thus predisposed to see anomalies in his experiments (Kuhn, 1970).

A conceptual problem has to do with the inadequacy of a theory on grounds other than its inability to account for observations. Among the differing sorts of conceptual problems is the discovery that a theory is self-contradictory. Theories can also fail to provide clearly the explanation that they seem to promise. Faraday's early model of electrical interaction did not eliminate the notion of actions at a distance as promised. Theories can be shown to be circular. Among them was the early kinetic molecular theory that postulated that gases are elastic because molecules are elastic.

Perhaps more important are conceptual problems that occur when one theory conflicts with another while both account for the relevant observations. In some cases, accepting one theory means the other cannot be accepted. Ptolemy's astronomy solved the problems of retrograde motion that had plagued the Greek astronomers but violated the idea that the motion of planets is perfect, that is, circular about the earth at a constant speed. In other cases, the problem is that two theories seem to account for a particular phenomenon nearly equally well. The behavior of light has been seen as explainable by references to either particles or waves. Other conflicts between theories occur when one makes the other unlikely (implausibility) or one seems to have offered no support or contradiction for another when it seems that being about the same class of phenomena they should be related.

Recommendation. Students should be taught that what counts as a scientific problem depends upon what theories or initial ideas about the natural world are being used; and that there are two primary types of scientific problems—empirical and theoretical. The study of both types of problems and their resolution needs to be a significant part of the science curriculum.

Form a Hypothesis — A Statement That Can be Tested

A primary consideration regarding the formulation and testing of hypotheses is that not all scientific problems can be resolved by observational testing of a hypothesis. Some scientific problems are conceptual in nature. These must be solved by the creation of new ideas. Every scientific field of study has examples of problems that were solved by a new idea and not by observations alone. Copernicus' placing the sun at the center of the universe in place of the earth is one of the most widely known. Wegner's theory of continental drift is another.

It is also necessary to remember that hypotheses are the result of the researcher's theory of the phenomena and are restrained by it. Scientists and people in general have some more or less well-formulated set of beliefs about the phenomena they encounter. Without them, thought would be impossible. The formulation of a hypothesis that actually will improve an understanding of the phenomena under study is a highly creative intellectual act. That creation is based on the quality of the researchers' initial theory, their knowledge of the methods available for testing the hypothesis, and their theory of what observations will be relevant. The formulation of the hypothesis about how some natural phenomenon works—that is, predicting what will happen if—is a major part of the fun and excitement of scientific inquiry, but the inquirer must be conceptually prepared to do so.

Recommendation. Students should learn that scientific problems can be solved by observations and by the creation of new theories.

Conduct an Experiment That Will Test the Hypothesis

Some scientists, philosophers of science, and science educators have developed the idea that the experiment, if not the one and only way of learning the truth about the natural world, is at least the primary and most certain way. E. Mayr proposed in 1997 that this outdated perception probably results from the "rigorous experiment" that was the primary tool of the physics of mechanics during the early part of the scientific revolution. However this has come about, it is absurd and contradicted by the methods that are in use in many fields of study.

Collecting and cataloging the natural objects and processes of the world have been considered scientific study for millennia. Biologists, geologists, astronomers, and chemists have been concerned with the question of what objects are there in this world and universe, their relationships to one another in time and space, and how humankind could organize its thinking about them. It is not too far from the truth to say that all sciences are founded on collecting and classifying. The collecting, moreover, has gone on in a myriad of ways that are not experimental. People have watched the sky using everything from the naked eye to arrays of radio telescopes. Geologists have mapped the earth beginning with little more than their legs, compasses, and notebooks and now use seismographs, satellite imagery, sonar, and radar. Modern chemistry was preceded by many years of alchemy during which a great deal was learned about the substances that made up the matter of the earth, air, and water. The description of phenomena has been the cornerstone of all sciences and therefore cannot be excluded from it.

In many circumstances, experimentation is impossible. Geologists do not formulate experiments as to the effects different earthquake intensities have on various geological features. The people living in the areas affected would object. Astronomers do not control and manage extraterrestrial events.

Ecologists seldom systematically poison ecosystems or remove keystone species such as the alligators in the Florida Everglades from an area to improve their ability to make predictions. In some cases, the development of physical and mathematical models and their testing are substituted. In others such as in much of astronomy, geology, paleoecology, paleontology, and climatology, studying the records of the past is the basis of predictions about the future. None of the techniques used in these inquiries are direct experimental tests, but they are certainly legitimate and essential ways of conducting scientific inquiry.

Presenting the experiment as the primary feature of scientific inquiry would surprise many great scientists in history. Most of the greatest scientific accomplishments have not been the "discovery" of some new idea by experimentation. The theories have not been discovered in the real world, but have been created by the human mind as ways of considering, observing, and accounting for experiences there. If experiments were at the core of science, then Copernicus, Lyell, and Darwin could not be called scientists.

A truly controlled experiment would have to accomplish the insurmountable task of determining all of the variables that may influence the results. What investigators actually do is to control for the variables that they have other reasons to believe do make a difference. Innumerable possible variables are unknown. Roentgen's discovery of x-rays is a good example. The discovery of x-rays was widely attacked because many, many experiments with cathode-ray tubes had to be reconsidered so as to account for the possible effect of this previously unknown energy source.

Nor are empirical problems always resolved by experimentation. Changes in theory can render them irrelevant. The Michaelson-Morley experiments were planned to determine the drag coefficients of bodies moving through an electromagnetic aether. At the advent of special relativity theory, all questions about the elasticity, density, and velocity disappeared.

Recommendation. Students should learn that there are more methods of inquiry than experimentation, each with its own demands for rigor and justification of the claims that are made about the results; limitations are substantial on what experiments can be conducted; and limits exist on the extent to which experimentation alone can solve scientific problems.

Collect, Organize, and Analyze All Relevant Data

Theories provide the ideas that determine what observations should be made. A high energy physicist dropped in the middle of a human genome project and asked to collect the relevant observations would probably be clueless as to what observations were even possible, let alone which ones would be relevant to the problem of unraveling the human genome. The physicist would be theoretically unprepared to participate. Even inquiry into relatively simple phenomena such as the thermal expansion of metal tubing depends on existing theory to tell what to observe. What is to be observed: the atmospheric pressure in the room, the time of day and year, the effects of noise in the laboratory, the length of the tube, the type of metal, the wall thickness, the original temperature of the tube, the type of heat source, the temperature of the heat source, where the heat source is applied to the tube, or how long the heat source is applied? Without the use of existing theories and the "laws" they encompass, then it might be necessary to consider each of these variables. If the theory is very fully developed, then only a few of these variables enter into the investigation.

Just as it is impossible to know whether all relevant variables have been controlled, it is logically impossible to tell whether all the necessary observations have been made. The observations that investigators chose to make are determined by their theory of what is relevant to the problem and no theory is or can be so complete that it tells all of what should be observed. Many times in the history of science, researchers have missed an observation when their theory did not cue them to its importance to understanding the phenomenon.

Implicit in the phrase "collect the data," moreover are a whole set of assumptions based in theory about what methods of data collection should be used and under what circumstances. Measuring temperature is a good example. It is kinetic molecular theory that provides an understanding of what has been observed indirectly—the average kinetic energy of the molecules in the system. Guidelines driven by theory define the precision required of the instrument chosen. In some cases the temperature must be knowable to a very small fraction of a degree and in other cases anything within a few hundred degrees will suffice. The same is true with respect to visual observations. Sometimes a glance at an object will be adequate for observing what is required. At other times the need is for a hand lens, an optical microscope or an electron microscope. The decision of which observational instrument to choose is dependent on what the initial theory declares to be required.

Organizing and analyzing data cannot be done in an absolutely objective way: in a way that is unbiased by previously held ideas. A student of igneous rocks who thinks grain size and the proportions of light and dark minerals are important may choose to classify the sample according to these characteristics. But a belief that mineralogical relationships may reveal what is sought will make for a different classification and conclusions. The researcher may even use the proportions of various elements and compounds. In plant and animal taxonomy, theory may determine whether comparisons should be morphological or by the genetic materials of organisms.

Similar arguments can be made with respect to the method of organizing the data that are collected. Simple tabulations, sketches, and drawings organized into sequences, maps, the construction of physical or mathematical models, and statistical testing are all possible types of data analysis. The proper methods of analysis are determined by the theory in use.

Recommendation. Students should learn that the decisions about what data to collect, how the data are to be collected, how much data are collected, and how they are analyzed and interpreted are all dependent on theory.

Form Your Conclusions—Which May Lead to Another Hypothesis

A component of the traditional concept of the scientific method is that deductive logic should prove or falsify a hypothesis. Philosophical studies have thrown doubt on this idea. The application of formal deductive logic cannot be used to "prove" a statement. This is well known but still presented to students. No matter how many confirming instances of a hypothesis are found, the possibility that the next instance will be disconfirming cannot be eliminated. Given the hypothesis that all ravens are black, there is no logical guarantee that the next raven will not be snow white. The law that all ravens are black therefore cannot be proven. Neither can a particular hypothesis be disproved. The difficulty is not in the relationship between evidence and the hypothesis. In truly axiomatic systems like mathematics and symbolic logic an outcome that is contrary to the one that is predicted disproves the hypothesis. The problem is that it is impossible to isolate a single hypothesis. Any hypothesis is intertwined with a complex of assumptions and ideas that constitute a theory. All that contradictory evidence can show is that there is something wrong someplace in the theoretical system in which the hypothesis is embedded. In a geologic study involving the use of x-ray diffraction a number of years ago, the results of the x-ray diffraction measurements were consistently and repeatedly contrary to what had been expected. After many days and nights of work that included examining alternative theoretical perspectives and new explanations, expected results were found but only in the early morning hours. The problem was in the x-ray diffraction equipment, which was responding to power surges generated by daytime demands. The evidence actually contradicted an assumption about the equipment and not the hypothesis under test. The expected data were evident only when most people were asleep and the demands on the electrical system were limited.

How to conduct scientific inquiry in a way that guarantees truth is not clear. Limitations on human perception and thought and ambiguities in data and instrumentation keep the process from being entirely logical. That does not mean that it is irrational. The decisions that are made in formulating conclusions are a matter of professional judgment in the light of all of what the researcher and the researcher's associated community of scholars know. The judgment is rational in being supported by complexes of reasons, some of them empirical and logical, others theoretical. Some are even determined by the culture in which the conclusions are drawn.

Recommendation. Students should learn that scientific truth is not absolute, but represents the best of collective thinking about natural phenomena when a full range of reasons are employed for understanding them—existing theories, competing theories, logic with all its limitations, past observations, new observations, and complexes of beliefs that too often are incorrectly considered as beyond the domain of science.

Present the Theory—A Hypothesis That Has Been Tested Again and Again by Many Scientists with Similar Results Each Time

This is perhaps one of the most misunderstood aspects of the scientific method. This last step of the method is presented as if everyone will somehow automatically accept the new theory just because a scientist or group of scientists claim to have followed the scientific method. Nothing could be further from the reality of scientific inquiry. It takes less than a few moments of reading any research journal or any newspaper to see challenges to new scientific ideas that range from gentle criticism to vicious personal attacks.

The history of science is replete with such confrontations. Scientists are challenged by claims to alternative theories, contradictory data, better methods, better analyses, and better reasoning. The scientists are challenged by assertions that the data are irrelevant and the methods invalid or poorly used. There are criticisms based on religious, social, political, and economic beliefs as well as criticisms of the researcher's scientific status, gender, morals, ethics, academic background, innate intelligence, and probably parentage. Charges of fraud and the theft of data, methods, and ideas abound. Many of the challenges are made as a way of investigating the validity of new ideas. Many are made to protect the competing ideas in which many practicing scientists have invested their lives, fortunes, and futures. Others are related to deeply personal animosities. In any case, few if any scientific theories are accepted simply because "a hypothesis…has been tested again and again by many scientists with similar results each time."

Recommendation. Students should learn that the presentation of a new idea is not the last step. In fact, it is usually the beginning of a long and often arduous sequence of discussions, arguments, replications, new investigations, and modifications of the new idea. This whole complex of events is demanded by the scientists' scientific community and often by many others from the larger society as well. Politicians, religious figures, special interest groups, the media, and the general public often become engaged. Observations, logic, proper methodology, and experimental replications may not in themselves put an end to a question.

A Last Consideration

Almost enough has been said about what is now known about the traditional scientific method. There is, however, one consideration that has not have been emphasized enough. Real people, with all their scientific knowledge, attitudes and biases, social and personal relationships, political, religious, and social beliefs, values, morals, and ethics, and limitations, conduct scientific inquiries. And because science is a human endeavor, no aspect of the scientific method is or can be made immune to being human. So whatever version or features of the scientific method are taught must account for the people in this fascinating, unique, powerful, and engaging enterprise. Our students need to meet the people who have investigated the natural world, learn about their theories, the associated problems, methods, observations, arguments, influences, and reasons for the claims they made about the natural world. It is at least as important that students learn something of the times and circumstances in which the inquiries were done. With this as background and consideration of how the social and cultural context influences every aspect of science, perhaps students will see and in some sense experience the excitement and satisfaction of following a curiosity about the natural world. There is a myriad of personal, social, and cultural factors that are ALWAYS critical.

REFERENCES

American Association for the Advancement of Science. 1993. *Benchmarks for science literacy*. New York: Oxford University Press.

Dewey, J. 1916. Method in science teaching. *General Science Quarterly* 1:3-9.

Kuhn, T. S. 1970. *The structure of scientific revolutions*. Chicago: University of Chicago Press.

Lauden, L. 1977. *Progress and problems: Towards a theory of scientific growth*. Berkeley, CA: University of California Press.

Mayr, E. 1997. *This is biology: The science of the living world*. Cambridge, MA: Harvard University Press.

National Research Council. 1996. *National science education standards*. Washington, DC: National Academy Press.

Skinner, B. F. 1971. *Beyond freedom and dignity*. New York: Alfred A. Knopf.

Suppe, F. 1977. *The structure of scientific theories*. Urbana, IL: University of Illinois Press.

Thorndyke, E. L. 1932. *The fundamentals of learning*. New York: Teachers College Press .

Watson, J. B. 1924. *Behaviorism*. New York: Norton.

Part 2

What Does Inquiry Look Like?

Science As Argument and Explanation: Exploring Concepts of Sound in Third Grade

Sandra K. Abell, Gail Anderson, and Janice Chezem

It is part of the educator's responsibility to see equally to two things: First, that the problem grows out of the conditions of the experience being had in the present, and that it is within the range of the capacity of students; and, secondly, that it is such that it arouses in the learner an active quest for information and for production of new ideas. (Dewey, 1938/1963, p. 79)

INTRODUCTION

The active quest for information and for production of new ideas characterizes inquiry-based science classrooms. Many elementary school science classrooms have moved beyond a didactic orientation where they present science content and test for understanding through recall questions (Anderson & Smith, 1987). They have adopted science curriculum materials that engage students in first-hand experiences with phenomena. However, these activity-driven approaches, as Anderson and Smith observed in 1987, typically involve students in activities but offer them few opportunities to develop conceptual understandings. The *National Science Education Standards* published by the National

Research Council (NRC) in 1996 assert that teachers must promote inquiry in science classrooms. In particular the *Standards* challenge teachers to provide less emphasis on "science as exploration and experiment" and more emphasis on "science as argument and explanation" (p. 113).

What does an inquiry-based classroom look like?   What are reasonable expectations for such classrooms?

J. J. Schwab addressed these issues in 1962:

> With classroom materials converted from a rhetoric of conclusions to an exhibition of the course of enquiry, conclusions alone will no longer be the major component. Instead, we will deal with units which consist of the statement of a scientific problem, a view of the data needed for its solution, an account of the interpretation of these data, and a statement of the conclusions forged by the interpretation. Such units as these will convey the wanted meta-lesson about the nature of enquiry. (pp. 52-53)

Schwab described a process of classroom inquiry that includes finding problems, collecting and interpreting data, and forging conclusions. This sounds very much like "science as exploration and experiment." Schwab's description fails to provide an adequate portrayal of inquiry in classrooms devoted to "science as argument and explanation." The purpose of this chapter is to provide stories from real classrooms that illustrate inquiry-based instruction emphasizing argument and explanation so that we can more completely understand what might be reasonable expectations for such classrooms.

SETTING THE CONTEXT

These stories take place in two different third-grade classrooms in two elementary schools not far from a large research university. In each classroom the teaching was shared by the classroom teacher, either Anderson or Chezem, and a university teacher educator, Abell, who had been released from her university teaching responsibilities to share the work of teaching elementary science in these two schools. Shared teaching meant that we—Chezem and Abell, or Anderson and Abell—collaborated over the course of several units of instruction in the planning and enactment of science instruction, including assessing students and reflecting on our teaching. (Other examples of shared science teaching can be found in Abell and Roth [1995] and Abell et al., [1996]).

The stories of our science teaching occurred during two separate teaching events, the first in the spring at Chezem's school, followed by another in the fall at Anderson's. Each teaching team had independently decided to concentrate on the topic of sound for a third-grade unit of science instruction. We deemed this topic age appropriate and offering many opportunities for inquiry. We planned to develop a community of inquiry in these classes by involving students in first-hand experiences supported by "scientists meetings" (Reardon, 1993), where teachers would help students think through, share, and compare their science ideas.

As we started planning our units on sound, we consulted the *Benchmarks for Science Literacy,* issued in 1993 by the American Association for the Advancement of Science (AAAS), and the *National Science Education Standards* (NRC, 1996). These documents declare that:

By the end of the second grade, students should know that:

▶ Things that make sound vibrate. (AAAS, 1993, p. 89)

As a result of the activities in grades K-4, all students should develop an understanding of:

▶ Position and motion of objects
▶ Sound is produced by vibrating objects. (NRC, 1996, pp. 123, 127)

Thus in both classes we began our inquiry into sound by examining the concept of vibration. The reform documents make teaching the concept seem quite clear cut; the chapter on "The Research Base" in the *Benchmarks* does not mention problems students might have in understanding it. The *Standards* did add one caveat: "Sounds are not intuitively associated with the characteristics of their source by younger K-4 students, but that association can be developed by investigating a variety of concrete phenomena toward the end of the K-4 level" (NRC, 1996, p. 126). Thus we proceeded on the assumption that exposing students to many sound phenomena would contribute to their conceptual understanding of sound and vibration.

Next came examining published science curricula that address the topic of sound. Many of the science curricula we examined seemed to agree that the relationship between sound and vibration could be developed by investigating a variety of phenomena. In the "Sounds" unit in *Science & Technology for Children,* a curriculum published by the National Science Resources Center in 1991, students explore vibrating rulers and vocal cords. The unstated assumption is that they will see a connection between sound and vibration. In the *Full Option Science System* "Physics of Sound" module developed by the Lawrence Hall of Science (1992), students observe sound originating from a variety of vibrating sources, including tuning forks, string phones, water bottles, and xylophones. In the *Insights* "Sound" module published by the Education Development Center in 1991, students are taken through a learning experience in which they explore vibrations with drums, tuning forks, and rubber bands. Accompanying the *Insights* investigations are suggestions for classroom discussions. In the culminating discussion the teacher is directed to "Explain that some vibrations are so small and/or fast that they can't be seen or felt" (p. 109). This was the only indication in any of the curriculum materials examined that the concept of vibrations might be a problem for students. The stories told here unfold against this backdrop of our preparations to teach the sound unit.

STORY #1: OPENING THE CLASSROOM TO ARGUMENT AND EXPLANATION

We (Chezem and Abell) introduced our unit on sound with a large group activity on feeling various parts of your body—lips, nose, throat—while you vocalize. Whereas some students used words like "moving" and "tickling" to describe the sensation, one student, Tyler, used the words "vibration" and "wave" in his answers. Next, students interacted at several exploratory stations at which they were presented with vibration phenomena: rubber bands, rulers, cans covered with balloons, and dancing rice. The rice proved most interesting to the students. At the station we placed rice on plastic wrap affixed to the top of a plastic bowl. When students banged the bottom of a metal pan, open end facing the bowl, the rice danced. During the scientists meeting that followed, Cindy explained that the sound traveled to the bowl and made the rice jump. We were surprised to hear this sophisticated reason so early in the lesson and wondered how other students were interpreting their observations.

The next lesson engaged students in explorations with tuning forks: "Try touching the tuning fork to your set of materials and see what happens." Students were excited when they placed a vibrating tuning fork on a plastic cup, water, and a ping pong ball. They observed the cup buzzing, the water jumping, and the ping pong ball bouncing. At the end of the lesson students came together to share their observations, which we summarized on a class chart.

Up to this point in our instruction, we had emphasized science as exploration. Students had explored various phenomena and we had summarized their observations. We knew we must now engage students in discussions in which they would invent explanations to account for their observations. Where such a discussion would lead us, we could only guess. At our next class meeting we synthesized student observations and asked for explanations: "Why do you think that happened?" we asked. John fixed his explanation on the vibration of the tuning fork and the transfer of the vibration to another object. When we asked students to find other instances of this transfer phenomenon, we noticed they could not give examples other than the observations from the tuning fork explorations.

To give students an opportunity to represent their ideas in another way, we instructed them, "Draw a picture of how something makes a sound and how you hear it." Several groups drew a textbook-looking "wave" picture to represent sound between the sound maker and the listener. We wondered where that idea came from. Cindy and Bobby drew a picture of two children talking on a string telephone. They drew a jagged line to represent the sound across the string. When asked what was going on in their picture, Cindy replied that the sound was moving along the string. Bobby corrected her, saying that it was the vibration in the string. Two other groups drew stereos with sound "lines" coming out of them. When John's group was probed about the lines, this conversation ensued:

Teacher: What are those lines?
John: Well, in my stereo there is a laser that plays CDs.

Teacher: What do you think is happening in the speaker? Have you ever seen the inside of a speaker?
John: There are lots of wires.

Teacher: Anything else?
John: A big round thing.

Teacher: What happens to that round thing when sound comes out of the speaker?
John: It moves in and out.

Later, during scientists meeting, John described how speakers move when sounds are being made. He embellished the description with a story about his uncle's truck, where the speakers do not have covers. According to John, not only do the speakers move when music is played, but they move more if the volume is turned up.

After explorations of sound traveling through different media—air, water, string, and wood—we asked students to represent their ideas in another way. We gave each team this assignment:

> Your job as a team is to write three sentences about sound that you can agree about. You can base your sentences on any of the investigations we have tried. Each team member will write one sentence. The others will help.

Their sentences mentioned vibrations, sound traveling, and hearing sounds. Some sentences referred to activities we had done in class, others to everyday life:

- Sound can travel through almost anything.
- The catcher [pinna] catches everything you hear such as a dog barking.
- If you put your ear against the ground you can hear vibration.
- Cars make different sounds.
- Sound doesn't travel very good in air.
- Sound is traveling through the air all the time.
- If you put your hand on your throat and talk you can feel vibration.
- Sound travels better through wood and [string] telephone than air.
- It vibrates when it goes through something.
- Sound can be loud or soft.
- Sound travels and vibrates when you sing or talk.
- If you put something close to your ear you could hear it very good.
- It's like a vibration.
- When you touch your Adam's apple and talk you can feel it vibrating.
- When something hits something it makes a sound or it vibrates and makes a sound.
- When you talk the sound vibrates and goes to your ears.

We consolidated the sentences into a set of ten statements and displayed them on a large chart. Our plan was to ask students to agree or disagree with each statement and give evidence for their thinking. We thought that everyone could readily agree upon the first statement: "Sound is made when a thing vibrates." After all, most of our explorations and interactions had demonstrated just this idea. What a surprise when a number of students disagreed with the statement! We asked for their evidence. They began to present what to them were disconfirming cases. Mandy said, "If you stomped on the ground, you would hear a sound but the ground would not vibrate." We asked Mandy and a few others to pretend they were buffalo scouts, ears to the floor, to find out whether they could feel the stampeding buffalo feet of the rest of us. They did, but were not convinced. Mark offered another example. "What about if two cars crashed? There would be a loud sound, but the cars would not vibrate." From their own experiences with car accidents, several students offered evidence contrary to Mark's statement. Mark seemed unconvinced. As teachers we left class that day wondering how the lesson had gotten so far off track. Or had it?

Reflection

In this story, we as teachers started from a stance of science as exploration and experiment. That is, we began by having students explore a series of vibration phenomena. Tyler's answer concerning vibrations in the first activity of the unit, and our early examination of curriculum materials and standards documents, brought us to assume that getting to the concept would be an easy journey. But when we at last opened the classroom to argument and explanation, we found out some things about the students' thinking that surprised us. We learned that students did not all readily agree that "Sound is produced by vibrating objects." Their collective theory more likely resembled this variation: "Some sounds are produced by vibrating objects."

Our original orientation to science as exploration and experiment did not help students grasp the concept we were addressing in our instruction, nor did it reveal to us anything about student understanding. When we opened the classroom to argument and explanation, things changed. We found that students had constructed a diversity of ideas about sound. Some had separated science class phenomena like tuning forks from real-world phenomena like stomping feet or crashing cars. By the end of the sound unit, many students still held two theories for sound, one for class phenomena and one for events in the real world. We also found out that, even when their theories did not match the scientifically accepted one, the third graders were capable of developing explanations that fit the evidence, of finding discrepant data, and of arguing for and against certain theories. What we learned would prove fruitful the next time we attempted teaching a unit about sound.

STORY #2: ASKING STUDENTS TO EXPLAIN
AND CHOOSE THEORIES

The following fall another teaching team (Anderson and Abell) in a different school used these new understandings to design and enact a unit on sound with another third-grade class. This time we wanted to begin the unit by bringing the real world into the science class, trying to avoid the dichotomy between real world and school science observed in the spring. We conducted a brainstorming session in which students created a class list of things that make sounds. Their list included barking dogs, crying babies, computers, CD players, and many more. We again wanted to provide experiences with sounds and vibrations that would lead to opportunities for argumentation and explanation. Thus in the second lesson, we engaged students in activities with rubber bands, drums, tuning forks and so forth. At the end of the session, we asked students to write a rule for sound based on their observations. Every student's rule mentioned vibration: "The sounds are made by vibrations"; "When someone or something makes a sound, it vibrates"; "Things vibrate."

The third lesson began with the request, "Think of the most unusual tuning fork experiment you did in team work yesterday, and we're going to ask you to share that and show us what happened." Students willingly shared their most interesting experiences with the tuning forks.

> "If you touch it right here [to water], it makes it um, kind of vibrate."
> "We're going to, Cody's going to hit the tuning fork and we're going to put it against the table and it will shock the table."
> "They're putting the tuning fork on Luke's glasses and they're making the glasses vibrate."

We then encouraged students to see patterns in the findings.

> **Teacher:** We saw a lot of different things. Did you notice anything the same about what your group tried? Cherril.
> **Cherril:** Almost all of them vibrated.

> **Teacher:** You know what I'm wondering about is maybe we should all agree on what "vibrate" means. Do we all mean the same thing when we say "vibrate?" What did you mean, Cherril?
> **Cherril:** (shrugs)

Teacher: Can you think of another word that you might use instead of "vibrate?" What about the rest of you? A lot of you have used this word the past couple of days, but what do you mean by that? Cherril?
Cherril: Movement.

Teacher: Movement. Timothy?
Timothy: Moving back and forth.

Teacher: Moving back and forth. So you say "movement" and he says "moving back and forth." What about you, Rachel?
Rachel: Moving fast

Teacher: OK, so moving fast. Cody?
Cody: It's not going very far.

Teacher: So, it's not going far but it's moving a little bit. Anybody else about vibrate? Ronny?
Ronny: Shaking like this.

Teacher: Shaking, that's nice. Sounds like you're all saying something about moving and something about back and forth like Timothy said. And maybe not moving a lot, but a little bit. Sounds like we agree on what "vibrating" means. OK, then let's go back to the tuning fork idea. Cherril said, tell us what you said again about....
Cherril: Almost all of them vibrated.

Teacher: So, almost all, that must mean that one of them didn't vibrate and I'm wondering if you can think of any instances that didn't vibrate. That didn't move or shake.

Although the class had established a shared meaning of "vibration," we remembered that not all students in the spring class had completely bought into the notion that sounds are produced by vibration. Cherril's final comment was a clue that we would need to probe a little farther. So we returned to the list of sounds we had made in the first activity and the rules for sound students had developed the day before.

Teacher: The other day your sound rules all mentioned that vibration causes sound. Now let's go back and look at our list of sounds. We've got over fifty sounds up there. Why don't you look through them and see if you can find anywhere you think the sound is not caused by vibrations.

Hands flew up. Every student had an opinion. Nicole said that singing birds would not vibrate. Rachel's choice was a crying baby and Luke's a computer. Several other candidates for lack of vibration were mentioned. We next asked students: "How could we test this to be sure that there is no vibration involved? How could we find out?" Lucy mentioned putting a CD in a player and then plugging it in to see whether it vibrated. Lon suggested trying the computer right there in the room. He popped up and approached the computer with his hand out. The classroom quieted down so that only the humming of the computer could be heard. Lon placed his hand on the computer and nodded, "It's vibrating all right." Then we gave the students some homework: "Here are some things that you don't think vibrate when they make a sound: birds, cats, a baby crying. Those would be some things that you could actually try and test and see what you think."

While students went home to test these possibly disconfirming cases, we teachers went home to think about what had happened and what to do next. One of us had attended a seminar about children's theory choice (cf. Samarapungavan, 1992), which challenged our thinking. Perhaps in settling on disconfirming cases as the main way to convince students to revise their theories, we were going down the wrong road. After all, the literature strongly supported the idea that anomalous data alone would be insufficient to help students change their theories (Chinn & Brewer, 1993; Tasker & Freyberg, 1985). And, we had seen this play out in a real classroom in the spring when, despite evidence to the contrary, students like Mandy and Mark did not agree that sounds are produced by vibration.

We decided to switch from a strategy of disconfirming to a theory-choice strategy. In the next lesson, we provided three theories among which students could select: all sound is caused by vibration; some sounds are caused by vibration and some are not; some sounds are caused by vibration and some are caused by something else. At our next class meeting, the three theories were posted on the board and students were asked to decide which one they thought best explained what they knew about sound. We asked students to vote by secret ballot for their favored theory. The first theory received three votes, the second five, and the last twelve. "Is this like the election earlier this month? Should the theory with the most votes be accepted as the winner?" we asked. Cody responded, "You have to prove it in science." We asked each student to turn to the person sitting in the next seat and present some evidence from one of our investigations that would support the theory the student selected. This was very difficult for the students. Only Josh was able to build a case in support of the first theory, citing evidence from the tuning fork investigations.

We retreated from the theory-choice strategy and refocused students on the next part of the lesson. We asked them to draw pictures of what they thought happens when a sound is made and heard. Three of the teams drew nothing in between the sound maker and the ear. Two teams drew the textbook version of sound waves radiating from the sound source. Five other teams drew something else between the source and the ear: horizontal squiggly lines, vertical squiggly lines, or, in one case, a tunnel from a radio to an ear labeled "sound." When asked to explain their lines, many students used the term "sound waves" but when probed about what a sound wave could be, they did not have a response. Except for Josh. He stated that the air was moving and that moving air was being passed along as a person talked.

The final first-hand experience of our sound unit concerned sound traveling through solids, liquids, and gases. In our scientists meeting we discussed two different explanations to account for how the sound gets through the material:

- Sound waves are vibrations of water or air or wood.
- Sound waves are pieces of sound going through water or air or wood.

Six of nine teams supported the first explanation, using evidence from many different class experiences. The three groups who were in favor of a particle theory of sound had trouble supporting their position with evidence. These ideas led us back to the question about sound and vibration and to the activity of the day before. To bring closure to our sound inquiries, we asked students to choose between two theories of sound:

- All sounds are caused by vibration.
- Some sounds are caused by vibration.

In the end there was still a split decision, although more students than the day before selected the first theory. As teachers we were left with the need to reflect upon our experiences as we prepared for our next unit of instruction.

Reflection

In planning and enacting this sound unit, we based several changes on what we had learned from the spring teaching team. We tried to incorporate real-world experiences throughout the unit, not relying only on the science class equipment and experiments in discussions. We also brought argumentation into play earlier in the unit, asking students to predict and test disconfirming cases. And we added to our teaching repertoire a theory-choice strategy on comparing alternative explanations.

When we gave students the freedom to argue and explain what made sense to them, as teachers we had to be willing and able to listen to their arguments and explanations and try to understand them. We also had to be willing to let those science conversations be a major driving force in planning and enacting our science lessons. When we turned our classroom over to argument and explanation, we lost some of our control over the instruction. We could not always predict where the lesson would end up. Perhaps most difficult of all was that we had to accept that not all students finally agreed to the scientific explanation of sound and vibration.

CONCLUSION

Our teaching stories include two classrooms moving from science as exploration and experiment to science as argument and explanation as they inquired into concepts of sound. Science knowledge, as J. J. Schwab observed in 1962, originates in the "united activities of the human mind and hand" (p. 102). In our classrooms, students built knowledge from both their first-hand experiences with phenomena and their discussions with other students and with the teachers, what in everyday parlance is referred to as hands-on and minds-on instruction.

Though not all of the students came away with the accepted scientific notions about sound, they did all have opportunities to have first-hand experiences with science phenomena and to talk about their evolving science ideas. According to Freyberg and Osborne in an essay of 1985, it is reasonable that we would have differentiated goals for a class of students. Specifically they suggest that:

> The aim of science education for children should be to ensure that they are *all* encouraged: (i) to continue to investigate things and explore how and why things behave as they do, and (ii) to continue to develop explanations that are sensible and useful to them. (p. 90)

We want *many* children:

> (iii) to recognize that scientists have sensible and useful ways of investigating things, many aspects of which apply not just to science, and (iv) to regard at least some scientific explanations as intelligible and plausible and as potentially useful to society, if not to the child personally. (p. 90)

We can expect *some* children:

> (v) to replace their own intuitive explanations with, or to evolve their own ideas towards, the accepted explanations of the scientific community, and (vi) to become committed to the endeavours of advancing scientific knowledge still further. (p. 90)

If we accept these goals, we should feel satisfied with the outcomes witnessed in the sound stories. In the end, not every student understood or believed that all sounds are made by vibrating objects, but they all had opportunities to investigate, to invent sensible explanations, and to develop arguments in support of their explanations. We hope that these students, for whom the process of argumentation and explanation in science class was new, learned something valuable about making sense of their world. We expect they will continue to use these processes to inquire into natural phenomena. As science teachers we learned to provide opportunities not only for first-hand investigations but also for classroom discussions that emphasize argument and explanation. We continue to learn how to accept our students' ideas, trusting that those ideas represent what makes sense to students at a given point in the development of their scientific thinking.

REFERENCES

Abell, S. K., M. Anderson, D. Ruth, and N. Sattler. 1996. What's the matter: Studying the concept of matter in middle school. *Science Scope* 20(1): 18-21.

Abell, S. K., and M. Roth. 1995. Reflections on a fifth-grade life science lesson: Making sense of children's understanding of scientific models. *International Journal of Science Education* 17(1):59-74.

American Association for the Advancement of Science. 1993. *Benchmarks for science literacy.* New York: Oxford University Press.

Anderson, C. W., and E.L. Smith. 1987. Teaching science. In *Educators' handbook: A research perspective,* edited by V. Richardson-Koehler, 84-111. New York: Longman.

Chinn, C. A., and W.F. Brewer. 1993. The role of anomalous data in knowledge acquisition: A theoretical framework and implications for science instruction. *Review of Educational Research* 63:1-49.

Dewey, J. 1938/1963. *Experience and education.* New York: Collier Books.

Education Development Center, Inc. 1991. Sound. Module in the *Insights* Elementary Science Curriculum. Newton, MA: Author.

Freyberg, P., and R. Osborne. 1985. Assumptions about teaching and learning. In *Learning in science: The implications of children's science,* edited by R. Osborne and P. Freyberg, 82-90. Portsmouth, NH: Heinemann Press.

Lawrence Hall of Science. 1992. Physics of sound. Module in *Full option science system.* Chicago: Encyclopedia Britannica Educational Corp.

National Research Council. 1996. *National science education standards.* Washington, DC: National Academy Press.

National Science Resources Center. 1991. Sounds. Module in *Science and technology for children.* Burlington, NC: Carolina Biological Supply Company.

Reardon, J. 1993. Developing a community of scientists. In *Science workshop: A whole language approach,* edited by W. Saul, A. Schmidt, C. Pearce, D. Blackwood, and M. D. Bird, 19-38. Portsmouth, NH: Heinemann Press.

Samarapungavan, A. 1992. Children's judgments in theory choice tasks: Scientific rationality in childhood. *Cognition* 45:1-32.

Schwab, J. J. 1962. *The teaching of science as enquiry.* Cambridge, MA: Harvard University Press.

Tasker, R., and P. Freyberg. 1985. Facing mismatches in the class-room. In *Learning in science: The implications of children's science,* edited by R. Osborne and P. Freyberg, 66-80. Portsmouth, NH: Heinemann Press.

Designing Classrooms That Support Inquiry[1]

Richard Lehrer, Susan Carpenter,
Leona Schauble, and Angie Putz

A continuing point of debate, among both developmental psychologists and science educators, is over the appropriateness of the metaphor of the child-as-scientist. This metaphor suggests that children seek knowledge about the world as scientists do, generating and exploring hypotheses about phenomena, and constructing consistent and coherent theories about the world (Brewer & Samarapungavan, 1991). On the other side are researchers (e.g., Kuhn et al., 1995) who emphasize the stark contrasts between the reasoning processes employed by scientists and those routinely used by laypeople. According to Kuhn et al., even adults if they lack scientific training routinely distort evidence to preserve favored theories and make a number of systematic errors in the generation and interpretation of evidence.

In considerations about the science education of young children, these positions are sometimes played out as extremes. Often the debates refer to the same source. The work of the Swiss epistemologist Jean Piaget on infants' early mental development and its roots in exploration and actions on the world (Piaget, 1952), for example, is sometimes used to justify a rather romantic trust in the power of children's curiosity. Yet Inhelder and Piaget's (1958) pioneering work on the development of logical thinking has been invoked to justify the restriction of early science education programs to the kinds of activities for which children are presumably ready at the moment. In practice, this has often meant limiting elementary school science to the mere introduction of facts and simple relationships, supplemented, perhaps, by domain-general exercises in reasoning—such as categorization and transitive reasoning in the early grades and process skills like observation and measurement in the later grades. As Metz (1995) argues convincingly, the notion that genuine inquiry should wait until children are "developmentally ready" often rests on misunderstandings about young children's capabilities and the nature of inquiry in the sciences. But neither is curiosity sufficient in itself. Although children's curiosity is certainly the foundation upon which good science instruction builds, it is equally important to understand the forms of support that teachers deploy to stretch initial interests into sustained and fruitful programs of inquiry.

Our own position, consistent with the *National Science Education Standards* (National Research Council, 1996), is that "science as inquiry is basic to science education and a controlling principle in the ultimate organization and selection of students' activities" (p. 105). In our view, the interesting question is not whether children have some developmental capability to engage in genuine inquiry. Rather than regarding children's capabilities as inherent and presumably fixed, we understand thinking and reasoning as grounded within contexts that are inherently social, not naturally occurring. Thus, thinking is brought into being and develops within contexts that are fashioned by people. Whether or not we are aware of it, these contexts include norms for the kinds of questions worth pursuing, the activities that are valued, the forms of argumentation deemed convincing, and the criteria for a satisfactory explanation. Recalling that learning environments are designed (Glaser, 1976; Lehrer & Schauble, in press; Simon, 1981) helps to turn our attention to the "design tools" that we have available for making classrooms and other learning contexts effective (Carpenter & Lehrer, in press). If teachers are the designers of learning environments, what kinds of design tools do they have at hand for fostering inquiry in the early grades?

These questions are the topic of collaborative investigation among a community of elementary grade teachers in the school district where we conduct our research. A rural and suburban district about fifteen miles from the state capital, it is undergoing extremely rapid growth in both the number and the diversity of its students. Teachers representing all five grades at the district's four elementary schools work with us as co-researchers on a project aimed toward improving mathematics and science instruction. Teachers cooperate in learning to teach such new forms of mathematics as geometry, data, measure, and probability and to understand the development of student thinking in these understudied topics. The other major initiative of the project is to learn how students can use these mathematical resources to understand science, especially through approaches that emphasize the development, evaluation, and revision of models.

We will begin our consideration of designing classrooms for inquiry by following one of these teachers as she orchestrates a long-term investigation in her first-grade class. The teacher used questions, forms of argumentation, and inscriptions to build on students' curiosity, turning their thinking toward important ideas like comparison, measure, and mechanism. In the second part of the paper, we consider ways that teachers of older elementary grade students can provide challenge and lift for students in grades three through five.

INQUIRY IN THE FIRST GRADE: DECOMPOSITION

We begin by summarizing the chain of inquiry conducted in one first-grade classroom, where children's curiosity about changes in the color of apples kicked off a year-long investigation into conditions for decomposition and explanations of it. Although this cycle of inquiry was initiated by children, it was sustained through the work of the teacher, Angie Putz. Over the course of the year, what started as a simple question led to opportunities to explore ideas related to experimentation, the role of models in scientific inquiry, and the importance of inscribing observations.

Investigating Ripening

As the class convened in the fall, Ms. Putz asked students to bring apples to class. As they inspected and described the variety of colors and shapes, someone pointed out that apples change color as they ripen. Ms. Putz asked children how they might account for this change, and a few children suggested that the sun might be the agent. Ms. Putz countered by asking children whether they could think of a way to find out more about how the sun affects fruit. We have found that this cycle of teacher questions following students' questions is quite common and important for modeling inquiry. The children in Ms. Putz's class proposed that they could investigate that idea by observing fruit in the sun. After some consideration, they agreed that bananas or tomatoes would be good candidates, because children knew that both of these fruits noticeably ripen.

The idea of a test. The next day, Ms. Putz brought green tomatoes to class and asked, "How do you think the sun helps in changing the colors?" Most students responded that the sun "gives light." The teacher asked children how they thought they might test their idea. Notice how her question raised the stakes, implying that beliefs about a phenomenon need to be justified by a particular form of argument—the test. Without such prodding, students rarely move beyond simple assertions. Valuing forms of argumentation and justification like these is a design tool that effective teachers like Ms. Putz employ for rendering inquiry productive.

In response to Ms. Putz's query, several children suggested placing one tomato on the window sill and another in the dark. After some discussion about what would count as "dark," the children settled on a spot in the classroom under a cover. They readily agreed that if they observed a tomato in each location, they would then know whether or not light mattered for effecting color change. However, one child, Ben, objected, "But the sun is hot. Does heat matter?"

Several children found Ben's suggestion compelling. Once again, their teacher reoriented this discussion into a consideration of evidence and argument: "How could you test the role of heat?" Children suggested placing one tomato in the refrigerator. At this point in the discussion, the students were assuming that the tomato on the window sill could serve as the case testing the role of warmth. Here, Ms. Putz stepped in again, to help children elaborate their thinking about the factors that might be regulating the changes in color. She began to probe in ways that might help them untangle their original confounding of warmth and light. She asked, "How do you know that the window is the warmest place in the room?" Some students claimed that since this spot was closest to the sun, it was obviously the warmest. Others, thinking about proximity to the window during a cold Wisconsin fall, weren't so sure. How could this disagreement be resolved? One child's proposal that the class use a thermometer sparked a long discussion about how this measurement device might work (a discussion that we do not recount here). As a result of their explorations with the thermometer, children eventually settled on four conditions for observation, affording comparisons of light and heat. Their observational conditions were: light and cool (the windowsill), light and warm (a location away from the windowsill but still receiving a lot of sunlight), dark and warm (the covered tomato), and dark and cold (inside the refrigerator).

From observation to inscription. Ms. Putz next encouraged children to move beyond observation toward inscription. We propose children's inscriptions as an important design tool of fundamental importance. As Latour has explained (1990), even though we think of scientists as observing the world, scientists do not do the bulk of their work with raw observations. Instead, they most often work with inscriptions, which may include records, drawings, mathematical formulae, various kinds of output from instruments, and more. Choices of inscription are partly choices about what to preserve—inscriptions select and enhance information that is vital and leave out other information deemed unimportant. This fixing of experience provides a means of making public what all know consensually and of holding steady what unaided memory will lose or distort. Here, children decided to use drawings, a decision that provoked discussions about what changes should be represented and how these changes should be displayed. In this case, children decided to preserve a record of changes in the tomatoes over time.

Over the course of a few weeks, many children began to note a progressive discoloration, discharge, and change in how the tomatoes felt (as one described it, they became "squishy"). Discoloration was relatively easy to represent in drawings, but changes resulting from discharge and corresponding loss of turgidity were more difficult. Children settled upon a convention of using shadings of color to represent regions of "squishiness." There was much discussion in the classroom about how to use detail to capture the changes observed. Ms. Putz and her students were surprised about the number of decisions that needed to be made to translate from the mind's eye to an inscription of change that carried shared meaning for the class. Ms. Putz, of course, also knew that all of this careful observation and detailed inscription would focus children on transitions over time. The inscriptions served as the basis for comparing the contrasting conditions of light and temperature. Note that inscriptions confer the additional bonus of rendering public, shareable, and inspectable, the private thinking of individuals. A teacher who is aware of how children think has a considerable advantage in crafting instruction that is tuned to children's understanding.

Children's initial conjectures about light were confirmed, but many were surprised by the role of temperature: "Hey, Ms. Putz, that's sort of a pattern! The tomato in the window was colder, so it took longer to change. The same thing happened to the tomato in the fridge." Margaret added to this idea, commenting on the tomato in the refrigerator and the one in the dark place in the room: "Of all the tomatoes, these two changed the slowest." Ms. Putz pushed further for explanation, asking, "Why do you think that?"

Margaret replied, "Because the other tomatoes changed faster because they had light."

Finding this statement somewhat ambiguous, Ms. Putz gently challenged Margaret's statement: "So does this mean that light is the only factor in the changing color?" This question led to a conversation during which children compared their four locations and decided that the fastest change was associated with both light and heat and the slowest in the refrigerator, where light and heat were absent.

Children noted intermediate changes in locations where only one of these factors was present, and thus concluded that heat and light both made contributions to a tomato's ripening. These conclusions were supported not only by appeals to the current appearance of the tomatoes themselves, but by arguments based on the cumulative drawing records describing the tomatoes in each of the test locations. This shift exemplifies a further value of inscriptions in science—they preserve information in a format that permits reasoning within the constraints and world of the inscription alone, without the need to resort back to the fleeting events of the world itself.

Identifying and inscribing new attributes. By the end of the month, children had settled on a new word to describe the changes they were observing: "rot." The odor was becoming noticeable, but children wanted to continue to observe change. They gathered their tomatoes and placed them in the school's courtyard for further observation. At this point, it was drawing close to Halloween, and the class moved on to a different seasonal theme, "pumpkin math." Each child brought a pumpkin to class, and children wondered whose pumpkin was the biggest. This question provided another opportunity to consider how to problematize, clarify, and inscribe what at first seems self-evident. "Biggest" posed interesting questions about what was meant by "big": Did it mean the pumpkin that was the biggest around? The tallest? The heaviest? Once again the children were faced with the problem of deciding attributes worthy of investigation. Reaching consensus about standards and units of measure occupied considerable debate. As children investigated whether pumpkin size was related to the number of its seeds, they had to reach consensus not only on how to measure size, but also on what would count as a seed (for example, what about the immature seeds)?

After exploring the mathematics of pumpkins, the children proceeded to carve them. Shortly thereafter, the telltale signs of rot once again became evident. Children drew upon their experience with tomatoes to predict what might happen with pumpkins. After further observation, they represented similarities and differences between pumpkin rot and tomato rot with the Venn diagram shown in Figure 1.

**FIGURE 1. FIRST-GRADE OBSERVATIONS OF SIMILARITIES AND
DIFFERENCES BETWEEN "TOMATO ROT" AND "PUMPKIN ROT"**

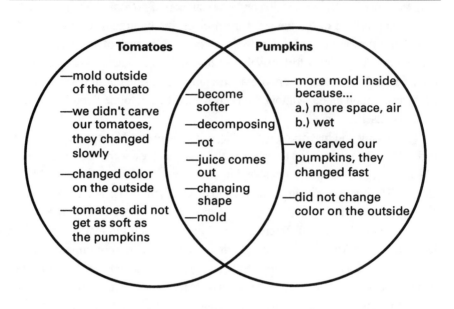

Children expressed surprise at the comparative rates of change of tomato and pumpkin and asked about potential causes. Ms. Putz cut pieces of pumpkin and placed them into dishes for ease of observation. Children began to notice additional changes: "There are different kinds of molds." Ms. Putz asked how they knew that, a question that led to further discussion about the relationships between the children's observations and their inferences from the observations. The class generated an additional question: "How does mold happen?" The students pointed out that the change observable in the pumpkin dishes was a lot like what they were observing outdoors in their pile of tomatoes, although the rates of "rot" differed substantially. But, the children were concerned that with the approach of winter, their rot experiment would be arrested by the cold (or at least, hidden by the snow). At this point, Ms. Putz introduced compost columns as a model for the outdoor rot process.

The Inquiry Shifts: Modeling Rot

Filling two-liter plastic soda bottles with decomposing material, the teacher began by raising the question whether compost columns might serve as models for the outdoor system, asking, "Why couldn't we just watch the pile that is already outside?" and "So what is it we are trying to do in our class with these columns?"

One child noted, "It is showing what something looks like," and "You see how something is made and you make one yourself." The emphasis in children's responses on the idea of "looks like" reflects a more general orientation that we have noted among children toward thinking of models initially as representations that copy or <u>resemble</u> the phenomena being modeled (Penner et al., 1997). As one child proposed that a compost column was like a model airplane, Ms. Putz asked, "Look around the classroom. Are there other models?" Children considered globes and maps to be models, suggesting that some models resemble the world but still are not identical to it. "A globe isn't the real same size or color, but it shows people what it looks like." Returning to the compost column as a model, one boy suggested, "We want to watch what happens to the tomatoes until spring, but we want to do it in our room."

<u>Constructing models</u>. Having come to convergence on the purpose of the compost columns, children proceeded to the problem of deciding which elements of the outdoor system should be replicated in the columns. Their choice was of objects that <u>looked like</u> the phenomena they had decided to track (rotting tomatoes outdoors in the dirt). Accordingly, they argued for inclusion of moldy tomatoes, dirt, leaves, gum wrappers, and a piece of foam. Once again, the teacher asked them to consider factors that might influence decomposition. Drawing upon their previous experience, children decided that one compost column would be kept warm and another cold (in the refrigerator). Other compost columns were constructed with pumpkin (to compare with those including tomato), and all were watered to mimic the effects of rain, which children felt would be important in the rotting process.

Children again observed change over time, drawing and noting what they observed: "I see mold, garbage, and leaves. The mold is white on the dirt." Many children noticed the increasing amounts of mold in the columns. Ms. Putz asked, "Why do you think there is more mold?" Most children explained that there were a lot of dead things in the columns. They did, however, notice that compost columns kept in the room apparently had more mold than the column kept in the refrigerator.

Ms. Putz pressed further: "When we started to make the columns, both column A—the one in the room—and B—that in the refrigerator—had the same things. So why does column A have more mold in it than column B?" Children suggested, "Because of the cold." When one child suggested that the mold might be growing, his remark was met with stunned silence. Ms. Putz reframed the conjecture: "Is the mold growing more in column A than in column B?" Most children objected that mold could not be alive. They seemed instead to associate mold with some unknown process connected with "dead things" and moderated (somehow) by cold. This notion that mold emanates from "dead things" is an example of a copy theory of generation that we sometimes observed: the belief, for example, that dead begets dead.

Attack of the fruit flies. In another compost column, one containing pumpkin, children noticed fruit flies, an observation prompting the question, "Where did the fruit flies come from?" A few children recalled that some of the pumpkins had worms, so they conjectured that the larvae had metamorphosed into fruit flies. Ms. Putz again pressed for evidence: "How do you know that the larvae turned into fruit flies?" Children returned to look at the column and were excited to find more larvae crawling around under some leaves. One boy noticed "bumps" on the wall of the container. "Ms. Putz, the larvae turn into bumps on the wall. Then they hatch into fruit flies! The fruit flies lay more eggs, and the eggs hatch more larvae!" Another boy added, "It's like a circle story!"—a reference to a story the class had read about the life cycle of insects.

The next inquiry came from questions asked by children in other classrooms and their teachers, with varying degrees of asperity: "Where are these fruit flies coming from?" Yes, the fruit flies had escaped into the school. Noting that the complaints were coming from all over the school, children sent envoys to each classroom to count the number of flies observed during a brief visit. They then represented their observations on a map of the school, displaying different ranges of counts by coloring classrooms accordingly (classrooms with the highest counts of fruit flies were colored green on the map).

As they reviewed the completed map together, Ms. Putz asked, "Why are there three green classrooms in a row?" One child replied, "That's easy, because they are next to our classroom, and our classroom has a lot of fruit flies."

"Then why," asked Ms. Putz, "are some of these classrooms in a different hallway green?"

A child excitedly replied, "I remember in all of those rooms the teachers said they saw the fruit flies around food places." Other children explored the implications of this conjecture by reexamining their display and ultimately connecting the higher and lower counts to what they knew about potential sources of food and water. The implications of these speculations about food and water sources were elicited again by Ms. Putz, who asked children to predict what might happen over time. Children suggested that when the food ran out, so too would the fruit flies. Subsequent observations confirmed this conjecture.

The fruit fly episode represents yet another important feature about inquiry, often brushed aside in the hustle of classroom schedules but valuable for helping children understand an important point about how scientific investigation proceeds. When things are cooking, either in the laboratory or in the elementary school classroom, investigations do not stop with the pursuit of isolated questions. Instead, they often stretch into a <u>chain of inquiry</u> from questions to investigation to new (and often more interesting) questions. Encouraging and helping children to extend that chain is something that excellent teachers like Ms. Putz do. This sends the important message that work conducted is not work completed. In these classrooms there is a continuing cycling of knowledge, questions, inscriptions, and data into new and more challenging next steps—the conceptual ante keeps rising, and children keep rising to meet it.

<u>Is it alive?</u> The children remained uncertain about the status of mold, so Ms. Putz decided to have them grow mold on wet bread. Using magnifying glasses and microscopes, the students observed change in greater detail. They viewed a "Magic School Bus" video about fungi, which served an important role in orienting them to the details of what to look for. Children noticed, for instance, that molds "...have stems like a plant and ball or a different shape on top for the flower or leaf part." Analogies like these convinced children that the change in the volume of mold was due to growth, not to some unspecified abiotic process. To help children map from the bread mold to the mold in the compost columns, Ms. Putz asked a further question: "How is the mold able to live in our compost columns?" The children volunteered several factors: "Because it has food—leaf litter, tomato, pumpkin—and we have water in the columns. It would die if we stopped putting water in."

As the character of the compost columns changed, the children's inquiry continued to evolve. Many children wondered about changes in the volume of material and the disappearance of pumpkin and tomato. Students developed conjectures about the potential roles of mold and fruit flies in eating the food sources, and wondered whether the fruit flies themselves were then eventually dying and going into the soil. It was noted that some food sources decomposed more rapidly than others. Ms. Putz again raised the stakes by asking children for an explanation. The responses indicated that children associated more rapid decomposition with observations of the actions of larvae and mold. Others noticed that the leaves were decomposing too, and Ms. Putz asked about their evidence for this claim. One girl explained, "When we first started, the leaves were crunchy and dry. Now they are wet, moldy, and smaller."

When spring finally rolled around (a late event in this Northern state), children resumed observation of their outdoor tomato pile. Now, however, their inquiry and observations were informed by their experiences with the compost columns. Over the course of the year, children had come to regard science as a cycle of inquiry, observation, and inscription. Each step in the cycle built on the previous one, and each drew its meaning from the whole. The teacher's questions introduced students to notions of conjecture and evidence, to considerations of models and modeling, and to the importance of comparison over time and among conditions. Although Ms. Putz certainly honored and worked with children's natural curiosity, curiosity alone would not have taken the students very far. Instead, she worked systematically to design a classroom in which children engaged in progressive cycles of inquiry and evidence. Her design tools included questions that pushed her students' questions further and acquainted them with norms of argumentation and evidence, the use of inscriptions and displays they devised, and evolving chains of inquiry. It was by refusing to rely on children's curiosity alone that Ms. Putz fostered it.

UPPING THE ANTE IN THE LATER GRADES

Motivating and sustaining the curiosity of young children seems to be a matter of hooking to their interests and building on them, but many teachers wonder where and how to direct this enthusiasm when they work with students in the upper elementary grades, especially when the demands of the subject increase. At these ages we suggest, teachers are able to turn children's reflection back upon their own inquiry, so that inquiry becomes more thoughtful and increasingly governed by a refined judgment about the questions worth pursuing. We also consider the advantages of helping students stretch their first inscriptions toward increasing mathematization. We are arguing not for letting science dissolve into computation, but for helping students develop a taste for the power of mathematical forms of argument.

Helping Students Pose Good Questions

One notable thing about Ms. Putz's teaching is how quick she is to pick up children's questions and push them forward in fruitful ways. Sometimes, especially in the later elementary grades, when teachers become primarily concerned that students acquire the knowledge and skills of a particular subject, they forget that genuine inquiry is rooted in questions. But how can children make meaning of an experiment or data collection that is not well anchored in a question that is real to the participants? The issue then on the table is: How does a teacher help students learn to pose good questions? In the older elementary grades, three through five, teachers will want to pay increasing attention to shifting students' attention beyond simply posing questions and toward reflecting about the potential and interest of the questions that are generated. Although many have encouraged teachers to welcome and listen carefully to children's questions and to let students discuss and investigate their own questions (Chaille & Britain, 1997; Gallas, 1995), little information is available to assist teachers at making these questions productive.

As Ms. Putz's experience shows, even the youngest students can generate a wide variety of questions about phenomena. But there are instructional practices that assist the process. We need to recognize that it takes time. Often, we rush right past question posing into data collection. Yet this is a time to slow down—explicitly opening up the process of generating questions and aiming toward the long-term goal of helping students develop criteria for what counts as a good question. Questions are generated readily when students work in groups, or when they can write a few questions individually and then contribute those to a group list. We find that students build on one another's ideas, especially if the teacher models appropriate criteria for evaluating questions as they are listed. One of our third graders, for example, asked of Wisconsin Fast Plants (a plant that completes its entire growth cycle within forty days), "How long will they grow?" The teacher asked which sense of "long" was intended. Was the student talking about size or time? The teacher also suggested the need to be sure we know to what "they" refers. As questions are generated and considered, the teacher will want increasingly to cede to the students themselves this process of evaluating questions.

Children generate more questions and more interesting questions when they are encouraged to build upon their own knowledge and experience about the phenomenon under investigation. Sometimes this means beginning with an extended conversation about what children know of a topic. Recall, for example, how the first graders' initial knowledge about ripening helped them generate the comparison conditions for their tomato experiment. Shared experience in the classroom with an interesting event or organism serves as an excellent prompt for further questioning. As their knowledge of a phenomenon grows, children ask increasingly interesting questions. So sometimes it is wise to devote at least two phases to inquiry: one to gaining familiarity with the phenomenon of interest, and the second to more concentrated investigation. Most questions that one class of third graders generated before planting seeds to grow Wisconsin Fast Plants looked to endpoints of growth: "How tall will they grow?" A few concerned timing of events in the life cycle. During its second round of growing Fast Plants, however, the class generated more subtle questions. Some were oriented toward function such as the role of petals and pollen, others toward development: for instance, the typical shape of the growth curve. Interval—"On what day?"— raised issues. Still others involved comparison: for example, the effects of different amounts of fertilizer. Over cycles of inquiry, questions became increasingly elaborated: "how long does it take" gave way to "how many more days" and then to "what day." From "flower buds" to "role of petals" and "what makes pollen," questions grew more specific. They also reflected increasing cognizances of variation: the words "usually," "normally," and "mainly" begin to be used to qualify statements. Students also turned from queries about endpoints to questions about change over time and rates of growth.

One way to begin examining and evaluating questions is to record them on index cards and ask small groups to arrange and rearrange them into categories. Then each group of students describes its category system, a process that encourages children to read and become familiar with the range and variety of questions, as well as to consider additional ways of categorizing. We have observed similarities in the ways that students in the third grade through the fifth (and groups of teachers!) categorize questions. Some categorize by words that appear in the question; for example, all questions containing the word "flower" are grouped together. Others group by concepts. Questions about living organisms may be sorted into groups labeled, "growing," "size," or "environment." Some groups classify questions into the familiar format who-what-when-where-why-how. Occasionally, a student suggests that the questions be sorted by the type of answer expected. This insight often helps students understand that many questions that can be answered by a simple "yes" or "no" are less interesting than queries that call for more complex answers. Students may separate from problems that they think unsolvable others that can be addressed by authorities such as books and experts or by investigation. It can also be useful to ask the class which questions are interesting or simple, and what makes them so. The class may consider which questions they could investigate within a given amount of time and which would take longer. Class discussions about how a question can be investigated are as important as later discussions about what has been learned from the investigation.

As students evaluate their questions, the teacher will also be considering which questions are most likely to be productive for extended class work. This will require attention to children's prior knowledge, the tasks and tools the question calls for, and the potential for developing reasoning and argument at both the planning stage and the resolution. Many hands-on science programs treat questions as givens, which invites students to regard science as the precise execution of prefabricated recipes of steps in pursuit of a solution to a question that nobody cares about. Time spent in helping students work at posing and revising questions also pays off in a deeper understanding of the results.

From Inscription to Mathematization

Recall that Ms. Putz's students' initial conjectures about whose pumpkin was biggest could not be answered definitively by merely inspecting the pumpkins. Instead, they had to achieve agreement about such attributes as height, width, and weight and about measurement: for young students, developing a firm understanding of measure is itself an accomplishment. In cases like these, children learn that their arguments and conclusions rest on firmer warrants if they can find ways to mathematize the world.

In another first-grade classroom, students explored the growth of flowering bulbs—amaryllis, hyacinth, and paperwhite narcissus. Although students could readily observe the bulbs growing, they required inscriptional resources to record, describe, and analyze the change. In addition to recording and drawing changes in their journals, students worked with cut out paper strips that preserved the height of each plant at each day of growth. The green color of the strips mapped easily onto the green of the stems. As is often the case with young students, these first graders found difficult the move from copy to representation. Eventually, however, they began to regard the strips not as copies of plants, but as representations of an attribute of the plants: height. The teacher assisted this move to representation by rearranging the strips in various ways to support conjectures raised by the children—for example, to compare the height of one plant as it grew and flowered, or to compare the height of different plants on the same day.

When the teacher introduced the question, "Which plant is growing <u>fastest</u>," children initially selected the plant that was currently the <u>tallest</u>, and supported their argument by referring to the longest strip for each of the plants being compared. But one child pointed out that it wasn't sufficient to refer just to the heights of individual strips; the class needed to consider <u>differences</u> in heights on successive days. On our segment of videotape, we can clearly see her considering the relevant height strips mounted side by side on a classroom chart. Using her thumb and forefinger, she marks off the successive differences from day to day so that the other students can see what she means when she talks about the difference. Children found this argument compelling, and went on to compare successive differences to answer the teacher's question about two different conditions in which bulbs were grown. The students readily concluded that while the paperwhite narcissus planted in soil had grown faster than the narcissus planted in water, eventually the bulb grown in water alone "catched up."

Although the strips were first employed mainly in side-by-side eyeball comparisons, they eventually inspired more sophisticated questions about linear measure. Issues of measurement were first raised when an adjacent classroom asked, "Whose amaryllis is growing bigger, ours or yours?" Students tried to answer by holding up pencils alongside their growth strips and reporting back to the partner class that their amaryllis was "three pencils tall." Then, a new question came back: "How big was your pencil?" The teachers found these conversations very fruitful for eliciting consideration of the need for standard units, iteration of units (with no spaces in between), and measuring from a common baseline, all important understandings for young children developing a theory of measure (Lehrer, Jenkins, & Osana, in press). Eventually, the children reinscribed the growth data in a table of measurements, and confirmed that the new data display could also be used to support their conjectures about endpoints, rates, and timing of growth.

Notice how closely intertwined in the first-graders' work are questions, inscriptions, and argument. As children's inscriptional resources became more sophisticated and mathematically powerful, the quality of their questions and arguments also expanded. Sustaining productive scientific reasoning requires going beyond exclusive reliance on observation and memory. Learning to describe events in ways that lend themselves to flexible and mobile forms of comparison is an important resource for moving classroom argument beyond appeals to what initially sticks in memory.

Similar issues come to the fore with older elementary students. In higher grades, however, teachers should be capitalizing on the increased leverage purchased by work in the earlier grades at developing children's inscriptional and mathematical resources, reflective criteria for evaluating their own inquiry processes, and internalization of mathematical and scientific argument.

A fifth-grade class's examination of insect growth for about six weeks was organized around a central question generated by the students: "Do [the larvae] grow better if they eat green pepper or 'recipe'?"—a standard formula developed by the University's Department of Entomology. As in the earlier grades, the class began by refining the question extensively. The students started with a discussion of the kinds and forms of data that could contribute to a satisfactory resolution. They eventually settled on rearing two groups of larvae, one fed exclusively on green pepper and the other fed on recipe. Students proposed that growing "better" might mean growing bigger or growing faster. As the investigation continued, additional senses of growing "better" were identified. Students pointed out that larval size and growth rate could be conceived by reference to length, width, height, or weight. Consensual procedures for finding values for each of these attributes were negotiated in the classroom. Several days into the experiment, a student proposed that growing "better" might mean "living longer," and this idea was also incorporated into data collection.

Their early data and observations led students to speculate that growth in length of larvae might be inversely related to growth in width and that larvae that grew faster and bigger might not live so long. This speculation contrasted with an earlier conjecture that growth rate and size would be directly related to mortality. These ideas were revisited later in light of data collected over the entire life cycle. Eventually, students compared growth rates for several attributes and debated which would count as growing better.

Toward the close of the unit, students developed frequency and line graphs to develop and justify conclusions about the growth of the two groups of larvae. These mathematical representations, in turn, inspired several new conceptions of growing better. One student proposed that larvae showing normal or typical growth might be growing better than one that was simply large, partly because larger organisms might be more susceptible to predation. Another student pointed out that the larvae feeding on green pepper lived longer in the caterpillar stage than the ones fed with recipe, which survived to pupate and eventually emerged as adults. This idea was extended by a student who thought that another sense of living longer (and hence "growing better") would include larvae that grew large over a longer time than others took.

A central point is that during their repeated reflections about their data displays, many of the most interesting questions students posed were inspired not by direct observation of the organisms, but by the emerging qualities of the displays themselves. Increasingly, these qualities included important mathematical ideas. Students inspected one chart displaying a bivariate frequency plot comparing the effect of recipe with that of pepper on body length of the larvae on different days of growth. They noted that the display showed not only differences between the two treatment groups in typical body length but also that the larvae fed recipe showed greater variability in body length. As one student put it, "The recipe got kind of spread out, and they're not really bunched up.... Green pepper doesn't really spread out much." Students speculated about possible causes of the discrepancies. Did it matter, for example, that the insects fed recipe tended to grow faster and therefore may have been moved to a larger container more quickly than the others? These conjectures, in turn, led to discussion about how the original experiment might be redesigned to eliminate possible confounds. This kind of repeated cycling between data and explanation, working to identify ways that each can illuminate the other and to seek alignment between them, is typical of the kind of scientific argumentation conducted by practicing scientists.

CONCLUSION

We might conclude that elementary school students are like child scientists because our experiences and those of others suggest that reasoning about the natural world can be provoked by inquiry not too far removed from children's curiosity and play. Sustaining and elaborating these initial efforts, however, requires attention to some important design features that are seldom articulated, features that teachers can orchestrate to help children build a chain of inquiry rather than a succession of fleeting interests.

Like scientists, students work most productively when in communities that embody and inculcate norms about interesting inquiry, good explanation, and argumentation based on evidence. Teachers like Angie Putz initiate and sustain chains of inquiry about the natural world by calling children's attention to events—what happens to apples; by questions that prompt consideration of explanation and evidence; and by efforts to help children reflect upon the history of their inquiry: comparing "How we used to think," for example, with "How we think now." Inquiry is not regarded primarily as exploration and experiment, in which the meaning of investigation is assumed to be self-evident. The work centers instead in argument and explanation, the negotiated and constructed nature of meaning and evidence. That approach is consistent with recommendations in the *National Science Education Standards* issued by the National Research Council in 1996.

Children, like scientists, mediate their inquiry by tools, inscriptions, and nota-tions. Although we certainly want children eventually to understand the powerful symbols and tools of scientific practice, the treading here must be light. It is usu-ally a mistake to give children solutions too soon to problems that they have not yet experienced as problems. Although one can simply teach children accepted procedures for collecting and representing data, we find it much more powerful to build from the inscriptions and displays that children invent on their own in the process of pursuing questions that they have helped pose and refine. This involves students themselves in considering what properties the display should feature. Even more important is that they are brought into the evaluation process that occurs when the first attempt is completed: Does the first try at a graph or drawing or diagram clearly communicate to someone who wasn't part of the data collection team? Does it really throw any light on the original question? Almost always, revisions are required, and students learn the important lesson that inscriptions need to be revised and retuned to the purposes at hand. They also understand that the tools of science were invented for a purpose.

Questions, inscriptions, and argument go hand in hand. Growth in one almost always leads to growth in the others. Yet in our experience, this kind of development does not spontaneously emerge. Representational displays and arguments are important for framing questions in fruitful ways, for answering them, and for provoking new questions that emerge out of the qualities of the inscription. In effective inquiry, teachers work to help students develop consen-sual criteria for what counts as a convincing argument and an interesting ques-tion. Learning to master the interplay between question, inscription, and argu-ment puts students on the road to becoming authors of scientific knowledge. It is in making students authors of knowledge rather than mere consumers that it is valuable to have them inquire about the growth of amaryllis, the spread of fruit flies, and the decomposition of tomatoes.

ENDNOTE

1. Correspondence concerning this article should be addressed to any of the authors at: Department of Educational Psychology, 1025 West Johnson, Madison, WI 53706. This research was supported by the James S. McDonnell Foundation's Cognitive Studies in Educational Practice Program and by the National Center for Improving Student Learning and Achievement in Mathematics and Science (NCISLA), which is administered by the Wisconsin Center for Education

Research, School of Education, University of Wisconsin-Madison. NCISLA is funded by the Educational Research and Development Centers Program, PR/Award number R305A60007, as administered by the Office of Educational Research and Improvement, U.S. Department of Education. The contents do not necessarily represent the position or policies of the funding agencies. We are grateful to the participating teachers and students and to other contributing research staff and students.

REFERENCES

Brewer, W., and A. Samarapungavan. 1991. Children's theories versus scientific theories: Differences in reasoning or differences in knowledge? In *Cognition and the symbolic processes: Applied and ecological perspectives*, edited by R. R. Hoffman and D. S. Palermo, 209-232. Hillsdale, NJ: Erlbaum.

Carpenter, T., and R. Lehrer. In press. Teaching and learning mathematics with understanding. In *Mathematics classrooms that promote understanding*, edited by E. Fennema and T. R. Romberg. Mahwah, NJ: Erlbaum.

Chaille, C., and L. Britain. 1997. *The young child as scientist: A constructivist approach to early childhood science education*. New York: Longman.

Gallas, K. 1995. *Talking their way into science*. New York: Teachers College Press.

Glaser, R. 1976. Components of a psychology of instruction: Toward a science of design. *Review of Educational Research* 46: 1-24.

Inhelder, B., and J. Piaget. 1958. *The growth of logical thinking from childhood to adolescence*. New York: Basic Books.

Kuhn, D., M. Garcia-Mila, A. Zohar, and C. Anderson. 1995. Strategies of knowledge acquisition. *Monographs of the Society for Research in Child Development* 245, 60, No. 3.

Latour, B. 1990. Drawing things together. In *Representation in scientific practice*, edited by M. Lynch and S. Woolgar, 19-68. Cambridge, MA: MIT Press.

Lehrer, R., M. Jenkins, and H. Osana. In press. Longitudinal study of children's reasoning about space and geometry. In *Designing learning environments for developing understanding of geometry and space*, edited by R. Lehrer and D. Chazan. Mahwah, NJ: Erlbaum.

Lehrer, R., and L. Schauble. In press. Modeling in mathematics and science. In *Advances in instructional psychology*, Vol. 5, edited by R. Glaser. Mahwah, NJ: Erlbaum.

National Research Council. 1996. *National science education standards*. Washington, DC: National Academy Press.

Metz, K. E. 1995. Reassessment of developmental constraints on children's science instruction. *Review of Educational Research* 65: 93-127.

Piaget, J. 1952. *The origins of intelligence in children*. New York: International University Press.

Penner, D. E., N.D. Giles, R. Lehrer, and L. Schauble. 1997. Building functional models: Designing an elbow. *Journal of Research in Science Teaching* 34:1-20.

Simon, H. 1981. *The sciences of the artificial*. Cambridge, MA: MIT Press.

Ways of Fostering Teachers' Inquiries into Science Learning and Teaching[1]

Emily H. van Zee

What does it mean to use inquiry approaches to learning and teaching? How do teachers who teach this way think about what they do? One way to find out is for researchers to study how such teachers teach. Another way is for the teachers themselves to study their own practices and to communicate their findings. This chapter reviews some of the relevant literature and then describes my efforts to foster such research.

WHY WOULD TEACHERS CHOOSE TO INQUIRE INTO THEIR OWN TEACHING PRACTICES?

Teachers typically have little time, resources, or encouragement to undertake inquiries into their own teaching practices. Why would they choose to do this? In *Teaching as Research*, Eleanor Duckworth articulates a vision of conducting research as an integral part of teaching:

> I am not proposing that school teachers single-handedly become published researchers in the development of human learning. Rather, I am proposing that teaching, understood as engaging learners in phenomena

and working to understand the sense they are making, might be the *sine qua non* of such research.

This kind of researcher would be a teacher in the sense of caring about some part of the world and how it works enough to want to make it accessible to others; he or she would be fascinated by the questions of how to engage people in it and how people make sense of it; would have time and resources to pursue these questions to the depth of his or her interest, to write what he or she learned, and to contribute to the theoretical and pedagogical discussions on the nature and development of human learning.

And then, I wonder—why should this be a separate research profession? There is no reason I can think of not to rearrange the resources available for education so that this description defines the job of a public school teacher. (1987, p. 168)

Through such inquiries, teachers can document the details of their students' thinking, deepen their own understanding of both science content and pedagogy, share their insights with colleagues, and contribute to knowledge about learning and teaching.

In *Doing What Scientists Do: Children Learn to Investigate Their World*, Ellen Doris provides many examples of teaching as researching. She writes:

We can, in effect, work as researchers in the classroom, observing children carefully, listening to what they say, noting when our responses seem to baffle them and when we help them to take a step forward. Science textbooks may offer us information to present or questions to raise, but only our careful attention to children will enable us to gather information about what they find interesting or puzzling, which ideas they understand and which ones confuse them. (1991, p. 11)

Doris shares insights and experiences by interpreting transcripts of conversations in her classroom, commenting about her students' drawings and writings, and telling stories about what happened as they explored together.

"Conducting formal and informal classroom-based research is a powerful means to improve practice," according to the National Research Council (NRC) in the *National Science Education Standards* (p.70). Listening closely to what students say, for example, can help teachers become more expert in diagnosing student thinking and in modifying instruction accordingly. A shift in emphasis from "teacher as consumer of knowledge about teaching" to "teacher as producer of knowledge about teaching" is desirable in designing in-service activities for teachers (p. 72). Professional Development Standard D states:

> Professional development for teachers of science requires building understanding and ability for lifelong learning. Professional development activities must...provide opportunities to learn and use the skills of research to generate new knowledge about science and the teaching and learning of science. (NRC, 1996, p. 68)

The premise is that teachers may find discussing data they have collected in their own classrooms more interesting and directly helpful than in-service sessions in which they hear experts talk about applying the results of formal research to teaching.

Teachers' inquiries also may yield insights and information that outside researchers would not be able to access or generate. The potential for teachers' inquiries to inform and reform educational practices motivated an initiative by the Spencer Foundation. In 1996, the Foundation launched a program to support research on ways to increase communication and mentoring among practitioner researchers because:

> ...research conducted in school sites by educational practitioners may offer specific and useful knowledge about education which can best be, perhaps only be, generated out of the experience of the practitioner. (1996, p. 32)

Funding can provide support for regular substitutes who engage students in on-going and coherent instructional experiences while the teachers work on their research. Such support can also enable teachers to present at conferences and to devote summers to analysis and writing.

Interest in teachers' accounts of their practices is not new. More than half a century ago, John Dewey recognized the potential of teachers to contribute to knowledge about teaching:

> This factor of reports and records does not exhaust, by any means, the role of practitioners in building up a scientific content in educational activity. A constant flow of less formal reports on special school affairs and results is needed.... It seems to me that the contributions that might come from classroom teachers are a comparatively neglected field; or, to change the metaphor, an almost unworked mine. (Wallace, 1997, pp. 26-27)

In particular, Dewey advocated such inquiries in progressive schools that emphasized self-initiated and self-conducted learning:

> The method of the teacher...becomes a matter of finding the conditions which call out self-educative activity, or learning, and of cooperating with the activities of the pupils so that they have learning as their consequence ...A series of constantly multiplying careful reports on conditions which experience has shown in actual cases to be favorable or unfavorable to learning would revolutionize the whole subject of method. (pp. 125-126)

Developing such case studies of student learning is a way that teachers can contribute to the knowledge base for improving instruction.

WHAT DO TEACHERS INQUIRE ABOUT THEIR OWN TEACHING PRACTICES?

Many teachers feel uneasy when first formulating research questions, particularly if they think of research as involving the treatment and control groups typical of traditional studies. "What are you curious about in your classroom?" may not seem appropriate. Most teacher research is interpretative, however; both the questions and the analyses evolve throughout the research process. In a preface to a special issue of *Teacher Research: The Journal of Classroom Inquiry*, the editors describe the beginning of such inquiries:

> It might be a question that darts into your mind and you take the time to ponder it, rather than brushing it aside with an "I'm too busy to think about it now" shrug. Or it might be your decision to look a little more closely at that kid who is driving you crazy in class and keeping you awake at night. It could be the first time you put pen to paper in a teaching journal, press the record button on a tape recorder, or sit down to talk with a teaching colleague about something you wonder about. But it begins somewhere. (1998, p. v)

In addition to papers by both beginning and experienced teachers, this journal includes a tool box section in which teacher researchers describe some of their research methods.

Some teachers have documented and interpreted changes that were occurring in their schools. Such studies are appropriate for publication in *Teaching and Change*, a publication of the National Education Association:

> *Teaching and Change* provides an open forum for reporting the experiences of classroom teachers as they learn how schools must change to make good practice possible. The journal is devoted to helping teachers as they work to strengthen their learning communities. Issues discussed in *Teaching and Change* include what is taught, how it is taught, and the different ways schools are organized. (1997)

Articles in this journal have traced the results of changes in classroom practice, examined ways of teaching particular processes, explored the effects of program restructuring, and investigated development of student understandings.

Sometimes teachers get started by participating in study groups in collaboration with an university researcher. In a chapter of a book co-edited by such a group, for example, Charles Pearce reported his thinking at the beginning of a research project:

> What if, I thought while driving home that day, I tried giving fifth-grade students boxes of materials with no directions, no packets of activity cards? There would be no hidden agendas, no regimented steps to follow, no expected outcomes. Would learning take place? Could that learning be assessed? Would this approach engender higher-level thinking and enable the students to monitor and evaluate thinking processes? Could the curriculum still be addressed if students were afforded a wide range of choices? What if.... (1993, pp. 53-54)

Pearce's chapter includes many examples of student work, such as questions students wrote on a Question Board, a student's entries on a Know, Wonder, Learn form for an activity on mealworms, a completed discovery log for a crayfish activity, a sign-up list for an inquiry period, a completed inquiry period log sheet, a completed tower workshop log sheet, several examples of completed "What I Accomplished" forms, the first page of a class Body of Knowledge booklet, and a story written by a student. Pearce's inquiries eventually developed into a book of his own published in 1999, *Nurturing Inquiry: Real Science for the Elementary Classroom*, in which he presents and discusses a wide variety of data from his students' explorations.

Teachers can gain information and inspiration from the work of their colleagues. Barbara Bourne, for example, reports an outgrowth of Pearce's "What If" thinking in a chapter in a second book by their teacher-researcher group, *Beyond the Science Kit: Inquiry in Action*, published in 1996. Bourne's chapter describes her students' experiences at a Kids' Inquiry Conference during which they shared findings from their investigations of topics of mutual interest. Reading this chapter inspired a fourth-grade teacher, Diantha Lay, to consider ways to shift from competitive science fairs to more collaborative contexts at her school. She discusses her experiences in her contribution to the collection here.

Some teachers have written books that focus upon particular aspects of their practices. Karen Gallas describes her inquiries, for example, in *Talking Their Way into Science: Hearing Children's Questions and Theories, Responding with Curriculum:*

> This book is about science. But it is also about a question. It is intended to be a very focused look at one aspect of science teaching and learning: Talk. Within the realm of talk, it focuses on a very particular kind of talk—that is, dialogue among children.... What I will describe in this book is how our practice of Science Talks developed in my primary classroom in response to my own question as a teacher researcher. My reflections will focus alternatively on what "real" science is, on the study of science in schools, on children as thinkers, on the role of theory in the science classroom, on the nature of collaboration and discussion, on different kinds of talk, on the acquisition of a discourse, on the teacher's role in science instruction, and on the social construction of learning. In this process, I will necessarily share the details of some of my work as a teacher researcher, and those details also will illuminate the ways in which the act of teaching and learning evolved in my classroom. (1995, p. 1)

Gallas' book includes many examples of students talking with one another and their teacher, along with complete transcripts of two talks in the Appendix. Gallas espouses many of the values that a university researcher, Jay Lemke, stressed in his study, *Talking Science: Language, Learning, and Values*, but she discusses these from the perspective of a teacher who is sharing with other teachers a deep knowledge of what has worked well for her, how, and why.

HOW DO TEACHERS INQUIRE INTO THEIR OWN
TEACHING PRACTICES?

Many teachers collect data as part of their usual ways of doing things, such as writing down their thoughts about a lesson and ways they might make changes next time. In reflecting upon her work, Peggy Groves comments upon the additional demands of research:

> The difference between my recent classroom research and my usual classroom practices is that for my research I kept notes about what I did, I looked more closely at what happened, I asked myself harder questions, and I wrote about it all. These differences took a lot of time, but I think I'm a better teacher for it. And maybe even a better writer. (Hubbard & Power, 1993, p. xv)

Such reflective writing can help develop questions that can guide further explorations. Talking with colleagues also can be useful. In an article about collaborative action research, for example, Allen Feldman (1996) describes ways in which a group of physics teachers generated and shared knowledge that enhanced their normal practices.

Systematic inquiries require time, resources, and a collegial milieu. The support of school administrators is critical. The *National Science Education Standards* articulated the kind of school support necessary:

> Program Standard F: Schools must work as communities that encourage, support, and sustain teachers as they implement an effective science program.... Schedules must be realigned, time provided, and human resources deployed such that teachers can come together regularly to discuss individual student learning needs and to reflect and conduct research on practice.... Time must be available for teachers to observe other classrooms, team teach, use external resources, attend conferences, and hold meetings during the school day.... For teachers to study their own teaching and their students' learning effectively and work constructively with their colleagues, they need tangible and moral support.... As communities of learners, schools should make available to teachers professional journals, books, and technologies that will help them advance their knowledge. These same materials support teachers as they use research and reflection to improve their teaching. (NRC, 1996, pp. 222-223)

Such standards can provide guidelines for both teachers and administrators who choose to initiate teacher researcher programs.

Extensive teacher researcher programs may include school-based study groups, district-wide seminars, newsletters, and conferences. Teachers in Fairfax County, Virginia, for example, initiated a teacher research network that publishes a quarterly newsletter, *The Networker*, and mounts an annual Teacher Research Conference where teachers can present their work.

One of the first books to provide guidance for teacher researchers was published in 1993, *Inside/Outside: Teacher Research and Knowledge* by Marilyn Cochran-Smith and Susan Lytle. Others include *Research and the Teacher: A Qualitative Introduction to School-Based Research* by Graham Hitchcock and David Hughes. The editors of the journal *Teacher Research*, Ruth Hubbard and Brenda Miller Power, have published two guides, *The Art of Classroom Inquiry* in 1993 and *Living the Questions: A Guide for Teacher Researchers* in 1999. The latter includes research strategies illustrated with examples, guidelines for setting up school-wide inquiry groups, ethical considerations, and advice from veteran teacher researchers on many topics.

Angelo Collins and Samuel Spiegel (1997) provide advice on doing action research as part of a collection of science teachers' studies in *Action Research: Perspectives from Teachers' Classrooms*. In *Probing Understanding*, Richard White and Richard Gunstone (1992) describe many techniques that teachers can use to diagnose their students' thinking. These include asking students to represent connections among ideas with a concept map or list of word associations; inviting students to justify predictions and then to reconcile their predictions with observations; interviewing students about instances, events, or concepts; interpreting students' drawings, line graphs, or relational diagrams; and assessing questions students produce in response to various prompts.

HOW CAN PROSPECTIVE TEACHERS LEARN HOW TO DO RESEARCH WHILE THEY LEARN TO TEACH?

As a new instructor of courses on methods of teaching science in elementary school (1998b), I wanted to prepare prospective teachers to do research as well as to teach. My vision of teachers as researchers reflects my experiences collaborating with the co-editor of this volume, Jim Minstrell, in trying to understand how he used questioning to guide student thinking (1997a,b). Minstrell (1989) had established a research site in his high school physics classroom where he, his students, several colleagues, and university researchers such as myself all collaborated on studies of learning and teaching. I also drew upon my experiences in collaborating with teachers on an investigation of questioning during conversations about science (van Zee et al., in press). In interpreting dialogue, we used methods derived from my graduate studies with an ethnographer of communication, Gerry Philipsen (1982, 1992). The collection here includes case studies of questioning developed by two primary teachers, Marletta Iwasyk and Akiko Kurose, an upper elementary school teacher, Judy Wild, and a high school physics teacher, Dorothy Simpson. I had met these teachers while assisting in physics programs at the University of Washington (McDermott, 1996).

My approach to teaching science teaching is similar to that of instructors such as Sandra Abell and Lynn Bryan (1997) who emphasize reflective practices. Like Cronin-Jones (1991), I use interpretive research methods in teaching teachers. Many of these are similar to ways that George Posner (1985) recommended to student teachers for reflecting upon their field experiences. Activities and assignments in my course include a joint analysis of factors that foster science learning, development of a personal framework for science teaching, a sustained inquiry into a natural phenomenon, research on learning and teaching, and formulation of a research question for the final.

Joint Analysis of Factors That Foster Science Learning

The prospective teachers begin learning to do research at the beginning of my course on methods of teaching science in elementary school. The opening activity is an example of eliciting experiences from a variety of individuals and identifying common themes. I ask the prospective teachers to think about experiences inside or outside of school in which they have enjoyed learning science. They draw pictures of these experiences, write captions, and identify factors that fostered their learning in these instances. Then members of each group make a poster with their drawings and jointly construct a list of factors that fostered science learning across their experiences. They introduce themselves to the class by showing their poster, describing their experiences, and stating the factors they identified. Then we construct a list of factors common to all the groups.

On the first day of the fall 1998 class, for example, one of the prospective teachers drew a picture of herself moving in front of a computer connected to a motion detector. She and her classmates constructed the following list of factors that foster science learning: hands-on, relating to real life, interesting and fun, different environments, working in groups, a sense of anticipation, trial and error, creativity, asking and answering questions, student-centered, self-discovery, curiosity. These were a good match to aspects of the teaching standards advocated in the *National Science Education Standards* (1996) but emerged from the prospective teachers' own analyses of their positive experiences in learning science. Although many thought they had not heard the phrase "inquiry approaches to teaching and learning" before, I interpret these findings to mean that these prospective teachers were entering my course with substantial prior knowledge on which to build.

When I asked the prospective teachers to raise their hands if the factors they had identified were typical of their science learning experiences, however, few hands went up. The positive experiences they had remembered and drawn had been unusual. Few seemed to remember studying much science of any kind in elementary school. Most had had negative experiences in high school and college science classes. We interpreted the results of this informal survey as evidence for the need for reform.

Development of a Personal Framework for Science Teaching

Throughout the semester, the prospective teachers continue reflecting upon factors that foster science learning. They write weekly journals that first describe science learning events they observe or experience themselves and then reflect upon factors that fostered science learning in these instances. One of them wrote a journal, for example, in which she described in more detail her experiences in the physics course that she had remembered on the first day of class:

> In our science methods class, I drew a picture of a science learning process where I was involved as a student. Last semester, I took a class called Physics for Elementary School Teachers. The class was taught dramatically different than any other science class I've taken. We designed our own experiments and created our own formulas. The constants in the formulas were values that were results of experiments we did in class. We also formulated our own definitions for scientific terms. (This can be harder than it sounds.) It was common knowledge that the professor and teaching assistants were not the sources of answers. If we asked a question, they would answer it with a question. If an experiment was necessary to answer the question, they would point us to the materials. One of my favorite experiences was one that my partner and I designed...(describes experiment). The

> most important part of this experiment was not the value of the initial temperature of liquid nitrogen. It was that we were able to design our own experiment and solve for the temperature using an equation we created. The class equipped us with a confidence in science that motivated us to persist until we found the answer. (First reflective journal, Spring 1998; emphasis added on the last day of class)

Thus week by week, the prospective teachers continue to reflect upon science learning in progress. At the end of the semester, they analyze these self-generated data for common themes and then use these themes to build personal frameworks for science teaching and learning. The prospective teacher who wrote the journal above, for example, underlined sentences in which she had stated factors that fostered learning, cut these out, and sorted them into a pile along with similar statements from later journals. These she taped together on a sheet of paper and wrote a summary statement at the top: "Students should learn to develop their own questions and design experiments to answer those questions." For the final, she used this and other themes to articulate recommendations for science teaching. For example, she wrote "Teachers should model scientific inquiry by encouraging students to develop their own questions and design experiments to answer their questions. This will increase the students confidence in science." As part of the final, the prospective teachers also present a lesson on a topic of their choice and describe how they would meet their recommendations in this context. Through this process, they have used a research technique to develop and elaborate their own principles for action as science teachers.

Sustained Inquiry

We also engage in a sustained inquiry about a natural phenomenon, the changing phases of the moon, in a manner similar to that described by Eleanor Duckworth (1987). I draw on *Where is the Moon?*, developed for students by *Elementary Science Study* (1966), and the astronomy section of *Physics by Inquiry*, developed for teachers by Lillian McDermott and the Physics Education Group at the University of Washington (1996).

At the beginning of the semester, the prospective teachers record their current knowledge about the moon, the nature of scientific explanations, and inquiry approaches to learning and teaching. Their assignment is to look at the sky daily, enjoy what they see, and record their observations if they see the moon. If they cannot see the moon, they record that too. In class, they share their observations with one another and generate questions to guide further observation. Eventually we go outside on a sunny day when the moon is visible and hold up balls so that the lit portion of the ball matches the shape of the lit portion of the moon. The prospective teachers move the balls so that the pattern of the changing shape of the lit portion of the ball matches the pattern of the changing phases of the moon that they have observed. We move inside to work with a bright light, ping pong balls, and themselves to model the sun, moon, and Earth. They then write papers that present their observations, articulate the explanatory model we developed in class, and reflect upon changes in their understandings of the phases of the moon, of the nature of scientific explanations, and of inquiry learning and teaching.

In reflecting upon this sustained inquiry about the phases of the moon, a prospective teacher wrote:

> I can now see that our moon project was specifically designed to model this for us. We students were active participants in the shaping of our own learning. Sure, we were directed to observe the moon on a daily basis, but at the same time we were allowed the freedom to interpret our findings in a way that made sense to us. I know that I personally, came up with many "whys" and "why nots" along the way and that this only served to further motivate me in ways that I never thought possible. Because of these questions that kept popping up as a result of my ongoing observations, I was excited to observe even further to find all of the answers. I think that this is what the "inquiry approach to learning and teaching" is all about. It's almost as if a sense of curiosity is aroused in the individual that can only be satisfied through further inquiry. It kind of builds upon itself. I for one enjoyed the whole experience and plan to take this approach with my own students some day!

My conviction is that prospective teachers need such experiences themselves in order to envision the approaches to science teaching and learning that I advocate in the course.

Research on Learning and Teaching

The major assignment for the semester is a research project that each prospective teacher conducts in the placement setting. They all are placed in schools with diverse populations of students, many from low-income immigrant families. The project involves exploring resources for teaching science in this setting, consulting with the mentor teacher to identify a science topic to teach later in the semester, examining ways various curricula present the topic, identifying relevant children's literature and technology resources, interviewing children to hear how they think about the topic before instruction, and designing a conversation about the topic. I use the phrase "conversation about science" to refer to the lesson in order to emphasize the importance of engaging students in discussing what they think. I use the term "design" because I require more than the usual components of a lesson plan. A design for a conversation about science also includes specifying questions to elicit student thinking, discussing accommodations for children with special needs, indicating ways to integrate across the disciplines, and making connections to district, state, and national standards. Each of the prospective teachers also formulates a research question to explore in this context. Before teaching and researching in their placement settings, the prospective teachers prepare by teaching and researching with peers in class. They collect data such as tape recordings of their lessons, copies of students' work, journal reflections, etc. The final product is a reflection about what happened both in the teaching and the researching.

Formulation of a Research Question for the Final

For the final, the prospective teachers write about ways in which they would teach lessons of their choice to meet the recommendations they developed as part of their personal framework for science teaching. They also formulate research questions that they can examine during their student teaching. Trisha Kagey, for example, elaborated her initial question:

The Role of Science Journals in a Science Inquiry Classroom

I want to explore the role of science journals in a science inquiry classroom. I believe that science should be integrated with other content areas, including language arts. Writing and drawing are methods of expression and communication. I am curious as to how students communicate their observations and results from investigations in their science journals. I also want to explore how science journals can be involved in the questioning, designing, experimenting, and communicating phases of scientific inquiry. I plan to discover the teacher's role in encouraging thoughtful responses in scientific journals.

My hope is that formulating a research question for the final will encourage prospective teachers to focus on an issue that interests them during their student teaching. Kagey, for example, was able to present her work in progress the following semester at a research festival that I had organized. She was able to continue her research during student teaching because of her placement with an earlier graduate of my course, Deborah Roberts. We are attempting to create a community of teacher researchers so that student teachers can be mentored in researching as well as teaching during their student teaching semester.

HOW CAN BEGINNING TEACHERS BUILD THEIR EXPERTISE IN RESEARCHING AS WELL AS IN TEACHING?

Forming or joining a teacher-researcher group can provide support for teachers interested in inquiring into their own teaching practices. The Science Inquiry Group (SING), for example, includes student teachers, beginning teachers, experienced teachers, and myself, a university instructor. With funding from the Spencer Foundation Program for Practitioner Research, we have been meeting monthly after school to share experiences and insights about science learning and teaching (van Zee, 1998a). Initially we met as one group in a participant's classroom. This year the original group is meeting at a more central location, in a local library, and a new group has formed within a participant's school. An on-site group has many advantages such as sharing equipment, ideas, and encouraging words on an on-going basis.

SING participants are developing case studies of science learning and teaching. We try to focus upon positive aspects of our practices: What are we doing that is working well, about which we might collect some data as we teach in order to understand better what is happening, so that we can communicate these successes to our colleagues? This paper, for example, is a case study of what I do as an instructor of courses on methods of teaching science and as the initiator and facilitator of a teacher researcher group.

Our case studies evolve through a complex process. We have found that a good way to start is to write an abstract. We prepare these abstracts for Research Festivals in which the teachers discuss their ongoing research with prospective teachers enrolled in my courses on methods of teaching science. Writing abstracts, collecting some data, and engaging prospective teachers in discussing these data seems to be a good way to begin making progress and to build self-confidence in doing research.

Some members of the group are interested in going to conferences and presenting their work in more formal settings. Presentations at the Research Festivals form the basis for writing proposals to do so. The Research Festivals also serve as rehearsals for our conference presentations. Typically we propose individual papers to be presented together in a group session. We begin by having presenters briefly introduce their case studies; then we divide into small groups so that the presenters can discuss their research in a non-threatening environment with a lot of interaction with participants in the session; we close with a whole group discussion based upon issues that have emerged in conversations within the small groups. Presenters usually hand out copies of their case studies as works in progress. Further refinement occurs as those interested prepare their case studies for publication. An example is "The Sky's the Limit: Parents and First-grade Students Watch the Sky" by Deborah Roberts (1999) published in *Science and Children.*

HOW CAN UNIVERSITY FACULTY FOSTER INTERACTIONS AMONG PROSPECTIVE AND PRACTICING TEACHER RESEARCHERS?

One of the most important opportunities that the Science Inquiry Group (SING) provides is for interaction among prospective and practicing teacher researchers. At the beginning of each semester, the SING teachers present their case studies at the Research Festival that we hold jointly with my course on methods of teaching science in elementary school. This enables me to put the undergraduates in direct contact with practicing teacher researchers. We meet after school in the classroom of a graduate of my fall 1995 class who is now a third-year teacher. At the spring 1999 Research Festival, presenters included an undergraduate student teacher, two first-year teachers, and several earlier graduates of my course as well as some experienced teachers. The graduates were able to show how the research questions that they had formulated in my courses have evolved into their current projects.

After the SING teachers discuss their own research, each helps a small group of the undergraduates plan a lesson to conduct in the SING teacher's classroom and also to formulate a research question to examine in this context. Each small group of undergraduates then visits their SING teacher's classroom to observe this teacher in action and to complete their collaborative planning. Next they try out their teaching and researching with peers in my class at the university. After doing the same with children in the SING classrooms, the small groups use my next class to reflect upon their experiences by developing posters that report their findings. The following class is again a joint meeting with the Science

Inquiry Group at which they discuss with the SING teachers their teaching and researching experiences in these classrooms. I intend this complex process to provide hands-on experiences in teaching through inquiry, in authentic contexts, in collaboration with practicing teachers who teach science this way, with opportunities to formulate and explore the prospective teachers' own questions about science teaching and learning. In other words, I am attempting to teach science teaching through inquiry.

This process requires five class sessions and two monthly meetings of the Science Inquiry Group. An evaluation of this process at the close of the Spring 1999 joint meeting indicated that both groups seemed to value this investment. The prospective teachers were pleased to have experience teaching in a realistic setting. One wrote, for example, "It was REAL!! We were working with students. This is much better than any lesson we could learn sitting and watching a teacher." They also appreciated the opportunity to observe and practice inquiry-based teaching. One commented, "As for the inquiry lesson, I learned a great deal. The children were active learners; this stuff really worked! I was really happy after teaching the lesson; I had this feeling of accomplishment." In addition, they enjoyed meeting with Science Inquiry Group teachers and talking with them about teaching and researching. One noted, "The SING meetings themselves were extremely valuable. I thoroughly enjoyed hearing the ideas and views of practicing SING teachers, as well as their input on what we are doing."

The prospective teachers also gained confidence in themselves and in teaching science. One wrote, "I really learned a lot from both the students and my colleagues but also from myself. This was the first time I ever stood in front of fourth graders. I was really surprised and pleased by my own confidence and performance." Another wrote, "I learned that I don't need to know everything about science in order to teach it. The students will come up with their own wonderful ideas as long as I facilitate and ask questions to keep them thinking." Several commented on the importance of having an opportunity to exchange ideas and views with others. One noted, "We worked in groups which exposed us to new ideas and views." Some indicated they had experienced a change in attitude toward science. One commented, "The investment of five class periods was very worthwhile and my experience in a science classroom has changed to very positive (since I wasn't very enthusiastic about science before)."

The Science Inquiry Group teachers felt they too had learned from the process. One of the SING teachers, for example, wrote about her experiences with the prospective teachers, "They continue to "stretch" me and force me to reflect on what I am doing. It gives me a chance to observe/critique their teaching, which is helpful in critiquing myself." One mentioned that the children also had gained by having guest teachers teach them a lesson.

ENDNOTE

1. The development of this manuscript was supported in part by a grant from the Spencer Foundation Practitioner Research: Mentoring and Communication Program. The data presented, statements made, and views expressed are solely the responsibility of the author. An earlier version of this paper was presented at the 20th Annual Ethnography in Education Research Forum at the University of Pennsylvania, Philadelphia in March 1999 under the title "Facilitating Reflective Research in School Settings: The Science Inquiry Group."

REFERENCES

Abell, S.K. and L.A. Bryan. 1997. Reconceptualizing the elementary science methods course using a reflection orientation. *Journal of Science Teacher Education* 8:153-166.

Bourne, B. 1996. The kids' inquiry conference: Not just another science fair. In *Beyond the science kit: Inquiry in action,* edited by W. Saul and J. Reardon, pp. 167-188. Portsmouth, NH: Heinemann Press.

Cochran-Smith, M., and S.L. Lytle. 1993. *Inside/outside: Teacher research and knowledge.* New York: Teachers College Press.

Collins, A., and S.A. Spiegel. 1997. So you want to do action research? In *Action research: Perspectives from teachers', classrooms*, edited by J.B. McDonald and P.J. Gilmer, pp. 117-127. Tallahassee, FL: SouthEastern Regional Vision for Education.

Cronin-Jones, L. 1991. Interpretive research methods as a tool for educating science teachers. In *Interpretive research in science education*, edited by J. Gallagher, 217-234. Monograph No. 4. National Association for Research in Science Teaching.

Dewey, J. 1928/1956/1959. Progressive education and the science of education. In *Dewey on education selections*, edited by M.S. Dworkin, 113-126. New York: Teachers College Press.

Doris, E. 1991. *Doing what scientists do: Children learn to investigate their world*. Portsmouth, NH: Heinemann Press.

Duckworth, E. 1987/1996. Teaching as research. In *The having of wonderful ideas and other essays on teaching and learning*, edited by E. Duckworth. 2nd ed. New York: Teachers College Press.

Education Development Center. 1966. *Elementary science study*. Nashua, NH: Delta Education.

Feldman, A. 1996. Enhancing the practice of physics teachers: Mechanisms for the generation and sharing of knowledge and understanding in collaborative action research. *Journal of Research in Science Teaching* 33: 512-540.

Gallas, K. 1995. *Talking their way into science: Hearing children's questions and theories, responding with curriculum*. New York: Teachers College Press.

Hitchcock, G. and D. Hughes. 1995. *Research and the teacher: A qualitative introduction to school-based research*. London, UK: Routledge.

Hubbard, R.S., and B.M. Power. 1993. *The art of classroom inquiry*. Portsmouth, NH: Heinemann Press.

Hubbard, R.S., and B.M. Power. 1999. *Living the questions: A guide for teacher-researchers*. York, ME: Stenhouse Publishers.

Lemke, J. 1990. *Talking Science: Language, learning and values*. New York: Ablex.

McDermott, L.C. 1996. *Physics by inquiry*. New York: Wiley.

Minstrell, J. 1989. Teaching science for understanding. In *Toward the thinking curriculum: Current cognitive research*, edited by L.B. Resnick and L.E. Klopfer, 131-149. Alexandria, VA: Association for Supervision and Curriculum Development.

National Education Association. 1997. Information for authors. *Teaching and Change*. 4(2).

National Research Council. 1996. *National science education standards*. Washington, DC: National Academy Press.

Pearce, C.R. 1993. What if...? In *Science workshop: A whole language approach*, edited by W. Saul, J. Reardon, A. Schmidt, C. Pearce, D. Blackwood, and M.D. Bird, 53-77. Portsmouth, NH: Heinemann Press.

Pearce, C. R. 1999. *Nurturing inquiry: Real science for the elementary classroom*. Portsmouth, NH: Heinemann Press.

Philipsen, G. 1982. The qualitative case study as a strategy in communication inquiry. *The Communicator* 12: 4-17.

Philipsen, G. 1992. *Speaking culturally: Explorations in social communication*. New York: SUNY Press.

Posner, G. J. 1985. *Field experience: A guide to reflective teaching*. New York: Longman.

Roberts, D. 1999. The sky's the limit: Parents and first-grade students observe the sky. *Science and Children* 37(1): 33-37.

Spencer Foundation. 1996. *Twenty-five years of grantmaking: 1996 annual report*. Chicago, IL: Author.

Staff. 1998. A note from the editors. *Teacher Research: The Journal of Classroom Inquiry.* 5(2): v.

van Zee, E.H. 1998a. Fostering elementary teachers' research on their science teaching practices. *Journal of Teacher Education* 49:245-254.

van Zee, E. H. 1998b. Preparing teachers as researchers in courses on methods of teaching science. *Journal of Research on Science Teaching* 35:791-809.

van Zee, E.H., M. Iwasyk, A. Kurose, D. Simpson, and J. Wild. In press. Student and teacher questioning during conversations about science. *Journal of Research in Science Teaching.*

van Zee, E. H., and J. Minstrell. 1997a. Reflective discourse: Developing shared understandings in a high school physics classroom. *International Journal of Science Education* 19: 209-228.

van Zee, E. H., and J. Minstrell. 1997b. Using questioning to guide student thinking. *The Journal of the Learning Sciences* 6: 229-271.

Wallace, J. 1997. A note from John Dewey on teacher research. *Teacher Research: The Journal of Classroom Inquiry* 5(1): 26-28.

White, R., and R. Gunstone. 1992. *Probing understanding.* New York: The Falmer Press.

Learning to Teach Science Through Inquiry: A New Teacher's Story[1]

Deborah L. Roberts

M y undergraduate courses introduced me to learning through inquiry and doing teacher research. That was the beginning of a journey I am still enjoying as a third-year teacher. The road is not always direct, the questions often change, but the learning is fulfilling, enlightening, and fun!

LEARNING TO INQUIRE

My first experience in learning through inquiry was in an undergraduate physics course for future elementary and middle school teachers taught by Dr. John Layman. Dr. Layman is an expert in inquiry teaching (Layman, Ochoa, & Heikkinen, 1996; National Research Council [NRC], 1996). In this class I learned a lot about physics, but I also learned that it can be exciting to learn! I vowed during this class that this was the method of teaching I would try to model in my future classroom.

My first experience in doing teacher research was in a course on methods of teaching science that was taught by Dr. Emily van Zee (1998b). As described elsewhere in this collection, she required weekly observations of "science learning in progress" in our placement classrooms or in our personal experience. The culminating assignment involved rereading all of our entries and looking for commonalties. We each came up with an individual list of the recurring themes in our observations. The last day of class, we shared these themes. I discovered four common themes among all of our lists: science learning needs to be hands on; science learning needs to follow the natural curiosities of the children; science learning needs to make real-world connections for children; and there needs to be a dedicated caring teacher who is able and willing to develop a sense of community in the classroom so that risk-taking will occur.

This discovery of what I considered to be the four essential conditions to foster science learning shaped my philosophy of teaching. All of these conditions had been evident in my physics class and my science teaching methods class. These discoveries also enabled me to do my student teaching according to my philosophy and to withstand criticism from my cooperating teacher and teaching supervisor, neither of whom appeared to have an understanding of inquiry teaching.

HELPING TO FORM A RESEARCH COMMUNITY:
THE SCIENCE INQUIRY GROUP

During my first year of teaching, I helped form a group of teachers who were interested in doing reflective research in collaboration with Dr. van Zee (1998a). This version of teacher research differs from action research in its focus on positive aspects of the teachers' practices. The purpose of collecting data is to communicate to others what is working well rather than to guide actions for improving problematic situations. We call ourselves the Science Inquiry Group or SING. Among our goals are building our capacities to do research, communicating our findings to others, and supporting one another in our efforts to do research. We meet once a month after school to share our experiences and ideas about teaching science in ways that the *National Science Education Standards* (NRC, 1996) suggest.

At our first meeting, we discussed what our research questions would be. Formulating a research question was very difficult for me. What could I research as a first-year teacher that would be beneficial to anyone? I had no idea that many experienced teachers feel this way too (Hubbard & Power, 1993). Encouraged by my colleagues, I came up with an idea. My very first attempt was called "Teaching Other Content Areas Through a Science Perspective." Although I was unsure of what exactly I would be researching, I wrote up an abstract to send off to an ethnography conference as a possible presentation of work in progress (Roberts, 1997). I have a strong belief that doing research should not be solely to improve teaching practice but also to share with other teachers who might be able to use what I have learned or experienced. The ultimate goal, of course, is to benefit students, particularly students like mine, many who come from low-income Central and South American immigrant families.

PARTICIPATING IN A RESEARCH FESTIVAL

Throughout my first semester of teaching, I had been making careful observations in the classroom. I tried to write my own reflections, as well as tape student discussions, make copies of student work I thought might shed light on this research, and from time to time, (when the school's video camera was available) videotape activities in the classroom. All of these are typical methods for teacher research that Hubbard and Power describe in their books (1993, 1999).

In December 1996, I participated in a Research Festival that Dr. van Zee (1998a,b) had organized to bring together students in her course on methods of teaching science and teacher researchers in the Science Inquiry Group. In preparing for this, I had been in a panic. There was no way I had anything that was defined enough to share with other teachers or methods students. I finally put something down on paper and shared it with some very eager, enthusiastic, and kind methods students, who appeared to be quite interested in what I had to say. All of the reviews of the Research Festival from the perspective of the methods students seemed to be favorable (surprise again!). One student wrote,

> It was valuable because we got the chance to see and hear about teachers (real ones) helping students think and learn in a way that is compatible with what we have been learning in school. It has been very frustrating to hear what we learn does not fit in with the way the real world works, and these teachers have shown us that it can.

It was very motivating to me to be able to provide a firsthand view of what I was doing and to help create a vision for the methods students so that they could see that doing research is valuable, even for a beginning teacher.

RESHAPING MY RESEARCH QUESTION

My initial research question gradually was honed to "In what ways can I integrate science to motivate students to do expository reading and writing?" This happened rather unintentionally. I had this great idea at the beginning of the school year to take my first graders on a nature walk. I wanted to find a way that I could integrate some expository reading, and felt that a nature walk might help provide motivation. As a result of the students' continual questions and observations about what was happening in the natural world outside of the building, we went on a walk to see what we could find that was nature. We had decided on a definition of what nature was after a lengthy class discussion, in which one student offered a fairly succinct definition—"Nature is anything that can't be made by people."

We each collected some nature items such as leaves, sticks, small rocks, acorns, wood chips, weeds, flowers, and moss. I had the children come back into the classroom, sit in a circle, and share what they had found. We then took turns gluing our own items to a piece of tag board. The children had generated many, many questions as a result of this activity. I made an experience chart by soliciting from each student one idea to share with the class.

Next, I modeled writing about what I had experienced on this walk. I asked the children to write about their experience, too. The children were eager to do just that. With a peer and then with me, they each shared what they had written. After some editing, and a little bit of research by one student about what kind of leaf he had found, each had one page to put into a book we made about nature walks.

I had not intended for this to be an ongoing activity. The students, however, began bringing things in from home and from recess. They were looking for books about trees and flowers, insects, squirrels, and seasons whenever they visited the school media center or browsed in our classroom library. They wrote in their daily journals about things they had observed on the nature walk, or their ideas on why things were the way they were. I was persuaded by the overwhelming demand of the students to keep up the nature walks on a monthly basis. They maintained their enthusiasm for reading and writing about nature. We ended up making five different class books.

After a few months of nature walks, a surprising thing happened. Because I was so focused (thanks to my research question) on this project, I started to notice that the children were no longer putting their collection items on the tag board randomly. They were beginning to become particular about where things went and what they were. Now I began paying attention to how they were putting their items on the tag board and why they put them where they did. I feel that their need to organize things sustained their motivation for expository reading because they wanted more information. Table 1 is a transcript of a conversation during the time we were putting nature items onto the tag board.

TABLE 1. CONVERSATION ABOUT THE NATURE COLLECTION

1	**Student 1:** Ms. Roberts, Ms. Roberts, Look what I got!
2	Ms. Roberts, what is it? Is it real nature?
3	**Student 2:** You got real nature, it looks like a bark to me.
4	**Teacher:** Let's sit down and make our nature collection.
5	**Student 3:** Where should we put the sticks?
6	**Student 1:** But what do I have? Is it a stick?
7	**Teacher:** Who has an idea about where things should go?
8	**Student 4:** Ms. Roberts, can we put the rocks at the top so they don't fall off and put the sticks at the bottom?
10	**Student 5:** I think we should do that, yeah, that's a good idea.
11	**Teacher:** Everyone OK with that?
12	**Student 6:** Who has brown leaves? I have brown leaves.
13	**Student 3:** I have sticks.
14	**Student 7:** Why are the leaves different colors? I have reddish brown ones.
15	**Student 8:** Because they are different trees.
16	**Student 9:** My mom said that they have different designs because they do.
17	**Student 10:** Can we get some more stuff? I didn't get any nuts.
18	**Teacher:** What nuts? Where do you see nuts?
19	**Student 10:** Like she has—those....
20	**Student 9:** These aren't nuts, these are acorns. Acorns with holes in 'em.
21	**Student 10:** Let me see the holes!
	Many students asking to see the holes.
22	**Teacher:** Let's pass that around so everyone can see.
23	**Student 10:** Why does it have holes? How did they get there? Can we crack it open?
24	**Student 9:** Don't crack mine!
25	**Teacher:** Maybe we can look for some others with holes.
26	Look on your way home from school today and see if you can find some with holes, and bring them tomorrow.

28   Let's start gluing things on the collection, ok?

29   **Student 2:** Can I put mine on first?

30   **Student 1:** Ms. Roberts, I don't know what I have. Do you know?

31   **Students 8 and 2:** Let me see it.

Teacher holds up object and asks for input.

32   **Teacher:** Does anyone have any ideas about what this might be?

33   **Student 2:** It looks like bark.

34   **Student 7:** It looks like a moon thing because it has bumps on it.

35   **Teacher:** What do you mean it has bumps on it?
     Why do bumps make it like the moon?

36   **Student 7:** In my library book, there are pictures of the moon.

37   and it has these. . . like bumps on it.

38   But that thing is like a part of a tree,

39   but it looks like the moon because it has bumps on it like the picture.

The students' enthusiasm for this activity was obviously high (lines 1-2, 17, 21, 23, 29). I was impressed that while we were making our collections, the students were answering one another's questions (lines 2-3, 8-10, 14-16) and not relying on me to be the central information source. They had also improved on the dynamics of conversation: they were able to take turns (lines 12-17), wait (at times) for one another to finish talking, and show they were listening by commenting on the responses they had been given (lines 21, 23, 24, 36).

Listening to the audiotapes of the children's conversations taught me many things. Often, I was able to hear conversations between two or three students who were talking at the same time the group was talking. Although they were not sharing with the class as a whole, they were obviously very much engaged in what was going on. I learned about myself, the ways I asked questions, the comments I attended to, and how many times I was talking when I should have been quiet.

PRESENTING AT NATIONAL CONFERENCES

In group sessions with my colleagues at two national conferences, I presented my findings as work in progress (Roberts, 1997; Roberts, van Zee, & Williams, 1997). In the evaluation of one of the sessions a participant wrote, "raised important issues about teaching, learning, and doing research. I was very impressed by the depth of engagement with children—respect for a child's thinking and work, their questions, and their multiple ways of documenting science learning, classroom as community, and teacher as inquirer." These comments made me realize that even novice teacher researchers may have something to share that others may value and be able to use.

COLLABORATING WITH COLLEAGUES AT MY SCHOOL

My research journey does not end here, nor do the surprises and pitfalls. I am continuing to be a research practitioner in my classroom. A wonderful surprise came when I was able to persuade several other teachers in my school to try reflective research as well. We have our own branch of SING now at our site and presented as a group at the Ethnography in Education Research Forum this spring (Crutchfield, 1999; Harris, 1999; Kagey, 1999; Roberts & Bentz, 1999).

REFLECTING ON THE PROCESS OF BECOMING
A TEACHER RESEARCHER

Another wonderful surprise came when I was able to bring my first-grade students back to the same physics lab I had been in as an undergraduate. We went to Dr. Layman's class and the first graders had wonderful experiences using motion detectors. The prospective teachers were excited and pleased to have had the experience of interacting with "real live" students. I was able to share with them how I am using what I learned in this course, and how it has shaped my teaching. At the 1998 American Educational Research Association Annual Meeting in San Diego, California, I reflected upon what I learned from this experience by presenting "Physics and First Graders: What a Good Match!" The classroom that had laid the groundwork for me to become a teacher researcher had become a place of research!

After several cycles of presenting and writing, I had a paper published in a journal for teachers, *Science and Children*. This paper had its beginnings when my classmates and I had watched the moon in Dr. van Zee's science teaching methods course in 1995. Then in 1997 I had collaborated with second- and third-grade teachers who were watching the moon with their students. We compared the questions our students asked and presented our findings in a paper together at an ethnography conference (Lay, Meyer, & Roberts, 1998). I had been surprised by the ways in which my students' parents became involved in the moon watching during this project. The parents' experiences became the focus of a paper I presented at the International Conference on Teacher Research in San Diego later in 1998. Then I refined this and submitted it for publication early in 1999. Seeing my work in print in *Science and Children* was exhilarating!

Doing research has been a valuable experience for me. I have learned a lot about myself as a teacher, about students and the ways they think and learn, and about ways other people I have met at conferences also do teacher research. Even though first-year teachers have a sometimes overwhelming burden to bear, the time and effort put into teacher research was extremely beneficial to me. I have learned, through conversations with colleagues and others, that many teachers would rather hear or read about the research of a fellow teacher than read what they term "university research." I would encourage all teachers, new and experienced, to take the opportunity to develop and pursue a research question that you have a burning desire to understand. It is more than worth the effort, although frustrating and nebulous at times.

Teachers who are considering doing teacher research might begin by attending conferences where teacher researchers are presenting. Encourage a colleague or two to try teacher research and support one another through the project. Read books written by teacher researchers, such as Karen Gallas' *Talking Their Way Into Science*, (1995) and realize that she started in the same place we all start. I have heard her admit that research is at times a very frustrating journey, but a journey well worth taking. I wholeheartedly agree!

ENDNOTE

1. This study was supported in part by a grant from the Spencer Foundation Program for Practitioner Research: Mentoring and Communication to Dr. Emily van Zee, Science Teaching Center, University of Maryland. Opinions expressed are those of the author and do not necessarily represent those of the funding agency.

REFERENCES

Crutchfield, M. 1999. Plant growth observation. Paper presented at the 20th Annual Ethnography in Education Research Forum, University of Pennsylvania, Philadelphia.

Gallas, K. 1995. *Talking their way into science: Hearing children's questions and theories, responding with curricula*. New York: Teachers College Press.

Harris, J. 1999. Comparison of literature-based and self-guided exploration approaches to science learning and teaching. Paper presented at the 20th Annual Ethnography in Education Research Forum, University of Pennsylvania, Philadelphia.

Hubbard, R. S., and B.M. Power. 1993. *The art of classroom inquiry: A handbook for teacher-researchers*. Portsmouth, NH: Heinemann Press.

Hubbard, R. S., and B. M. Power. 1999. *Living the questions: A guide for teacher-reseachers*. York, ME: Stenhouse Publishers.

Kagey, T. 1999. Fourth-grade scientists share their electric circuit investigations at a scientists' conference. Paper presented at the 20th Annual Ethnography in Education Research Forum, University of Pennsylvania, Philadelphia.

Lay, D., S. Meyer, and D. Roberts. 1998. Watching the sky with primary students. Paper presented at the Ethnography in Education Research Forum, University of Pennsylvania, Philadelphia.

Layman, J., G. Ochoa, and H. Heikkinen. 1996. *Inquiry and learning: Realizing science standards in the classroom*. New York: The College Entrance Examination Board.

National Research Council. 1996. *National science education standards*. Washington, DC: National Academy Press.

Roberts, D. 1997. Teaching other content areas through a science perspective. Presentation at the Ethnography in Education Research Forum, University of Pennsylvania, Philadelphia.

Roberts, D. 1998. Physics and first graders: What a good match! Paper presented at the annual meeting of the American Educational Research Association, San Diego.

Roberts, D. 1999. The sky's the limit: Parents and first-grade students observe the sky. *Science and Children* 37(1): 33-37.

Roberts, D., and J. Bentz. 1999. Using motion detectors with elementary teachers: A study in physics and inquiry. Paper presented at the 20th Annual Ethnography in Education Research Forum, University of Pennsylvania, Philadelphia.

Roberts, D., E.H. van Zee, and S. Williams. 1997. Research on learning and teaching science in elementary schools: The Science Inquiry Group. Presentation at the 4th Annual International Conference on Teacher Research, Evanston, IL.

van Zee, E.H. 1998a. Fostering elementary teachers' research on their science teaching practices. *Journal of Teacher Education* 49(4): 245-254.

van Zee, E. H. 1998b. Preparing teachers as researchers in courses on methods of teaching science. *Journal of Research in Science Teaching* 35: 791-809.

Kids Questioning Kids: "Experts" Sharing[1,2]

Marletta Iwasyk

A s a kindergarten/first-grade teacher in an alternative school, I have much latitude in curriculum development and instructional methods. Questioning and dialog are an integral part of my teaching. The National Research Council states, "Inquiry into authentic questions generated from student experiences is the central strategy for teaching science" (1996, p. 31). I believe that children are capable of being teachers and while engaged in the teaching process, they reinforce and solidify their own learning.

To examine how this happens in my classroom, I conducted a case study to show children using questioning and communication skills during conversations about science (van Zee et al., 1997). The case study involved analysis of transcripts of students discussing the subject of shadows in which two students became the "teachers" or "leaders" and the rest asked questions for clarification or gave input of their own.

To document the discussions, I used a tape recorder with a micophone placed on a desk near the seated children. I also placed a video camera high in an unobtrusive corner. The camera was trained on the seats in the middle of the circle where I placed the "leaders" of the discussion. If other students had something to contribute, I asked them to step to the middle of the circle where I knew they would be visible to the camera. The object of the case study was to record the children's talk while they discussed the subject of shadows to see if there was any carryover from the teacher-directed questioning/discussion activities. The focus of the study was not only the facts being presented, but also how the discussions took place.

EMPHASIS ON COMMUNICATION

From the first day of school, I model questioning and communication skills that I hope the children will emulate as the year progresses. Groups and pairs of children are allowed to converse a great deal during the day and so are comfortable speaking with one another.

The emphasis this year for the entire school is using kind and respectful words. I also stress the role of a respectful listener, which provides an environment in which children feel safe to risk speaking and sharing their ideas. As a class, we practice listening and speaking skills in many subject areas.

During "show and tell" time each week, the class president facilitates the sharing period, and the person called upon to share, in turn, asks for questions and comments from the rest of the class. As children respond to one another, I help them analyze and decide whether they are making a direct comment about what they heard. In the beginning, many children will add to the sharing rather than commenting on what was said (e.g., I have a dog too.). Also, their questions are usually too specific when asking for more information about what was shared (e.g., Is your dog's name Rover or Fido?). We talk about a better way to ask the question (e.g., What is your dog's name?).

Another opportunity for modeling questions comes when the class president is interviewed for stories that are written by each individual for the "president's book." Again, early in the school year, the questions tend to be very specific and limited (e.g., Do you like spaghetti? Do you like apples?). The children are encouraged to think in terms of general questions instead (e.g., What is your favorite food or fruit? What do you like to do on the weekends?"). This clarification helps children to differentiate between types of responses and also determine appropriate times to make them. These skills carry over into other areas of study.

EMPHASIS ON SCIENCE

Science is wonder—that feeling of awe and excitement you have when you see a golden harvest moon, experience the power of the wind, see a rainbow, watch a salmon hatching, and experience the miracles of nature that take place around you. My school, which has an environmental and art focus, greatly values this view of science and nurtures it in every child's heart and mind. It is then a natural step to go from this "wonderland" of experience and appreciation of nature to the world of discovery and the desire to find out the "why and how."

And so, science is wondering, wondering about our physical and biological world—and young children wonder most of all! They have a natural, eager intellectual curiosity about the world around them and want to find out all they can, as evidenced by the many questions they ask. The dilemma for a teacher of young children is how to keep this natural curiosity alive within the confines of the classroom. Time, lack of materials or space, or other obstacles—such as a feeling of inadequacy in science knowledge—may limit the amount of experience the teacher provides in the area of science.

One successful activity that I have used to keep this connection to the world outside the classroom is the study of light and shadows, using, of course, the most visible object in the children's sphere of reference—the sun. Learning about objects in the sky, including observing the sun and its movement, is one of the science benchmarks for kindergarten through grade two (American Association for the Advancement of Science, 1993). Much of my knowledge about the sun was gained in a physics program for teachers at the University of Washington (McDermott, 1996).

SHADOW DISCOVERY

On the very first sunny day of school in the fall, we begin our study of shadows. This is a natural and easy way not only to nurture curiosity and wondering, but also to help the children develop the skills and attitudes that will make them successful lifelong scientists whether or not they go on to choose a career in a scientific field.

Figure 1 lists some suggestions for various shadow activities. During these shadow activities, many observations and recordings are made and compared throughout the year. Many questions arise, such as, "Why does my shadow change shape, length, and direction?

FIGURE 1. ACTIVITY SUGGESTIONS FOR SUN PLOTS AND SHADOWS: OBSERVING SHADOWS, RECORDING DATA, AND COMPARING

▶ Take a walk on the playground and observe shadows of poles, trees, and walls.

▶ Take a walk with your shadow and observe how it follows wherever you go and does whatever you do.

▶ Have your class stand in a circle and discuss shadows they see—some are in front, some are behind. Discuss orientation if children want to make a claim about the shadow position. (If students say, "The shadows are in front of us," ask, "Are all the shadows in front?" If the children say no, then ask, "How can we make that happen?") Ask questions that help children see that everyone needs to be facing the same direction to see their shadow in front, behind, and so on. This can be facilitated by the use of a "shadow line" (discussed below) that is close to where your class lines up everyday. I have my class line up on the "shadow line" after every recess whether it is sunny or not. If it is sunny, then we can quickly make some observations, think about some questions, and have a short discussion before coming in—very efficient and easy to do. (At this time, I do not say whether a conclusion is right or wrong; I ask them to think about it.)

▶ Line up on a North-South line (if possible) on the playground. This orientation helps children see the shortest shadow pointing North at local noon. Give directions to your class, such as the following: Stand so that your shadow is in front of you (or behind, beside on the left, right). Which way is your shadow pointing? Toward the building or away from it? Is it long or short? Longer or shorter than this morning? Last week? There are many questions you can ask to promote thinking and stimulate observations.

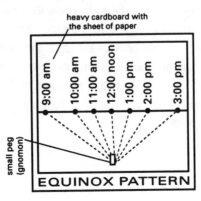

▶ Make a sunplot/shadow board (see above) for children to study shadows independently at home. To make the sunplot/shadow board, use heavy cardboard and stick a small nail (or other similar object) into the center. A sunplot/shadow board is placed in the same spot throughout the day. Children mark the end of the shadow (noting the time and date) using a piece of heavy cardboard with a sheet of white paper. Students can then observe the pattern the dots make, if connected.

▶ Train two children to use a sunplot/shadow board. Discuss the use of a gnomon, which is a straight object (peg, stick, rod) used to cast a shadow, and how to record data (length, time of day). Take as many readings as possible each day. The first team then trains the next team (a classroom job). If inside, shine a flashlight or bulb into the sunplot board to make "artificial" shadows (peg boards work well). This is a good exploration activity.

▶ On equinoxes and solstices, record the end of the shadow throughout the day on the playground with chalk and then use paint to make it permanent for future reference. A tall pole on a sturdy base works well as the gnomon for this (mark where the base goes).

▶ Transfer information from daily records to overlays for use on an overhead to compare fall, winter, and spring shadows (length, shape of line connecting dots, sun's position). Overlays are good for end-of-year discussions.

▶ Enrich the activity with shadow puppets, poems, journal writing, and literature.

In the beginning, I do not answer any of the questions; instead, I ask the children to think about the questions and discover how they can find the answers for themselves. If they make early conclusions about what they observe, I do not acknowledge any answer as right or wrong. It isn't until after the winter solstice, when the shadows are becoming short again, that we have an in-depth discussion of what we have learned about the sun and shadows, with the children facilitating as much as possible. To make this happen, much groundwork has been laid during the year to this point.

DISCUSSION OF SHADOWS

Teaching Standard B in the *National Science Education Standards* (National Research Council, 1996) states, "Teachers of science guide and facilitate learning. In doing this, teachers orchestrate discourse among students about scientific ideas" (p. 32). See Table 1 for an example of dialog that took place during a discussion of shadows, which typically lasts anywhere from 15 to 30 minutes, depending on interest and focus.

TABLE 1: STUDENT CONVERSATIONS ABOUT SHADOWS ON FEBRUARY 4, 1997

(Responses are noted by initials: T = teacher, * = male student, and ** = female student.)

L*: I think I know how they are made.

T: The shadows?

L*: Uh-huh.

T: Would you like to come on up here and be our second "scientist" then?

L*: (After positioning himself in the middle of the circle.) If there was a bright, bright light up here, and it goes like you were talking about (responding to the information C*, the other facilitator, had previously shared), and then you could be right here and you're covering part of the ground and you could be however you want.

C*: I know. That's what I said.

L*: Like if we were outside you can almost always see it on grass.

T: OK. Have a seat there and you can answer any questions these people have.

C*: (Calls on R*)

R*: How, I mean like.... Why does(n't) it have the color that you have on?

C*: It doesn't.

R*: But why doesn't it?

T: (Clarifying question) Oh, so, why doesn't it have the color that you have on?

C*: It's not really you, its just....

R*: A part of you.

C*: Yeah, it's just a reflection of you.

R*: Oh, okay.

C*: Black on the ground.

L*: Like the sun, like you're, you're, like you're a dark black cloud.

T: Ooo, we'll have to write a poem about that!

M**: I see a shadow in the room.

T: Oh, you're looking at your shadows in the room?

Everyone sees shadows on the shades with sun shining through the windows. There is great excitement and everyone is talking at once.

At the beginning of our discussion, the children posed questions they had thought about in connection with shadows, and these were listed on a KWHL chart as shown in Table 2 (e.g., Where do shadows come from? Can they see?). In the course of the discussion, two male students who had a lot to share became the facilitators for the discussions, calling on others for questions or comments. In doing this, I turned a possible negative (two male students dominating the conversation) into a positive by asking the two "leaders" to explain some of their statements. The two student leaders became quite humble at some points saying "I don't know" when asked a question.

TABLE 2. A KWHL CHART.

K	W	H	L
What do I Know about _____?	What do I Want to know about _____?	How can I find out about _____?	What did I Learn about _____?
(Prior knowledge or preconceptions. All ideas are listed.)	(Questions that students have.)	(Books, Internet, asking others.)	(Facts learned may be different from those listed under K.)

At first, the student leaders called on their friends (also male). The naturally quiet students were not as involved at the beginning, but as time went on, they joined in when called on or when they just wanted to speak. Soon, everyone was participating, both males and females. For the final discussion, when I facilitated, all but one female shared.

One of the questions asked and discussed was "Why doesn't it (the shadow) have the color you have on? This student wondered why the shadow wasn't the same color as skin. Throughout the discussions, many moments of spontaneous dialog occurred among the children. They were all respectful and involved listeners. Because of the accepting attitude of the group, no laughing or put-downs occurred, even if an idea seemed far-fetched. Some disagreed with statements but were willing to suspend judgment and try to find out for themselves. Also, spontaneous moments might be seen by some to be negative, but these were some of the more positive moments in my view. They showed that the students were really involved. It was enjoyable to just sit and listen to them as the children tried to explain their thoughts and communicate their ideas to the group, asking questions of each other for clarification. I plan on having more discussions during the rest of the year, with other students being the leaders.

LEARNING FOR ALL

Questioning techniques can be used by students to learn how to ask questions of themselves or of others to investigate or explore a topic of interest. Questions allow a child to become the leader or teacher as he or she enlarges or guides the discussion in a specific area, whether they are the ones asking or being asked. I firmly believe that as one teaches, one also learns; thus, children grow in their own skills as they teach others.

Just as questions can help children clarify their own thinking, the teacher can learn much about the students by listening to their discussions. It was very enlightening for me to observe thinking processes as the children gave explanations. I also gained insight into class dynamics. During the shadow discussion, the original leaders were male, but in many subsequent discussions, the females took the lead. I will continue to heighten my awareness of participants in discussions, making a special effort to draw in the quiet ones and encourage student leaders to do the same. My goal is to empower the students to have a role in their own education!

ENDNOTES

1. This article is reprinted with permission from NSTA Publications, copyright 1997 from *Science and Children*, National Science Teachers Association, 1840 Wilson Blvd., Arlington, VA 22201-3000.
2. Development of this case study was partially supported by a grant from the National Science Foundation (MDR-9155726) to Dr. Emily H. van Zee. Opinions expressed are those of the author and do not necessarily represent those of the funding agency.

REFERENCES

American Association for the Advancement of Science. 1993. *Benchmarks for science literacy*. New York: Oxford University Press.

McDermott, L.C. 1996. *Physics by inquiry*. New York: J. Wiley.

National Research Council. 1996. *National science education standards*. Washington, DC: National Academy Press.

van Zee, E.H, M. Iwasyk, A. Kurose, D. Simpson, and J. Wild. 1997. Teachers as researchers: Studies of student and teacher questions during inquiry-based science instruction. Workshop presented at the annual meeting of the American Association for the Advancement of Science, Seattle.

Eyes on Science: Asking Questions About the Moon on the Playground, in Class, and at Home[1,2]

Akiko Kurose

I stop, it stops too. It goes when I do. Over my shoulder I can see. The moon is taking a walk with me. (L. Moore, 1974)

Observing the moon gives children experience in making discoveries in a cooperative way that allows them to discover that everybody on planet earth is sharing the same phenomena. While you can't grab the moon, push it around, or change its position, you can see it, draw pictures of it, and talk about it. Table 1 shows some questions I ask my students throughout the school year. Since I learned about the moon in a special physics program for teachers at the University of Washington (McDermott, 1996), noticing the sky has been a favorite activity with my students.

TABLE 1. QUESTIONS THAT ARE THE BASIS FOR CONVERSATIONS ABOUT THE MOON

INITIAL QUESTIONS:

- What can you tell me about the moon?
- Please draw a picture of the moon.

- Where can you see the moon?
- When can you see the moon?

ONGOING QUESTIONS:

- Where did you see the moon?
- How high in the sky was it?
- When did you see it?
- What kind of moon did you see?
- Please draw what you saw.
- Are all the drawings the same?
- Why do you think some are different?

OBSERVING THE MOON

The *National Science Education Standards* (National Research Council, 1996) suggest that children from kindergarten through second grade should learn about objects in the sky by observation. My first graders and I frequently go out and look at the moon at different times of the day so the children can view the moon in a variety of phases. In the afternoon, the southern sky sometimes gives students an opportunity to see the waxing crescent moon. In the morning's western sky, we sometimes observe the waning gibbous moon. During our viewings, neither the students nor I use many words to explain what we see; we simply observe. So that they become aware of the position of the moon, I have them draw the moon in relationship to the horizon. (See Figure 1.) As the children gain experience they draw the sun on the same side as the lit part of the moon.

Engaging in this type of observation enables students to illustrate the different phases of the moon, keeping in mind its relationship to objects on the horizon. I encourage them to make predictions as to when and what the next phases will be. I also invite them to reflect on the phases of the moon previous to the moon they have illustrated. Eventually, students are able to seriate the different phases of the moon as well as identify which ones will set in the evening and which in the morning. (See Figure 2.)

**FIGURE 1. STUDENT'S DRAWING OF THE MOON DURING RECESS EARLY IN
THE SCHOOL YEAR**

**FIGURE 2. EXAMPLE OF STUDENT'S SERIATION OF THE MOON EARLY
AND LATE IN THE SCHOOL YEAR**

Students then express themselves creatively, demonstrating their knowledge as well as their appreciation of the moon by writing prose and poetry relating to its phases. (See Figure 3.) I also read stories about the moon and sun from other cultures such as *The Truth About the Moon* by Clayton Bess (1983).

FIGURE 3. STUDENT'S POEM AND DRAWING OF THE MOON

full moon

full moon full moon Shine so bright won't you guide my way Tonight.

I like the moon and stars

Watching the moon can easily be transferred into the child's home. Initially, parents say, "We don't let our kids stay up that late. And what is this, telling them to observe the moon at night?" But parents delight in learning from their young. Time and time again excited parents come to school, using the terminology their own children have taught them: "waxing, waning, and gibbous" and admitting that they "didn't know what those things were." As their children point out to them the waxing moon setting, the parents recognize that the moon is visible during suppertime, at sunset, and also when the children go home from school.

Similarly, they discover that the waning gibbous moon is observable in the mornings and into the class time. Parents have reported that when their children get up in the morning, they are so excited when they see the moon. Students are also able to observe the waning moon, which they illustrate during morning class time.

STUDENT'S QUESTIONS ABOUT THE MOON

Some colleagues and I have been developing case studies of questioning during conversations about science. Below are examples of questioning by my students.

When Stella went on a month's visit to Australia with her family, her assignment had been to keep a record of the phases of the moon while she was there, using the same format as her classmates in Seattle. When she returned, we made a comparison of her recordings in Australia with those of her classmates. Figure 4a shows her observations in that country, where the waxing crescent moon appeared to be lit on the left. Figure 4b shows her classmates' observations in Seattle, where the waxing crescent moon appeared to be lit on the right. The children and I asked her to share her observations as well as explain why this phenomenon occurred. For discussion, we used a globe and a map on the floor. Australia is in the southern and Seattle in the northern hemisphere. Some of the children lay on the floor and realized that if they are upside down something lit on the right looks as if it is lit on the left. One child asked what it would look like on the equator. See a transcript of a conversation about the observations of the waxing crescent moon in Australia and Seattle in Table 2 .

At other times, I ask the children to write down their own questions to share and compare with one another and try to come up with some answers. "What would the moon look like on the sun?" and "Why is there a moon?" are examples. During circle time, I ask students for more questions about the moon and tell them to think about the whole planet earth and not just Seattle. These are some of the questions that they develop: "How did the moon turn into different phases?" "This morning I saw a very thin moon. What kind of crescent moon is that?" "How can the moon be rising in the morning?" "Why does the moon follow me?" "Why does the moon look different in Hawaii?" "Does the moon look different in Africa?" "What does earth look like from the moon?" "Do they see different phases or do they see the whole earth?" "Why does the moon look the opposite in the mirror?" "What does the moon look like on the equator?" I also ask the students how they would get information about the moon and they suggest the dictionary, encyclopedia, a trip to the library, and looking at the moon.

FIGURE 4A. STUDENT'S OBSERVATIONS OF THE WAXING CRESCENT MOON WHILE VISITING IN AUSTRALIA

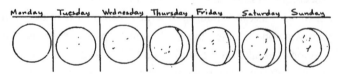

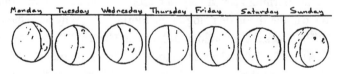

FIGURE 4B. STUDENT'S OBSERVATIONS OF THE WAXING CRESCENT MOON IN SEATTLE

TABLE 2. TRANSCRIPT OF CONVERSATION ABOUT THE OBSERVATIONS OF THE WAXING CRESCENT MOON IN AUSTRALIA AND SEATTLE

We compared the observations of the moon in Australia by a student, Stella, with her classmates' observations of the moon in Seattle.

Teacher: Stella, do you want to tell us about your observations in Australia?
Stella: The waxing crescent moon was facing the other way.

Teacher: The waxing crescent moon was facing the other way? Why do you think it was facing the other way?
Stella: It was on the other side of the world.

The children and I found Australia and Seattle on a globe. Then Stella drew her observation on the board, a crescent moon curved on the left.

Student: It's like the waning crescent.
Student: I thought it was like this! (The child turned upside down.)
Teacher: That's right. When he lies down on the ground he sees it the opposite way.

Then we discussed how the waxing crescent moon would look in Canada.

Student: Right side up.

Teacher: What do you mean right side up?
Student: The way we see it.

Teacher: Do you mean the way we see it in Seattle?
Student: Uh, huh.

Teacher: In Australia, do you think they would think they're upside down?
Student: No. Maybe. Stella thought so.

Teacher: For them, they would think that they're right side up, wouldn't they? It's just that they see it from a different perspective.

We also discussed how the moon would look in Florida and other places.

Student: My dad's been to Brazil.
Teacher: How do you think you would see the moon there? Remember where the equator is. Make a hypothesis.
Student: The way Stella saw it in Australia.

Teacher: The way that Stella saw it in Australia.... So it depends upon where you are on earth.
Student: How does it look if you're on the equator?

INTERPRETING FIRST-GRADERS' QUESTIONS ABOUT THE MOON

Studying the phases of the moon at a first-grade level is appropriate and challenging to students. Not only do they become enthusiastic about the learning process, their feelings are sustained.

My students look forward to the observations because they empower them to know what's going on in the sky. When students reach the point at which they are able to comprehend what they are seeing, they celebrate their discoveries and take ownership of them. The process of observing the moon becomes part of their daily experience and becomes cooperative in nature as they are able to share with their families and fellow students what they have learned and observed.

The children's experiences in our daily moon gazings over a period of several months also nurture their abstract thinking and questioning skills. In one conversation about observations, a child casually inquired, "how does the moon look if you're on the equator?" This question was neither prompted nor expected. I do not make it a practice to feed questions. Because of the manner in which we study the phenomena of the moon, children's queries are real. And because their curiosity is piqued, the answers have meaning. As a child at that age, I never entertained such questions because I had not been given the opportunity to study the moon in an organized way through observations, and it would not have occurred to me to think about what the phases of the moon would be like from any part of the world, including my own home. In discussing their observations, the children sometimes relate a shared experience—the feeling that the moon is following them. This experienced relationship with the moon leads naturally to inquiry. Other children have become interested in learning about the way the moon looks from different places on the earth and away from the earth such as on the sun. Several children were also curious about how the earth appears from the moon. These experiences engender thoughts about the moon from different places, granting the students the gift of a global perspective. Engagement in this type of abstract thinking and questioning has become part of our class culture, in which virtually all of the children participate.

Teaching about the moon is accessible to all teachers. It is truly an "eyes-on," "minds-on" curriculum. This kind of study is spontaneous and exciting for students, parents, and teachers alike. The students experience physical as well as mental freedom as they observe, cooperatively discuss their questions, and work with one another. Questioning strategies are important in encouraging and inspiring students to pursue their interests and engagement in dialogue about their experiences and realities. This curriculum integrates all the disciplines in a natural setting: science, mathematics, multicultural studies, reading, writing, language arts, music, and art all fit into the theme.

ENDNOTES

1. Development of this case study was partially funded by a grant from the National Science Foundation (MDR-9155726) to Dr. Emily H. van Zee. Opinions expressed are those of the author and do not necessarily represent those of the funding agency.
2. This paper was presented at a workshop, "Teachers as Researchers: Studies of Student and Teacher Questions During Inquiry-Based Science Instructions," at the 1997 annual meeting of the American Association for the Advancement of Science in Seattle.

REFERENCES

Bess, C. 1983. *The truth about the moon*. Boston: Houghton Mifflin.

McDermott, L. 1996. *Physics by inquiry*. New York: J. Wiley

Moore, L. 1974. In *Poems to remember,* edited by L. B. Hopkins. New York: Scholastic Book Services.

National Research Council. 1996. *National science education standards*. Washington, DC: National Academy Press.

Tapping Into Children's Curiosity[1]

Rebecca Kwan

uriosity drives the quest for knowledge. Curiosity brings wonders and questions. It then generates actions to answer questions. Curiosity is the basic spirit of science learning. A child's curiosity can be ignored because of the teacher's agenda.

In this case study, I followed up on a question one of my first-grade students had asked in social studies: "How come rain won't go through the roof?" His question arose while I was reading to the class from a page on thatched houses in Japan in *Houses and Homes Around the World* (Karavasil, 1986). I decided to develop a lesson on constructions based on this child's question and my county's science curriculum (Westley, 1988).

As a way of learning science integrated with the county curriculum for social studies, I had my students start to figure out "How can I make the roof with straw so that it won't leak?" I thought trying to figure out this question was a wonderful spin-off topic for my students while I was teaching topics such as building materials, shapes, and the motion of shapes. This lesson would also be a way to assess students, which would fit nicely with the requirements in the Maryland School Performance Assessment Program. Students were to determine ways to make a roof with straw so that it wouldn't leak. The lessons took two sessions for exploration and testing.

We started the exploration session by brainstorming various possibilities for making a waterproof roof. The students came up with four different ideas: to braid the straw tightly, to tape the straw on, to glue it on, and to weave it on. I provided students with these materials to explore: backyard grass for straw, green plastic strawberry baskets from the grocery store for a triangular roof frame, a science research work sheet to record their findings, and glue, tape, and scissors. From the beginning of this exploration session, the students were excited and highly stimulated.

The students investigated the several ways that had been suggested of handling the straw and basket. Some of them tried to braid the straw but the grass was too dry and broke as soon as they tried to twist it. They tried hard to have the glue stay on the plastic lines of the baskets. At first most of the glue dripped down between the plastic lines and the students did not know they needed to wait for the glue to dry before the straw would stick there. They struggled to weave the dry grass in and out of the narrow openings and their fine motor skills weren't sufficient. For some, it was the first time ever to attempt to bend tape around to form a loop in order to stick one side on the plastic and grass on the other side. There were a couple of students who had the small motor skills to weave away merrily. I was in awe with how the students' assimilated these skills. Some of the frustrated students turned their energy to imitating the ones who had succeeded in weaving and did a great job of it. By the time the second session started, most of my first graders were happily doing their weaving, gluing, and taping.

The second session ended with the testing. By this time, some of the roofs were quite appealing yet not finished. I took a gallon bucket full of water and a paper cup to test our theories. I poured water down the roofs and asked the students to observe. The students were fascinated by the water movement and excited to see water not passing through the part with more grass while passing through only the part with a couple of strands of grass. Some of the students took their roofs to test under the faucet in the sink. I handed out papers for them to record the results and asked them what they had learned. The conclusion was unanimous that they need to fill in the holes with more grass on the roof.

Teaching according to the students' curiosity requires a radical change in perspective from that which comes most easily to instructors. The waterproof thatch roof lesson was a modification of a lesson from the previous year, which had emphasized the effect of the direction of thatch on the water run-off. Since leaking was not the issue, I had used cardboard paper as the roof frame. But, when my student asked the question, "How come the roof doesn't leak?" I needed to come up with an activity to fulfill that curiosity. Modification of materials and method was needed. This is an example of teaching for understanding so that, as the standards proposed in 1996 by the National Research Council suggest, "activities and strategies are continuously adapted and refined to address topics arising from student inquiries and experiences"(p. 30). I was glad the students were enjoying their lessons and I took pleasure in watching them learn through their own discovery and gain physical skills along the way.

ENDNOTE

1. Development of this case study was partially supported by a grant from the Spencer Foundation Practitioner Research: Mentoring and Communication Program to Dr. Emily H. van Zee. Opinions expressed are those of the author and do not necessarily represent those of the funding agency.

REFERENCES

Karavasil, J. 1986. *Houses and homes around the world*. Dillon Press.

National Research Council. 1996. *National science education standards*. Washington, DC: National Academy Press.

Westley, J. 1988. *Windows on science: Construction*. Creative Publications: Sunnyvale, CA.

Giving Children a Chance to Investigate According to Their Own Interests[1]

Constance Nissley

As a science teacher in an independent school, I work with children from kindergarten through fourth grade. I have found that children enjoy having a chance to explore with materials in an unstructured time that allows them to shape their own investigations. In addition to class time under my direction, we schedule from twenty to thirty minutes each week for every class to have what we call Choice Time. During this time, students choose their activities, get out the materials, investigate, and then clean up.

Many of the activities are described in Table 1. The materials are on open shelves so that they are easily available to the students. Students may also request materials for additional activities they would like to explore. Sometimes these are extensions of other class activities, such as the electric circuits. At other times students have been continuing observations after a demonstration by a student or me: for example, working with siphons, putting on goggles and testing the popping power of ingredients that form carbon dioxide in vials with lids, or investigating electroplating.

TABLE 1. MATERIALS AND ACTIVITIES AVAILABLE FOR CHOICE TIME

Visiting and observing live animals. We have guinea pigs, hamsters, a rabbit, finches, quails, a frog, a turtle, a lizard, fish and snails in fresh water, and crabs and snails in salt water. The mammals can be held or placed in mazes made on the floor using blocks. All the animals can be fed and their behavior observed. Occasionally one of the animals has had babies, so that we have had opportunities to observe life cycles and development.

Mixing simple chemistry. Dropper bottle of diluted vinegar, baking soda solution, bromthymol blue solution and red salt water can be mixed in trays with small wells or in vials. Medicine droppers can be used to transfer mixtures. This selection of solutions works well because they form bubbles, the acid/base indicator changes from blue to green to yellow, and colored layers can form if added carefully. When mixing is done thoughtfully, children can make a variety of colors. At times cabbage juice can be added for additional colors.

Simple electric circuits. Circuits can be created by use of batteries with holders, wires with alligator clips, bulbs in holders, buzzers, and switches. The battery holders lend themselves to connections for both parallel and series circuits.

Magnets. Magnets of various strengths, sizes, and shapes can be manipulated with various materials—metals and non-metals—for testing interactions.

Computer. The favorite activity is the coral reef ecosystem game, Odell Down Under. The challenge is for various fish to keep alive by finding their food and avoiding predators.

Capsela. This motorized set can be constructed to have its motor simply make a fan propeller go or to make it a more complex wheeled vehicle that can go forward or backward with a switch.

Light and color. A box from the *Elementary Science Study* curriculum unit, "Optics," with a two-hundred watt light bulb shining through openings, is used with such supplies as colored filters, mirrors, prisms, and diffraction grating glasses.

Examining and comparing rocks, fossils, bone or shell samples.

Observing frog development. In the spring, we observe development of wood frog eggs to tadpoles and then frogs. These eggs are collected from our pond and returned as they mature into tiny frogs.

Certain activities like the following may be assigned during part of Choice Time to give each child a responsibility for a task or project shared by the class:

Animal care. Checking and filling water bottles and food dishes for the animals is an activity that I assign to students in rotation, unless children choose the activity. Some students like to change the chips in the cages, but it is not required.

Temperature graph. When a class is studying weather or a country in social studies, we may keep a record of daily temperatures. Placing a strip of masking tape along a marked line of a laminated sheet, the student colors in a bar of an appropriate length for the temperature observed on an outdoor thermometer. This strip is taped in place on the ongoing graph being created each day that the class is in the science room. When we are studying a country, we create a parallel graph. Each day we obtain from the Internet the temperature in the country's capital.

In this case study, I am interested in what amount and quality of journal writing is appropriate in this context. I have been investigating how to encourage the students to communicate their observations and understanding of their investigations. By their own choice, the students sometimes repeat activities in many variations. They find this more interesting and easier, however, than explaining what they have done and observed.

The usual format has been for students to record comments in their Science Log booklets. Originally I had asked them to write about what they had learned from their investigations or something new that they had observed. For the first time in our science classes, they were asked to write about self-chosen and self-directed activities. So the comments also had to be independently created. Although they have had experiences of journal writing in Language Arts and have recorded observations in science investigations directed by a teacher, many students wrote very little. In addition to being very brief, what they wrote was vague, such as "Chemistry is fun" or "I like Pumpkin" (the hamster). I have tried several ways of encouraging more extensive descriptions or more analytical comments. For example, I have asked students questions such as "What about chemistry is fun?" and then "What did you need to mix to get the bubbles?" and "What other powder or liquid might you add to get bubbles?"

Among group experiences that have helped students have an idea of how to write about their own investigations has been to start the class with a demonstration, such as placing a small bottle of blue hot water in a larger container of cold water, and asking students to write a description of the observations and an explanation. This has given us a chance to discuss observations made by individuals and what makes a good record of these.

Another approach is to follow a demonstration with a class discussion that gives students a chance as a group to share observations and build an explanation for their observations. For example, I did this after folding three pieces of paper into a cylindrical shape, a triangular column, and a rectangular column and testing which would best support a book. This produced a lively discussion among the students of why the cylindrical shape was the best support, getting to the idea in their own words that the circular top more evenly distributed the weight. One difficulty in the discussion, of course, was that a few students were most anxious to talk about their ideas and others didn't like to share in this setting. The students also needed reassurance that their ideas were valid and didn't need to be referred to the teacher's authority all the time. The discussion provided a way that was not very threatening for a student to question another's idea that differences in amounts of scotch tape were the reason for differences among their observations. They could make further observations and interpret them.

Another way to encourage both careful manipulation of materials and clear explanations of what happened is to ask each student to prepare a demonstration to present to the class. The presentation includes leading a discussion of the outcome and possible explanations.

We feel that all of these experiences have allowed students to follow their own interests more than would be provided otherwise in the curriculum. Also they focus individually on some analysis of their investigations.

Table 2 shows examples of the development of log writing for three third graders over a semester. These logs show increases not only in the detail of observations but in evidence of the thinking that was involved in their investigations. The last entry for each student includes brief statements of interpretations of these observations.

TABLE 2. EXAMPLES OF THE DEVELOPMENT OF SCIENCE LOG ENTRIES BY THIRD GRADERS DURING A SEMESTER

D.H.

9/16: I did siphons. I made different colors.

10/7: We cleaned lots of cages. They smelled really bad.

10/28: I was at electricity. Me and C. made bulbs light up. We made the buzzer sound.

1/20: Today M. and I did chemistry. We made explosions out of soda and vinegar. Some of them were big and some were small. The big ones we had to do outside. We used little tubes to put the soda and vinegar in. It was fun.

J.F.

9/16: I did Capsela.

10/7: We cleaned the animal cages. It was a lot of fun. The cages stink!

10/28: I did chemistry. I made a secret ??

1/20: In electricity, I made a huge circuit. I made a switch out of a light bulb. (Oral explanation later: The buzzer went on when the bulb was screwed in but the bulb didn't light.) The circuits were very interesting. We made the buzzer very loud.

L.Y.

9/16: I did siphons.

10/7: You have to shake the bubble out of the tube in siphons.

10/28: I did animal care. It was smelly because of all the pee.

1/20: I did mammals with L., and we played with Kuby and fed him carrots and celery. He preferred celery. I also did the computer for the ice cube experiment. I found out that mine was alive for 7½ hours.

The planning of this program relates to the teaching standards specified in the *National Science Education Standards* (National Research Council, 1996). This is an example of planning and development of a science program to guide and facilitate inquiry in an environment that provides students with time, space, and resources needed for learning. Choice Time and many of our other activities relate to Content Standard A in the *Standards*: All students should develop the abilities to do scientific inquiry and understand it (p. 121). The activities are related to Content Standards C, D, E, and F: physical science, life science, earth and space science, and science and technology.

ENDNOTE

1. Development of this case study was partially supported by a grant from the Spencer Foundation Practitioner Research: Mentoring and Communication Program to Dr. Emily H. van Zee. Opinions expressed are those of the author and do not necessarily represent those of the funding agency.

REFERENCES

National Research Council. 1996. *National science education standards*. Washington, DC: National Academy Press.

How Does a Teacher Facilitate Conceptual Development in the Intermediate Classroom?[1,2]

Judy Wild

As children investigate and explore the world around them in their preschool and primary years, the curiosity and creativity of a scientist seem natural to them. Continuing to nurture that enthusiasm within a structured science curriculum that enables students to develop scientific concepts is an awesome task for teachers. Studies in recent years report that students, especially at high school and college levels, hold misconceptions or preconceptions about natural phenomena (Driver, 1985; McDermott, 1990). How can teachers, especially in elementary science classes, facilitate conceptual development so that the knowledge the students construct is aligned with that of the scientific community?

Conceptual understanding is developmental; students must therefore be engaged in activities that are of interest to them and at the appropriate stage of their growth. Concrete activities, such as interacting directly with materials, help students to have a basis for shaping concepts and abstract reasoning. Adequate time is critical so that students may begin developing ideas and refining them.

An inquiry-oriented approach to science has been advocated by science educators for many years and is a major goal in both the *Benchmarks for Science Literacy* published by the American Association for the Advancement of Science in 1993 and the *National Science Education Standards*, which the

National Research Council issued in 1996. As students learn to question what they observe and search for answers, schools can provide a framework for their curiosity. Inquiry as a way of thinking and learning can enable students to develop knowledge and skills, to identify and solve problems, and to make informed decisions. Such a framework can extend learning beyond the science curriculum and beyond the confines of the classroom.

In inquiry applied to the study of electrical energy at the intermediate level (Educational Development Center, 1966; Lawrence Hall of Science, 1993; McDermott, 1996), students cultivate an understanding of ideas such as simple circuits, conductors, non-conductors, short circuits, series circuits, and parallel circuits. How can teachers assess their conceptual development and assess how the understanding that the students are arriving at compares with that of the scientific community?

Students need opportunities to communicate their ideas as they develop concepts. Discussing these with a partner, a small group, or the entire class, as well as writing about their learning, helps them to formulate and refine their understanding and makes their thinking known to others. Asking students "why" is important in assessing their understanding. Questions such as these are helpful: What do you observe? What do you infer? Why do you believe...? What is the evidence for...? How would you operationally define...?

In asking fourth-grade students to explain the reasoning for their observations and beliefs during lessons on batteries and bulbs, I discovered that at times it was difficult for some students to expand or apply their understanding of electrical circuits. Revising lesson plans to provide additional activities was necessary in these instances to help the students with conceptual development. Here are some examples of lessons in which students needed additional experience to help expand or apply their understanding of electrical circuits.

After one or two lessons, most students were able to identify correctly the two ends of the wire, the two ends of the battery, and the two ends of the bulb and to state that these ends needed to be included to make a circuit. But some students could not apply this generalization when predicting whether new configurations would light. (See Figure 1.) Whether the bulb was touching the battery directly was a factor in their predictions. Students needed to experience this idea in a variety of ways. Four or five lessons were necessary for some students to predict correctly circuits that would light in all or most of the configurations they drew or were given.

FIGURE 1. EXAMPLES OF CONFIGURATIONS WHERE THE BULB WAS NOT TOUCHING THE BATTERY DIRECTLY AND STUDENTS RESPONDED INCORRECTLY THAT THE BULB WOULD NOT LIGHT

Students had difficulty recognizing that connecting a wire to a battery that was already part of a completed circuit with a wire and a bulb resulted in a short circuit. Setting up a short circuit and connecting and disconnecting the wire without the bulb while observing the light going off and on helped students to identify a short circuit correctly on later evaluations. (See Figure 2.)

FIGURE 2. EXAMPLE OF SHORT CIRCUIT

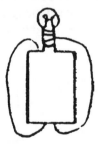

Some students who could identify short circuits correctly when given drawings of batteries and bulbs with wires were unable to add wires to drawings of batteries and bulbs only so that the bulbs would light. They drew additional wires that resulted in short circuits. (See Figure 3.) Setting up the circuits with batteries, bulbs, and wires as they had drawn them helped students to identify which circuits were short circuits and on later evaluations these students could add wires to drawings of batteries and bulbs that did not cause short circuits.

**FIGURE 3. EXAMPLES OF DRAWINGS BY STUDENTS THAT
RESULTED IN SHORT CIRCUITS**

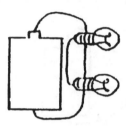

Some students were able to identify series and parallel circuits when two batteries were touching each other but not when the batteries were connected by
wires. Additional activities setting up circuits with batteries, bulbs, and wires,
making and checking predictions, and recording and discussing results helped
students on later evaluations to identify series and parallel circuits when the batteries were not directly touching. (See Figure 4.)

**FIGURE 4. EXAMPLES OF CONFIGURATIONS WHERE BATTERIES ARE
CONNECTED BY WIRES AND STUDENTS RESPONDED
INCORRECTLY THAT THE BULB WOULD NOT LIGHT**

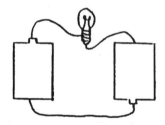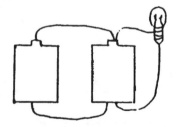

As lessons progressed, students could develop ideas about the flow of electrical current by using the brightness of the bulbs as an indicator of the amount of
current flowing. Questions such as these were to elicit thinking while the students
developed an electrical current model: How can you tell whether electrical current is flowing in a circuit? How does the brightness of bulbs compare in series

and parallel circuits? What happens to the flow of electrical current if you take a bulb out of parallel or series circuits? Which would wear out first—a battery in a series circuit or a battery in parallel circuits?

Because students differed in their ideas about the flow of electrical current, making their thinking known and listening to the reasoning of others was important in forming a model. For example, when asked to predict which battery would wear out first—one in a series circuit with three bulbs or one in parallel circuits with three bulbs—students did not agree in their predictions. Some were confident in their choices and were anxious to explain their reasoning while others were unsure. By the end of discussion, most students agreed correctly that the battery in parallel circuits would wear out first. Here is part of a class discussion representing the viewpoints and reasoning of the students.

Teacher: Which do you think would wear out first?
Student 1: Probably the series one.

Teacher: Why did you choose the series one?
Student 1: Because one circuit is going through three light bulbs and trying to get power to all three light bulbs with one battery.

Teacher: I see a lot of hands up. What do you think, Student 2?
Student 2: The parallel ones because it's brighter and it's charging more energy.

Teacher: So you think this one will wear out first (pointing to parallel circuits). And what do you think Student 3?
Student 3: I think it's parallel because it has to share enough for everybody and it has to go in all these different wires and some of the wires use up a little bit more energy just to make them so they can work—like water: sometimes it gets stuck on rocks; it doesn't always keep on going—it sometimes gets stuck on a rock when the river dries up.

Teacher: And how are you comparing that to the light bulb, Student 3?
Student 3: Some of the energy might wear off because it's going in all these different directions. And with the series it just goes all in one complete motion instead of splitting like in parallel.
Student 4: The parallel one would wear out faster because the battery only has so much energy and it gives it out to three wires.
Student 5: In series it only has to go one place and in parallel it has to go to three different places so I think parallel would wear out first.

Whether the ideas students are developing are about electrical energy or in another context, opportunities for students to communicate their viewpoints and reasoning are important to their conceptual development. These opportunities also provide windows to the thinking of students and help teachers to assess how the understanding that the students are developing compares with that of the scientific community.

ENDNOTES

1. Development of this case study was partially funded by a grant from the National Science Foundation (MDR-9155726) to Dr. Emily H. van Zee. Opinions expressed are those of the author and do not necessarily represent those of the funding agency.
2. This paper was presented at a workshop, "Teachers as Researchers: Studies of Student and Teacher Questions During Inquiry-Based Science Instruction," at the 1997 annual meeting of the American Association for the Advancement of Science in Seattle.

REFERENCES

American Association for the Advancement of Science. 1993. *Benchmarks for science literacy*. New York: Oxford University Press.

Driver, R., E. Guesne, and A. Tiberghien. 1985. *Children's ideas in science*. Philadelphia: Open University Press.

Educational Development Center. 1966. *Elementary science study: Teacher's guide for batteries and bulbs*. Nashua, NH: Delta Education.

Lawrence Hall of Science. 1993. Magnetism and electricity. Module in *Full option science system*. Chicago: Encyclopedia Britannica Education Corporation.

McDermott, L.C. 1990. Millikan Lecture 1990: What we teach and what is learned— Closing the gap. *American Journal of Physics* 59: 301-315.

McDermott, L.C. 1996. *Physics by inquiry*. New York: Wiley.

National Research Council. 1996. *National science education standards*. Washington, DC: National Academy Press.

Science Inquiry Conference—A Better Way![1]

Diantha Lay

"Inquiry into authentic questions generated from student experiences is the central strategy for teaching science," according to the *National Science Education Standards* (National Research Council, 1996, p. 31). As a fourth-grade teacher, I want my students to generate their own questions about some aspect of science and develop projects based on these questions. Then they discuss their projects with other students in our school during our Science Share and with students from other schools during an inquiry conference (Pearce, 1993; Bourne, 1996). Inquiry conferences are a better way to celebrate student learning because they involve the students in talking with one another about what they did rather than being judged by adults as is typical in traditional science fairs.

Preparing for an inquiry conference is a long-term process. Throughout the year, we keep a class list of things we wonder about as we move through the various science units. Among the questions my students have generated this year are "Why do big fish sometimes eat little fish?" "Why are minerals valuable and rocks aren't?" "Is lightning electricity?" "Are kids' reflexes faster than adults'?"

When the time comes to develop projects, we begin by looking at our list of questions. It is important to me that the students develop projects in which they have a vested interest. Research has suggested that projects are more meaningful if they are developed out of the interest of the children, rather than the design of the teacher (Katz & Chard, 1989).

The projects provide an excellent opportunity for an extended writing activity that takes the students through the entire process of writing, editing, rewriting, and publishing. First I ask the students to write to me about questions that they have. The questions should be based on something they already know that allows for further exploration. One student, for example, wrote about the purpose of a study of bubbles, "To find out what holds bubbles together by experimenting with a variety of solutions. I want to find out what causes the colors in bubbles. I also want to find out which solution makes the biggest and most colorful bubbles." Then we write back and forth in journals about the questions. I wrote back to this student, for example, "Can you take pictures? Can you compare colors? How? Sounds good!"

We also have conferences about the students' questions. I try to direct their questions to more meaningful topics beyond what we have done in class. One of the questions that a fourth grader wanted to explore, for example, was "Will oil and water mix?" As I remember our conversation, I asked, "Do you know the answer to that question?" The student said, "It's something I wonder about." I responded, "What do you wonder?" The student said, "Why don't oil and water mix?" and I said, "OK but do they mix?" "No, they don't mix" was the answer and I replied, "So your question is really not 'do they mix?' but 'why don't they mix?'" Sometimes my students need help getting focused on what their questions really are. They know they have some kind of question about a topic but they need to figure out what they are really wondering.

The students keep journals of the progress they are making. A progress report contains information regarding their research up to that point. I want to be sure that all students are working on their projects and to follow up on students who may be falling behind. After the students have formulated their questions, the next task is to list all of the materials they need for the research. They also have to explain how they plan to find the answers to their questions. Here is an example progress report:

> This week I got all my materials ready. For the past few days I was thinking when I would start my project, and I decided Thursday night. Today I started. I was unable to get strawberry juice so I decided to use coconut juice. Tonight I tested punch and orange. First, I got both juice in a cup and then two shiny pennies (keep on reading to see why). After that I went outside to my front yard and there is a faucet. So, I turned

on the faucet and let the water run on the dirt. Then I put dirt on both pennies. After I did that I put the dirty pennies in the punch and orange juice (one in each cup). I let sit for a while. When I looked at it and moved it with a toothpick and got the pennies out with my hand! Disgusting! Then I dried them. So far I think punch cleans a penny the most. But I have two more juices to test! I mean lemonade! What a long progress report!!

I responded to this student, "Great job! Leave it there for a few days and see what happens! What do you predict?"

The next step is to experiment and explore the students' questions to see if they can come up with answers. Then they draw conclusions based on what they find with their experiments. The students edit their own reports. Then I edit the writing pieces and return them to the students. After they write the final reports, they present their projects to the class.

The students also have an opportunity to assess their own work in the form of a rubric. They develop the rubric with my help and assess their own work and one another's. The rubric for the science projects had several parts: a point for the scientific process; two points for the scientific process and what you learned; three points for the scientific process, what you learned, and a comparison of your results with what you thought was going to happen (your predictions). To get four points, you had to include all that plus a reflective piece on what you would do differently if you were doing this over again—what went right, what went wrong, and if you were going to do the project again, what changes you would make.

Traditionally the big emphasis has been on the scientific process and making sure each of those steps is labeled properly on a display board. I wanted to change that. Students can go through the scientific process and still not really be engaged in reflective thinking about what they have done. They can have results and draw conclusions but not really think about the meaning of what they have done. So many think that they can run a test one time and that gives the answer. By presenting their projects to one another, the students learned to become more critical thinkers, not just with their own project but in evaluating their peers' work as well. The students gave their rubric sheet to one another, not to me. There was a place on the rubric sheet for comments and even the kids who got two's were thrilled to have a whole pile of papers with comments. They sat there and read them. It was a learning tool for them. This was private, for their own information, for them to learn and to give them some ideas about what they could do.

Someone might say, for example, "I wish you had spoken louder because you had good things to say."

At the Science Share at my school this year, projects were set up in individual rooms instead of one large place. There was a certain period of time when students would be at their projects to answer questions as people came through. No judging occurred but everyone who participated received a certificate of participation.

After reading Barbara Bourne's description of a Kid's Inquiry Conference (1996), I became interested in organizing one. Also I attended a seminar about these conferences and saw their value and benefits. As a participant in the Science Inquiry Group, I was able to convince its sponsor, Emily van Zee (1998a), and two of the other teachers to work with me. After a great deal of planning and preparation, sixty-five students in third through fifth grade came to the University of Maryland to participate in the conference.

The students discussed their projects with one another in small groups and then shared what they had learned in a large group discussion. They also evaluated the conference. They said that it was very helpful having as much time as you needed to talk about your project and to share with other people and be able to answer questions. At school, they had had to tell everything in a minute and a half. They thought it was helpful to be involved in talking with students besides those in their own schools. They said that presenting to people who do not know you is different because sometimes people you are with every day do not take you seriously. As part of this closing discussion, I asked what the students would say to teachers about learning science. They said that teachers hand out too many worksheets, don't do enough science, don't wait and let the students find out the answers, and tell the answers before class is over, and that the hands-on part of science is the best part.

The inquiry conference was one of the highlights of my teaching career. It was so powerful to be a part of all of the sharing and to watch the students talk about their projects, to have the confidence and ability to stand up and talk about what they were doing—just a remarkable growth experience for those students, something they'll remember for the rest of their lives. As we were leaving, two college girls got mixed in with our group. One said to the other, "Did you ever visit a college when you were in elementary school?" The other said, "No way! What are these kids doing here? Isn't that cool?" I answered, "Because they had fabulous science questions that they came here to share with the world!"

ENDNOTE

1. Development of this case study was partially supported by a grant from the Spencer Foundation Practitioner Research: Mentoring and Communication Program to Dr. Emily H. van Zee. Opinions expressed are those of the author and do not necessarily represent those of the funding agency.

REFERENCES

Bourne, B. 1996. The kids' inquiry conference: Not just another science fair. In *Beyond the science kit: Inquiry in action,* edited by W. Saul and J. Reardon, 167-188. Portsmouth, NH: Heinemann Press.

Katz, L., and S. Chard. 1989. *Engaging children's minds: The project approach.* Norwood, NJ: Ablex Publishing.

National Research Council. 1996. *National science education standards.* Washington, DC: National Academy Press.

Pearce, C. 1993. What if...? In *Science workshop: A whole language approach,* edited by W. Saul, J. Reardon, A. Schmidt, C. Pearce, D. Blackwood, and M. D. Bird, 53-77. Portsmouth, NH: Heinemann Press.

van Zee, E. H. 1998. Fostering elementary teachers' research on their science teaching practices. *Journal of Teacher Education* 49: 245-254.

Science Beyond Labeling[1]

Rhonda Hawkins

S cience, reading, and writing are nearly inseparable. In my sixth-grade classroom, I have three students who are considered resource students. They
carry labels that have reinforced their conviction that they are poor readers, writers, and students in general. As the science teacher, I have seen these students
conduct experiments with their peers and do just as well. When the hands-on
activity is completed, however, and the reading and writing portions of the
experiment must be tackled, the labels resume the mystical tyranny over their
abilities. In this case study, I have documented their conversations during the
experimental process and provided accommodations in reading and writing. In
the course of closing the disparity between their ability to participate in the
hands-on task and their ability to translate that into print, I have explored ways
in which these students can use the field of science to piece together the web of
reading, writing, activity, and understanding. Data collection included audiotapes of conversations within the students'groups during an experiment and
experimental write-up with accommodations that included textual readings by
the teacher or among the three, summarization of main points, questionings by
the teacher, and her transcription of answers dictated by the students.

My school is in a suburban area mostly of white-collar workers. There are
thirty-three students in my sixth-grade class, eighteen boys and fifteen girls, most

of whom are reading on a fifth- or sixth-grade level. The two boys and one girl in this study have not been held back but each receives services in reading as part of an Individualized Education Plan (IEP). The names are pseudonyms.

Noah is the most outspoken of the three students. An African American with diverse interests, Noah is not afraid to take risks. When asked to read or write in front of the class he does not hesitate. He makes an attempt and is willing to accept help from his classmates and his teacher. In describing how he had worked on a task in a small group, for example, he pointed out one of the parts that he had read and at what point he had someone else read for him. From his own admission, he is a poor reader but his ability to comprehend the processes of science and have a plethora of questions is not affected. The difficulty comes into play when he is unable to make basic connections between his mental processing of information and what is in printed text. When he does not understand something, he will ask questions of his teacher and classmates.

Caleb is an African-American student, often troubled by something. As a member of the class, he is quiet most of the time, hoping to go unnoticed. When he is called upon, he responds sheepishly, yet with my help he will come up with an answer. When working with his classmates, Caleb relies heavily upon them to do all of the work and to explain to him what is being done. I use discretion in pushing him so that he will remain a part of the working group and maintain some responsibility and ownership for the task being completed. Believing that he cannot do anything well, he attempts to do as little as possible. One-on-one Caleb is like another person. He works hard at solving problems and is attentive to what you say. A part of him remains withdrawn but he is willing to work in a small setting.

Michaila is an African-American female who is very outspoken and confident. She prefers to do things on her own and often gets offended if you make her feel as though she cannot read or do something her peers are doing. Even though she may need assistance, she will not admit it. As a working member of her cooperative group, Michaila is just as aggressive as the other students. When she is wrong, she plays it down and picks up in another part of the task. Because of the pull-out demands of her special education program, she is very sensitive to being singled out. When given the choice, Michaila prefers to stay with her peers and fit in. In private sessions Michaila is responsive as long as she does not feel as though she is in a remedial setting.

The teacher is an African-American female with five years of teaching experience. I am the teacher and researcher referred to in this study. You may notice that in expressing personal experience I refer to myself in the first person while in speaking objectively I employ the third. For four of the five years that I have been teaching, I have taught science for at least part of the year.

My strengths are discipline and organization. In order for me to have an organized classroom setting, my students must be comfortable and in agreement with how our class is run. One of the major difficulties for me is breaking organization for the sake of student inquiry. In my day-to-day assessment of how I am doing in relation to my students' needs, adapting and adjusting to their demands is a big consideration for me. One thing I do well is the dramatics that I engage in when teaching and interacting with the students. This manner of reacting and relating to my students establishes a rapport that transcends textbooks and traditional relationships between teacher and student. As a learner, I am always emerging into another state of comprehending. When I am teaching I learn as much as my students do. The biggest difference between my students and me is that I am responsible for organizing and presenting material in a way that will benefit all of them. I enjoy teaching, learning, and interacting with the students but I would be remiss if I did not admit to great frustration with constraints such as large class size, disjointed curriculum demands, administrative overloads, and unrealistic time allotments.

In my sixth-grade classroom that includes Michaila, Noah, and Caleb, inquiry learning and teaching centers on questioning by students and teacher. The students' search is within the context of science as it relates to their natural interests and prior knowledge. Mine is within the context of finding the best means of helping my students learn the content of science and apply the skills of scientific investigation to other areas of their learning. The teacher must also make sure that the students' inquiry is not hindered but capitalized upon, while at the same time guiding their inquiry so that it is directed enough to meet the requirements mandated by the state, county, and whoever else has some say as to what the students must know. By context, I refer to where the teacher or the student makes meaning of circumstances and conditions. Within the classroom, students are naturally inquisitive and this inquisitiveness is grounded within their context. Inquisitiveness can often lead to further exploration and enthusiasm in science investigations. Allowing students actively to engage in questioning, exploring, and refining their inquiry through reexamining their original question makes for both learning and teaching (National Research Council, 1996, p 31). The result is contrary to much that emerges within the school curriculum and traditional classroom instruction. That applies to my three special students along with the rest of the class.

All Maryland elementary schools put a great emphasis upon the Maryland School Performance Assessment Program (MSPAP), which drives instruction toward specific outcomes. In science, this program fosters hands-on science; there are tasks designed specifically for meeting the science outcomes and performance standards as designed by the Maryland State Board of Education. Using these guidelines, my students and I continually work on tasks that explore a topic in one of the science disciplines. These tasks not only concentrate on the content but are embedded with the skills needed to conduct scientific investigations (graphing, labeling, diagramming, identifying variables, predicting, and so forth).

Very often my students are so curious about the content of a MSPAP task that they attempt to rush through conducting a scientific investigation. For example, my students worked on a task involving chemical reactions and fair testing. Using Alka-Seltzer as the catalyst, the students went through two initial investigations to explore the basic components of chemical reactions and to expose them to going through the steps of a scientific investigation as a means of answering a question. The culminating task was to design their own investigation from hypothesis to conclusion, utilizing all the materials on a designated list. In all three classes, each with eight groups, the students were so fascinated with mixing chemicals that they completely disregarded controlling variables. At the end of the investigation, we had cups of liquids mixed with Alka-Seltzer and reaction times that varied with no apparent reason. After scoring the task as a class, we discussed the findings of the investigations and the implications upon conducting the fair tests.

Throughout this investigation much more was going on than a search and findings. For my students, there was an excitement that accompanied the freedom to investigate chemical reactions. Even though they had agreed to collect data on specific chemical reactions, every group failed to resist the temptation to mix several chemicals and then collect data on the reaction time and observed reactions. All of my groups collected data and formulated some way to show these through graphing or charts. Their inquisitiveness and the excitement of having the freedom to investigate caused them to formulate several questions at one time. It is at this height of excitement that I, as the facilitator, stepped in and attempted to guide their questions in some type of logical pattern to answer the questions they had put forth as well as those to meet the task requirements.

For the teacher, completing a task of this sort with students can be arduous. Their questions and the speed at which they generate them is enough to tire even the most hearty of science teachers. But the questions are the catalyst of opportunity to foster science attitudes and investigations that lead to a lifetime of learning. I often find myself choosing which to address, content or process.

As a means of meeting the demands of MSPAP, I continually assess and choose what needs to be emphasized and what can be revisited in other investigations. When my students had completed their investigations with chemical reactions, for example, we had such a mix of uncontrolled data that I chose to emphasize how they record the information that they collect. The one thing that is not negotiable is the factor of student inquiry. If they do not engage in inquiry through the questioning and challenging of what is being presented, then I do not have strong basis for holding and refining their attention to whatever task we are doing.

The written work of my three special needs students during the Alka-Seltzer investigation gave very little evidence of comprehension of the directions and the demands of the task. The written data reflected the students' lack of ability to translate text into their own written text. Students who have difficulty in this will meet with little success when science tasks rely heavily upon reading and not doing (Mastropieri & Scruggs, 1994; Scruggs et al., 1993). The portions of the task that the three completed had been with the help of group members. The pattern of their writing followed the ease with which a group member could commit time to assisting them with directions or writing.

Before repeating the first portion of the task, I explained to them that to understand what they did and did not understand, I needed their help. We also talked about how they felt about reading and writing. We looked at the text included in the task and they agreed that it was a lot and they expressed relief at not having to read all of it. When I explained that they would not be writing answers, they expressed their appreciation with a smile and even more enthusiasm towards completing the task. To make it as equitable as possible I allowed them to call out answers. When one person dominated the conversation or answered most of the questions first, I would call on one or the other to respond first so that answers would not be influenced by what others had said.

The responses to the task demands were very accurate and insightful, and would have resulted in a satisfactory if the three had been able to write them as well as they could say them. An example of their dialogue is shown in Table 1. They knew the purpose of the experiment (line 2), reflected upon the results (lines 7-8, 10-11), formulated new questions (line 12), and made predictions (lines 14-15). As for specific task behaviors, when it came to controlling variables in later parts of the investigation, these students reacted the same way the rest of the class had. They were so eager to mix chemicals that controlling the experiment was the last thing on their minds. Inquiry was winning over scientific process.

TABLE 1. DIALOGUE ABOUT THE ALKA-SELTZER EXPERIMENT

Before the students conduct the tests:

1. **Teacher:** When we conduct this experiment what are we looking for?
2. **Michaila:** How long it takes for the Alka-Seltzer to melt.

3. **Teacher:** Okay. What is another scientific way to say melt?
4. **Caleb:** Dissolve.

5. **Teacher:** Great. Timekeeper, are you ready? Don't forget to record your data on the data table.

After they conduct the tests:

6. **Caleb:** I told you the hot water would melt it quicker.
7. **Michaila:** The hot water dissolved it quicker.

8. **Teacher:** What are you talking about?
9. **Noah:** The hot water made the Alka-Seltzer dissolve faster because it was so hot.
10. **Michaila:** The cold water was the slowest because it slows down the reaction time.
11. **Caleb:** I wonder what would happen if you used hot water to dissolve other stuff?

12. **Teacher:** If we changed the size of the tablet will that affect the dissolving time?
13. **Noah:** Yeah. The less there is to melt, the faster it will melt. Dissolve.
14. **Michaila:** It would probably take a long time for a whole box of tablets to dissolve.

15. **Teacher:** When we change things on purpose in an experiment what is that called?
16. **Noah:** Variables?
17. **Caleb:** Independent variables. We changed the temperature of the water on purpose.

I plan to experiment further with having these students tape their answers. Then as a small group we can transcribe them with word processors. Approaching tasks and their learning this way may enable me to hold their attention and begin to bridge the gap between their facility in experimentation and their abilities in reading, writing, and processing text. When we worked in a small group with other students on or near their level, they appeared to be more comfortable and willing to take some risks. To improve in reading and writing, they have to take a lot of risks.

My three students need many things, some basic some complex. The basic needs include personal attention directly to address their problems in learning, compassion and understanding for their current level of accomplishment, and time to grow and process things that we take for granted other students know at this level of learning.

ENDNOTE

1. Development of this case study was partially supported by a grant from the Spencer Foundation Practitioner Research: Mentoring and Communication Program to Dr. Emily H. van Zee. Opinions expressed are those of the author and do not necessarily represent those of the funding agency.

REFERENCES

Mastropieri, M., and T. Scruggs. 1994. Text versus hands-on science curriculum. *Remedial & Special Education* 15: 72-85.

National Research Council. 1996. *National science education standards*. Washington, DC: National Academy Press.

Scruggs, T., M. Mastropieri, J. Bakken, and F. Brigham. 1993. Reading versus doing: The relative effects of textbook-based and inquiry-based approaches to science learning in special classrooms. *The Journal of Special Education* 27:1-15.

Collaborative Conversations: Strategies for Engaging Students in Productive Dialogues[1,2,3]

Dorothy Simpson

Many years ago the idea of discussing concepts with students in a mathematics context was presented at a workshop I attended. Since then I have continually tried to improve upon the process of letting my physics students develop their own ideas to reach a logical conclusion.

I believe this is a great method for helping students talk about their ideas, develop them logically, and reach some understanding of the process of science and the way real scientists work. These classroom discussions certainly fit the requirements for focusing and supporting inquiries, orchestrating discourse, challenging students, and encouraging and modeling skills of scientific inquiry as listed in Teaching Standard B of the *National Science Education Standards* (National Research Council [NRC], 1996, p. 32).

The teacher's role in orchestrating discourse, according to the *Standards*, "is to listen, encourage broad participation, and judge how to guide discussion—determining ideas to follow, ideas to question, information to provide, and connections to make" (NRC, 1996, p. 36).

Such discourse is very different from a lecture in which a teacher explains physics principles and demonstrates ways to solve physics problems. Such discourse also differs from a recitation in which a teacher asks *students* to explain physics principles and demonstrate ways to solve physics problems. Such discourse involves reflection, not only on what one knows but also how one knows something and why one believes that to be the case (Minstrell, 1989; van Zee & Minstrell, 1997).

IDEAS FOR DIALOGUE

The following presents my strategies for encouraging dialogue with students, from eliciting preconceptions to bringing closure to a unit. These strategies might provide a point of departure for teachers embarking on this approach for the first time or some insights for teachers more experienced with this approach to teaching.

The structure of a unit is shown in Table 1. For students to develop "big ideas" through dialogue, they need to start with a familiar situation to which they can relate. The situation should take the form of a pre-instructional exploration activity that encourages students to explore relevant physics ideas.

TABLE 1. STRUCTURE OF UNIT

Pre-instruction (exploration) activity
   Present situation from which to elicit student preconceptions

Dialogue about exploration
   Open out dialogue—many possibilities
   Develop hypothesis

Development activities
   Look for questions about situation
   Look for inferences to support observations

Dialogue about development activities
   Start closing
   Reexamine pre-instruction ideas
   Analyze for logic, consistency, validity, and reasonableness

Dialogue for closure
   Foster logical evolution of ideas toward thinking of physicist

---

A dialogue about the exploration activity encourages further thought and curiosity about the physics involved. After this dialogue, students work on developing the concepts through demonstrations and activities in a logical sequence. Small-group discussions provide the basis for class development of the big ideas related to the unit. The small groups present their ideas to the rest of the class, and then there is a dialogue about the development activities in which students introduce evidence to support or refute each idea. The class narrows the list as the students reach logical conclusions about each idea.

The teacher-mediated discussions lead to closure as the students use logical reasoning about the observations to establish inferences that physicists would consider acceptable. During discussions I use the strategies presented in Table 2. I specifically discuss with students my expectation that they will improve or develop the skill of asking good questions by considering the implications of the points listed in Table 3.

TABLE 2. STRATEGIES FOR TEACHERS DURING A DIALOGUE

Are you inviting all students to speak without judging their comments?

When opening a discussion, are you refraining from commenting on student ideas and remaining ambiguous regarding your own ideas?

Have you listed ideas on a board or overhead projector before discussion?

Do you ask for supporting evidence for each comment after all are elicited?

Do you ask questions to help the student construct a logical conclusion?

Do you refrain from "telling" the student the pieces he or she is struggling to construct?

Do you paraphrase each comment?

Do you validate each speaker with an acknowledging comment?

Do you provide "wait time" after a question, before allowing comments?

Do you ask for counterarguments for each idea after all ideas are elicited?

Do you ask questions that direct a student's thinking to a conclusion showing the fallacy of the argument?

TABLE 3. STUDENT STRATEGIES DURING A DIALOGUE

ACTIVE LISTENING

Do you listen carefully to what the speaker is saying?

Do you listen from the point of view of the speaker?

Do you actively consider the ideas presented?

Do you try to find a pattern in the observations and ideas of other students?

Do you mentally paraphrase what the speaker said?

Do you think of questions you could ask the speaker to clarify?

Do you think about the observations and look for missing pieces?

CONTRIBUTING

Do you indicate your desire to speak without interrupting the speaker?

Do you make comments that further the discussion about the ideas just presented?

Do you ask questions about what the speaker said?

Do you ask for clarification of what the speaker said?

Do you challenge what the speaker said, based on your evidence?

Do you refer to the ideas presented rather than to the person who introduced the ideas?

Do you try to present arguments and counterarguments to the ideas presented?

PRE-INSTRUCTIONAL EXPLORATION ACTIVITY

Pre-instructional activities create a context for student thinking by presenting some familiar situations related to the concepts in the unit. A possible exploration activity for a unit about forces might be students drawing force vectors on a diagram of the motion of a tossed ball. Discussion of the ball's motion provides a natural motivation for students to attempt to come to consensus about the forces in the situation.

I open the dialogue about the exploration activity with a discussion about any preconceived notions of the physics concepts involved. I then place ideas on the board or overhead projector for students to consider, and I invite students to give reasons from real-life observations as to why they think as they do. An example of a student idea might be that there needs to be a force on the ball in the direction of motion (McDermott, 1984); in this case students often propose that such a force would be provided by the hand. Our goal is to help students apply their understanding that hands cannot exert a pushing force if they are not touching the object.

Possible questions for the dialogue about the exploration activity include:

- What do you think might happen?
- What experiences have you had to support your idea?
- Does that always happen?
- What might be some reasons why _____ would not happen?
- What other possibilities might you suggest?
- Who has a different idea about what might be happening?

As students discuss their ideas, they try to create logical arguments and develop tentative hypotheses. Helping students focus on what they believe stimulates their thinking and gives them a starting point from which to compare observations and final inferences. During this discussion students consider evidence for counterarguments. They also revisit concepts such as forces touching and forces at a distance.

DEVELOPMENT ACTIVITIES

Development activities engage students in examining their ideas further with demonstrations or hands-on experiments. Students record and discuss observations within lab groups. During these activities they ask questions about what is happening and continue to consider their predictions. They might think of related situations and form additional "What if..." questions and then develop their own experiments to test their hypotheses. During this phase of the unit, students develop ideas to present as suggested "big ideas" from their group.

During the dialogue about the development activities, students elaborate and develop their ideas by reviewing the observations from the development activities. It is important that the observations be accurate enough that students can make valid deductions about the concepts. For example, students can move dynamics carts on a frictionless track and observe that they tend to continue at a constant velocity. At least one student will remember Newton's law of inertia and apply it to this situation, which is similar to that in which a ball is thrown horizontally. This result is in contrast to the students' initial hypothesis. By discussing the ideas from all lab groups more information is presented for finding patterns.

In a dialogue about the remembered ideas and the pattern that is formed, students develop their ideas and follow their reasoning to a logical inference. This discussion helps students move from observations to inferences using logical arguments to reach logical conclusions. They consider how valid their inferences are and the reasonableness of their conclusions. The comparison of the predictions from the exploration activity with the experimental observations is a critical part of the process of leading students to understand the concepts. The contrast between what students expect and what they observe or conclude is often an "aha" experience that helps them to mesh their original ideas with the logic of the conclusions.

Possible questions for dialogue about the development activity include:

▶ What is your evidence for that idea?
▶ What was your observation?
▶ What might you infer from that observation?

DIALOGUE FOR CLOSURE

I move toward closure by asking a variety of checking questions and carefully observing students' nonverbal expressions. Many ideas have been tossed about, and some students may have become confused with all the possibilities. Some students may have tuned out from active thinking about the inferences.

As the dialogue about the observations and inferences reaches a conclusion, it is important for the teacher to ask if the students agree about the conclusion. The agreed upon conclusion needs to be repeated several times in several ways so all students understand it and can mentally agree with it. If any students still have questions, it might be necessary to reopen the discussion with some pertinent points of the logical sequence from observation to big idea to help the doubting students follow the logic.

GENERAL COMMENTS ABOUT QUESTIONS

When teachers start using this process to help students think critically about the predictions, observations, and inferences, the dialogues usually tend to be dominated by teacher questions with student responses. As students become accustomed to the process and as they practice logical thinking and mental debate of the ideas, they will start asking questions. The student questions need to be answered with a teacher question that helps them move along in the logical reasoning.

As the process evolves, other students will become involved and some of the best reasoning dialogues will involve students only, bypassing the teacher entirely. This is a very valuable learning situation in which students talk and reason with each other. They are actively thinking and have matured in their ability to reason logically so they can move to conclusions without the crutch that the teacher questions provide.

To negotiate a dialogue of logical thinking leading to physics big ideas, the teacher needs to use a variety of techniques. Guiding the discussion requires a great deal of focus to listen and process the comments. Before a dialogue of this kind can occur, a safe atmosphere must have been created so each student feels free to take a risk and make a statement without fear that it might be wrong. Students should be reminded that there are no right or wrong statements in these dialogues.

Choices must be made by the teacher as to the direction of the dialogue. Is this a comment that requires more probing? Is this a comment that helps the group move to the big ideas? Is this a comment that leads to a tangential idea and needs to be postponed for another day? If this is a comment that might lead the class astray or add confusion, what question will help direct the conversation back to the big idea or help a student redirect his or her thinking? Is this a dialogue to open up the student ideas, a dialogue to consider observations and inferences, or a dialogue for closure, culminating with the big ideas?

Negotiating dialogues in this way requires much work by the teacher, who always strives to maintain the direction of the discussion, but the payoff is understanding gained by students.

ENDNOTES

1. This article is reprinted with permission from NSTA publications, copyright 1997 from *The Science Teacher*, National Science Teachers Association, 1840 Wilson Blvd., Arlington, VA 22201-3000.
2. Preparation of this manuscript was partially funded by grants to Jim Minstrell and Earl Hunt by the James S. McDonnell Foundation and to Emily van Zee by the National Science Foundation (MDR 91-55726). Opinions expressed are those of the author and do not necessarily represent those of the funding agencies
3. This paper was presented at a workshop, "Teachers as Researchers: Studies of Student and Teacher Questions During Inquiry-Based Science Instruction," at the 1997 annual meeting of the American Association for the Advancement of Science in Seattle.

REFERENCES

McDermott, L.C. 1984. Research on conceptual understanding in mechanics. *Physics Today* 37:24-32.

Minstrell, J. 1989. Teaching science for understanding. In *Toward the thinking curriculum: Current cognitive research,* edited by L.B. Resnick and L.E. Klopfer, 131-149. Alexandria, VA: Association for Supervision and Curriculum Development.

National Research Council. 1996. *National science education standards.* Washington, DC: National Academy Press.

van Zee, E.H., and J. Minstrell, 1997. Reflective discourse: Developing shared understandings in a physics classroom. *International Journal of Science Education* 19:209-28.

Teacher Inquiry[1]

David Hammer

INTRODUCTION

To think of inquiry in the classroom is almost always to think of student inquiry. One does not often associate inquiry with the teacher's role, other than with respect to questions that come up within the discipline, science questions for a science teacher, for which the teacher does not have an immediate answer. My first objective in this chapter is to promote a view of inquiry as central to the teacher's role, particularly inquiry into student understanding, participation, and learning.

Although it is becoming more common to think of teaching as inquiry,[2] the emphasis in education reform remains on methods, materials, and standards. Meanwhile, the progressive agenda of promoting student inquiry, along with the need to coordinate that agenda with the traditional one of "covering the content," places substantial intellectual demands on teachers. If these demands are not considered and addressed, the progressive agenda is unlikely to succeed. In other words, to pursue science education reform through the development of new curricula, new materials, or new standards is not sufficient. To promote student inquiry, we must do more to understand and support teacher inquiry.

To begin, teachers spend a significant portion of their days taking in and inter-preting information about their students. Much of this data gathering is deliber-ate and explicit: taking attendance, collecting homework assignments and labo-ratory reports, giving quizzes and exams. Other information arrives on its own, in a nearly continuous stream, in the questions students ask and comments they make as well as in their facial expressions, body language, tone of voice. It is an enormous amount of information.

How teachers interpret that information, what they perceive in their students (whether the students are paying attention, confused, interested, frustrated, etc.) can dramatically influence how they choose to proceed (pose a challenging question, provide information, continue to new material, digress to pursue a stu-dent's idea). Most of this interpretation happens—must happen—automatically. In this respect teachers are like other practitioners, from professional chess play-ers to doctors, whose judgments are and must be largely tacit.[3]

Regarding chess players and doctors, however, there is a general awareness that this perception and judgment is taking place, that it is intellectually demand-ing, and that its betterment is central to professional education. It is both possible and expected for chess players and doctors to make at least some of their reason-ing explicit, as a matter of professional practice and development, and they do so in the contexts of specific games and cases. For teachers, in contrast, it is rare to have the opportunity, let alone the expectation, to present information from their classes to others, to make explicit their interpretations, or to consider alternatives.

CONVERSATIONS AMONG TEACHERS

This chapter describes work from a project designed to engage teachers in pre-cisely this sort of conversation, centered on their ongoing experiences in the classroom. From March 1995 through June 1998, a group of physics teachers and I met every other week of the school year for two hours, to talk about students and teaching. Our conversations, recorded on videotape for transcription and analysis, focused on snippets from the teachers' classes, small samplings of the information they were taking in about their students, in the form of transcripts, video or audiotape recordings, or samples of students' written work. Reading, watching, and listening to these snippets, we talked about what there was to see in the students' participation, exploring a range of possible interpretations. The teachers in the group and their school affiliations during the 1996-1997 school year were: Elisabeth Angus, Winchester High School; Hilda Bachrach, Dana Hall School; Edmund Hazzard, Bromfield School; Bruce Novak, Watertown High

School; John Samp, Cambridge Rindge and Latin High School; and Robert Stern, Brookline High School.[4]

This chapter is organized around excerpts from three of our conversations along with the classroom snippets they concerned that were contributed by Robert, Hilda, and Bruce. These conversations, from three consecutive meetings in the fall of 1996, are representative of the substance and tenor of our work. They also reflect a range of physics topics and types of snippet. I will use these examples to advance three objectives:

1) **Teacher perception and judgment.** The first, as I noted above, is to promote greater appreciation for the role and demands of teachers' inquiry into their students' understanding, participation, and learning.

2) **A language of action.** The second is to offer an insight that has emerged from our work regarding the language teachers use to express what they find in that inquiry. In our conversations, the teachers often experienced and communicated their interpretations in a language of action, in other words as ideas for what to do in the given circumstance, rather than in an explicit language of diagnosis. For a simple example, a teacher may express an interpretation ("The students have forgotten what they learned about inertia") by suggesting an action ("I would review the concept of inertia").

3) **A role for education research.** My third objective is to propose a view of the role of education research in instructional practice. Specifically, I will suggest that its primary role is to contribute to teacher inquiry, to teachers' perceptions of their students and judgments for how to proceed, rather than to prescribe effective methods. The conversation between teachers and researchers should therefore be understood to take place mainly at the level of their respective interpretations of students' understanding and participation. This conversation, however, may be difficult to recognize and to facilitate, owing largely to differences in the language by which researchers and teachers experience and communicate their interpretations.

INTERPRETING A CLASS DISCUSSION ABOUT FREE FALL: TEACHER INQUIRY INTO STUDENT UNDERSTANDING AND PARTICIPATION

The first snippet we discussed in our meeting on November 18, 1996 was a transcript Robert had prepared of a discussion in his college prep class[5] about the forces on a skydiver. Robert wrote in the snippet that his goal for this activity had been "to reinforce the idea of the net force as the driving engine for acceleration." Here is roughly half of the transcript.

Teacher: What forces act on the skydiver when he first jumps out?
Student 1: He accelerates down; he goes faster.
Student 2: But the air slows him down so he can't fall faster.
Student 3: But he doesn't slow down so something must be getting bigger.

Teacher: Someone come up to the board and draw the forces acting on him.
Student 4: There's the gravity that pulls him down. (Student draws a vertical arrow down.)

Teacher: What's the common English word for force of gravity?
All Students: Weight.

Teacher: (to student 4) Add the letter W to your diagram. Now what?
Student 4: Then there's the air resistance. (He draws a vertical arrow up, but not connected to the weight arrow. Long silence.)
Student 5: You have to put the arrows together.

Teacher: Why?
Student 5: Because they're both pulling on the person.
Student 4: Yeah, that's right. (He draws both arrows connected to the same point.)

Teacher: How are the two arrows related? Are they the same? Is one bigger?
Student 4: Well, the weight is bigger because it's pulling down.

Teacher: Does everyone agree? (Calls on a student.)
Student 6: No, it can't be right because the speed is increasing. The force of gravity is getting bigger.

Teacher: What's the common word for force of gravity?
Student 6: Weight.

Teacher: So what are you saying? The person gets heavier as he falls?
Student 6: (smiling) No, but something is wrong. He keeps going faster as he falls, doesn't he?
Student 4: Sure he does but it's the gravity that pulls him down.
Student 7: But doesn't the air resistance get bigger?

Teacher: You have some good ideas, but there is confusion here. The difficulty, as I see it, is that you're confusing the MOTION with the FORCES. Remember that you started the year with learning how to describe motion (kinematics). All the graphs and equations you did. Now you're looking at FORCES (dynamics). It's the forces which make things move and we've got to separate these two effects. Let's concentrate on just the forces; then we'll connect them to the motion.

Excerpts From Our Conversation

Bruce started our conversation with the suggestion that Student 6's comments revealed a common misconception. Robert's response showed that he too considered Student 6's contribution significant, but for different reasons.[6]

Bruce: [Student 6 showed] a misconception, that we've talked about before. That the speed is proportional to the force (reading Student 6's comment from the snippet): "That can't be right because the speed is increasing. The force of gravity is getting bigger."
Robert: [Student 6] is usually very, very slow in reaching any sort of [original idea], so for her to say what she did.... She said it so immediately, she knew the speed was changing, but in all of the year it's the first time I've ever seen her, you know, come up with something herself. [There] must be something else, another force, another factor. It was nice to see her do that. She couldn't quite get it, and I'm not sure whether that's [important]. I thought it was a turning point in the whole discussion.

After a brief exchange to help others locate Student 6's comments in the transcript, I turned the conversation back to what Robert had been saying.

David: And that was a turning point and the student who said that was somebody who—
Robert: Who normally doesn't see things very intuitively. She's very methodical, she's very good at memorizing stuff.... But for original

ideas, no. This is the first time that I saw that with her. Which was that you can see that somewhere what we had is not enough. There needs to be something else. But you didn't know what it was.

Shortly later, Robert elaborated on what he had intended in this conversation and what he saw happening at this juncture:

> **Robert:** I've never done this one before.... I'm using a new textbook this year and I looked around, I thought that might be a good way to tie up some of the ideas, let the students talk. Instead of doing [lots of] problems today, we'll spend a while, whatever we need, just talking about [one] problem. And it just was so enlightening to me to see that, just what you're saying, [they came up with the] idea, there needs to be another "force." That's the key item: There needs to be something else to make it accelerate. It doesn't have to be an increase in the force, but it needs to be something.

Turning back to the misconception he saw in Student 6's comment, Bruce commented on Robert's response at the end of the excerpt.

> **Bruce:** You may reinforce [the misconception] with what you say, "it's the forces which make things move." Which makes it sound like you need the force to have the motion. Which is something a lot of us say, [although] we don't mean it that way.

This reminded Robert of a related difficulty: students' reluctance to accept a velocity as an initial condition of an object, a problem he agreed his language may aggravate.

> **Robert:** Typically the thing that comes up, now that you mention it is, even when you have problems with things moving at a constant velocity, there are always a handful of kids, you know, they want to get that acceleration in the beginning, [thinking] "you gotta get it going," and I say "OK, now it's going."...Well, maybe I contribute to that.

Bruce recalled a suggestion John had made the previous year of a strategy for responding to this difficulty: Start with the room lights off and then turn them on after setting a ball in motion. The idea is to help students distinguish between the concept of *velocity* and that of *force* by focusing their attention on the ball's initial *motion*—when the lights come on the ball is moving—and away from any prior, initiating *force*.

Hilda reminded us that the students had been talking about a skydiver, who had no initial downward velocity. In this case, she noted, the students' reasoning may have been appropriate, because "there had to be a force, otherwise [the skydiver] wouldn't come down." Robert maintained, nevertheless, that the students had not distinguished force as causing velocity from force as causing *acceleration*.

After a brief digression on the sensitivity of students' understanding to particular wording, I asked Robert to say more about the snippet. He reflected on his impressions of the discussion, reiterating his pleasure and surprise at how it had gone, and recounted more of what had happened in the class after the segment he transcribed.

> **Robert:** I thought it was a great class. The class ended, the kids didn't want to go! . . . I had no idea it would turn out this way. I started out with, here's this problem, let's look at the different forces, maybe get to the idea of seeing that the net force would keep changing.
> **Lis:** Were you drawing on the board at all?
>
> **Robert.** Very little, I did very little.
> **Hilda:** The kids did [draw on the board].
>
> **Robert:** The kids did most of it. At the very end, when this one student wanted to know how—we finally got the idea that the net force is changing—he wanted to know how does the net force change? I asked "what do you think would happen?" and [he drew] a set of axes with force and time. And he stood there a while, and eventually he drew a straight line decreasing to zero. Which was, I thought, a very good first step, because the kids have never done this before.

The student was correct that the net force on the skydiver would decrease to zero: As the skydiver's velocity increases, the force of the air resistance increases as well, until the force of air resistance (upward) equals that of the earth's gravity (downward). The straight line was not correct: The net force would approach zero asymptotically, not as a linear function. Robert was impressed by the student's having made the first realization; he was not worried that the explanation should be fully correct at this point when the students were first considering the question.

Teacher Perceptions of Students' Understanding and Participation

The snippet continued further, as did our conversation; we spent roughly half an hour talking about it, the amount of time we typically allocate for a snippet. Our conversation was also typical in the range of perceptions it reflected in the snippet's author and in the rest of the group. Among their interpretations of the students' understanding and participation, Robert and the other teachers noted:

▶ **a misconception, on the part of Student 6, "that the speed is proportional to the force."** Bruce mentioned this in our conversation, but Robert had evidently seen something similar in Student 6's contribution, since, a moment later in the snippet, he tells the students, "The difficulty, as I see it, is that you're confusing the MOTION with the FORCES."

▶ **an original contribution by a student, Student 6, who was more inclined to memorization.** Robert recognized the same misconception Bruce did, but he perceived Student 6's idea in several other ways as well. He saw Student 6 as participating in a way that was new for her, a perception not available to the rest of us from the snippet itself, since it depended on Robert's experience from the start of the year.

▶ **a valid insight in Student 6's idea that "there needs to be something else," and a productive turning point in the class discussion.** In addition to seeing Student 6's reasoning as reflecting a misconception—that is, a conception inconsistent with the Newtonian understanding he wanted students to develop—Robert saw it as containing an insight that could help her and the class progress toward that understanding.

▶ **the misconception as possibly reinforced, inadvertently, by the explanation, "it's the forces which make things move."** This was not, directly, a perception of the students' understanding, although indirectly it attends to how they might reasonably interpret a statement by the teacher.

▶ **students' difficulty with the idea of an initial velocity.** Bruce and Robert had been talking about the students' confusing the concept of motion with that of force, both with respect to this particular situation and as a more general misconception. Here, Robert connected their reasoning to a related pattern he had seen, that of students' difficulty thinking of an object as having an initial velocity.

▶ **students' interest, engagement.** Robert was enthusiastic about the outcome of the discussion, both for the students' engagement ("The class ended; the kids didn't want to go!") and for the substantive progress they initiated ("The kids did most of it.").

To be clear, the point here is not these particular perceptions, and I am not claiming they are correct. To be sure, I expect other teachers would offer different interpretations, as happened routinely in our conversations. My point is that these perceptions represent multiple dimensions of teacher awareness concerning the students' conceptions of forces and motion, their modes of reasoning and participation, the level of their interest and engagement. It is awareness of individual students and of the class as a whole, in general, over the school year, and in particular moments.

In fact, this list of teacher perceptions is incomplete, as it reflects only those Robert and the rest of the group made explicit. Much, it is clear, always goes unsaid in our conversations about snippets. For instance, Robert saw something in the students' reasoning that led him to press them with respect to vocabulary: "What's the common English word for 'force of gravity'?" It is a reasonable guess that he saw the students' distinguishing as two ideas (the *weight* of an object and the *force of gravity* on that object by the earth) what a physicist considers one idea. By insisting on their use of the "common English word," he was insisting that they apply their everyday understanding of weight —in particular that the weight of an object is independent of its motion—to their reasoning about "force of gravity." From Robert's comments on other occasions, it is likely as well that he perceived and hoped to address a general pattern of students' treating physics as disconnected from their everyday experience. Moreover, this is only an excerpt of Robert's snippet, which itself represents only a fraction of what transpired in a single period from a single school day.

Here, then, is a first illustration of this chapter's opening premise: Teachers take in and process an enormous amount of information about their students' understanding and participation. Most of this inquiry is and must be tacit: There is more information than explicit thought could accommodate. No teacher could articulate every perception and intention.

Still, it seems to be both possible and productive for teachers to articulate *some* of their perceptions and intentions. Nevertheless, at least in the United States, it is rare for this to occur. Teachers seldom have the opportunity or occasion to show others their "data," to present their interpretations, and to have those interpretations challenged with alternatives. Because of this, teachers are mostly left to themselves to develop the intellectual resources they need to meet the demands of interpreting their students' understanding and participation, diagnosing their strengths and needs, and making judgments for how to proceed.

We did not pretend that our conversations captured more than a fraction of teacher thinking. However, by capturing that fraction, they allowed the teachers to exchange and compare not only methods and materials but perceptions of students in particular moments of instruction. Our conversations, grounded in specific instances from the teachers' classes, provided not only ideas for instructional strategies, but also new diagnostic possibilities, an exchange of resources to support the intellectual work of teaching.

In this respect, in their ongoing inquiry into students' understanding and participation, teachers have much in common with education researchers, specifically those who conduct research on learning. They study essentially the same phenomena, that is student learning, although in different ways, and it is reasonable to expect that teachers and researchers could support each other in their efforts. The central purpose of this project was to explore how that may happen, particularly how perspectives from education research may contribute to teacher inquiry. I will discuss this further in the section below, "A Role for Education Research."

Before that, however, I will present another example of a snippet conversation. This will serve further to substantiate the view I am promoting of teaching as inquiry and of teacher expertise as involving intellectual resources for engaging in that inquiry. The main purpose of the section, however, is to reflect on the language by which the teachers articulated their interpretations: The teachers often experienced and expressed what they perceived in students as ideas for how to proceed in instruction.

INTERPRETING LAB REPORTS ON SIMPLE CIRCUITS: DESCRIBING PERCEPTIONS IN A LANGUAGE OF ACTION

Our meeting on December 12, 1996 opened with a snippet from Lis, a videotape produced by two of her college prep students as part of an optional project. They had performed two experiments in projectile motion. First, they fired a BB-gun across a field at a target, measuring the distance the BB fell in its trajectory below the horizontal, and showing that this distance, ten inches, was consistent with a calculation from kinematics equations they had learned in class. Their second experiment was to throw two pumpkins from a cliff, launching them horizontally at different speeds. They measured the times the pumpkins took to fall and the distances they fell outward from the cliff, again to compare with the theoretical predictions.

There was much to discuss about this tape, including the students' investment in their work, the validity of their reasoning and measurements, the value of their "seeing" the BB fall ten inches in its trajectory, as well as the students' campy humor. The conversation digressed often from the details of the videotape to talk in general about the motivational and conceptual value of real-world and open-ended projects, together with strategies for assigning and assessing them.

Here I will focus on the second snippet, which Hilda assembled by use of student reports and her observations of their much more traditional laboratory work. In one lab, the students systematically varied the voltage and resistance in a simple series circuit by changing the resistor or the number of batteries. They measured the voltage and current using three different resistors for each of at least five values of voltage, and they plotted their results on graphs to confirm Ohm's Law.[7]

Most but not all of the students' lab reports were in line with what Hilda had intended. For her snippet, Hilda collected some of the aberrant responses to questions in the results section, including "What happens to the current when the voltage is increased (R constant)?" and "What happens to current when the resistance is increased (V constant)?" The following is quoted from Hilda's snippet.

> Despite a discussion of cause/effect, there were still those who answered:

> "The current would decrease, as would the voltage, as the resistance gets greater, it allows less electrons to pass through the circuit at a given time." Then, in her conclusion, she said: "We could also see that when the resistance stays the same and the current increases, the voltage would increase in proportion to it. This could be proven by $R=V/I$."

> Even though they answered the questions correctly, in the conclusion where they are required to sum up, there were those who said:

> "We discovered that as current decreases the voltage decreases and the resistance increases."
> "Because the R must remain constant with each circuit set up, if the current is decreased, then the volts must be increased to compensate. This satisfies the equation and makes sense because in order to compensate for a lower current due to a higher resistance, the volts must be higher in order to push the electrons through."

"When the current flowing increases, the circuit voltage increases."
"As the current increases, the potential difference increases."

Included in "sources of error" were:

"The batteries: as we used them they lost energy."

At the end of the snippet, Hilda added a comment about work by one group on a previous experiment that had impressed her favorably:

> I'd like to mention a really interesting way that one student saved her group's experiment that was measuring with a tangent galvanometer the dependence of its magnetic field on the strength of the current. In this PSSC experiment, a light bulb is used in the circuit to limit the current and to show that there is current. About halfway through the process of winding on coils one at a time the light bulb blew! After changing the bulb they saw that the compass needle deflection increased by a much larger amount than expected for a single coil increment. One of the girls recognized that it must be a different bulb letting more current flow in the circuit—this was before doing the Ohm's Law experiment! She was able to select a bulb like the one that burned out, and they were back to similar increments.

Our Conversation

John opened our conversation about Hilda's snippet with an interpretation of the difficulty some of the students had on the Ohm's Law lab and his suggestion of a way to address it.

> **John:** I look at this, and my thought is, one of the toughest concepts that over the years I have had to try to teach is what electric potential or voltage is in the first place. Students come into class, and they've talked about volts . . . all their lives, [but] essentially nobody knows what it is. And about five years ago somewhere I got my hands on a piece of shareware called "Circuit Vision." I can bring in copies. It runs on a Macintosh.

This software, John described, allows the user to build a virtual circuit, made up of batteries, resistors, and wires. The program then enacts a mechanical analogy of that circuit, showing current as the motion of little balls. Little escalators carry the balls from lower to higher levels, analogous to batteries lifting charge to greater voltage, and the balls push paddle wheels as they fall back down, analogous to charge expending energy as it moves through a resistor to a lower voltage. In this way, the program visually presents an analogy between electric potential (or voltage), meaning the *electric potential* energy per unit of charge, and *height*, which can be understood as the gravitational potential energy per unit of mass.

> **John:** And I think, as a result of that, students get a better concept sooner of just what voltage is. And some of the questions, I mean, this one question that somebody made in the middle [reading from Hilda's snippet]: "Because R must be constant with each circuit setup, if the current is decreased then the volts must be increased to compensate," as if somehow voltage and power are different measures of the same thing—one goes up, the other one's gotta go down.
> **Hilda:** [Agrees.]

> **John:** Others are just kind of looking at it as a mathematical equation. In some cases they're getting it right, getting the mathematical equation right, but I still get the feeling they have no idea what they're talking about.
> **Hilda:** No, that's the thing. That's right. In other words, the inverse proportion is there and the mathematical equation, but it's not there in terms of the concepts.

Hilda elaborated, in a tone of amused exasperation, her perception of how the students were using the equation:

> **Hilda:** [They were] using the equation as though it were pure numbers and not [as though it was] a measurement of anything that had significance. So, when I talked about it, I talked about it as a cause and effect idea. Or, sometimes I'd say to them, "You know, we put the cart before the horse. You've got things not in sequence. What's controlling what? We often talk about an independent variable, [and] a dependent variable. What's the control here?"

That is, Hilda was noting, the students had manipulated the voltage by changing the number of batteries they connected in series. Several students nevertheless described their observations as though the change in voltage resulted from the change in current ("as current decreases the voltage decreases and the resistance increases"). In this way, Hilda was saying, they were not making a meaningful connection between the equation ($V = IR$) and their measurements in the lab.

We also spent some time talking about the group of students who had discovered they were using the wrong bulb. Hilda recounted a similarly impressive episode in which a group of students, working on the Ohm's Law lab, had found that their measurements did not correspond to the markings on a resistor, ultimately to decide it was mismarked. Hilda described what impressed her about these cases:

> **Hilda:** I thought they did a really good thinking job there. Where they weren't going to just write down this number and say, "I've got 200% error" or something like that. [They] came over to say, "You know, we really think that this one's [mismarked]."
> **David:** So that's another example sort of analogous to this one [in the snippet].

> **Hilda:** Yeah, yeah, where they are showing greater sense. . . . that something that's different isn't, "uh-oh, we've got some errors in our experiment," but they looked for what could make this happen so that they could talk about it, that's what actually they did in their report. [In contrast to] one girl [who] just reported 200% error and didn't bat an eyelid. . . .
> **David:** So . . . they found some discrepancy and they were committed enough to the ideas to deal with it.

> **Hilda:** *Right.* Exactly. I can't even, sometimes they go through a whole experiment and they don't even notice if they've got some really anomalous data that just doesn't fit.

For Hilda, this problem went beyond this particular experiment. She proceeded to describe another example, in which students had somehow misread a scale to find that it took more force to pull a cart up a shallow than a steep incline.

> **Hilda:** But they don't notice [the mistake] until you look at their numbers and ask them, "What went wrong here?" . . . They're doing exactly what they were told to do and they don't really see, is it good data or is it not good data.

This reminded John of students' failing to catch absurd answers on their calculators, specifically in finding trigonometric functions of angles measured in radians when the calculators are set to measure angles in degrees. Robert saw this as a general liability of their inordinate faith in calculators, which can lead them to accept results such as that "a person's height [is] 43.5 meters." John noted that he "had that problem before calculators," and everyone agreed calculators were not the root cause.

Our conversation turned to this topic in general, of students who do not notice absurdities in measurements or in calculations. Referring back to Hilda's examples of those who did notice and resolve inconsistent results, I asked why other students do not do this and whether it is something they could be taught.

> **Hilda:** I had a discussion about that one time and [the students said] they figured I was doing something to trick them. That if I'm giving them problems on a test, the numbers don't have to be real numbers and so I could make it come out like a person can be 43.5 meters tall. I got into this mode then, of telling them.... "This is a real problem. There's no tricks. The numbers should be the order of magnitude of what you would expect."

Ed referred back to Lis's snippet as an example of an instructional approach that might help.

> **Ed:** One answer to your question, to the teachable-ness of this, is to give them a BB-gun, take them out in the field and have them make a video, and see whether the ten inches is [real]. They even did the conversion to meters—that was very impressive. I wonder, is that a way to make them [think of the results as meaningful]?

Ed's comment about the students' having converted to meters prompted an exchange about the prevalence of unfamiliar units in introductory physics. Lis remarked that "we didn't grow up with kilograms. And I think that they don't really know what [it means]." John agreed, "Except for seconds, pretty much everything we deal with in physics is not real to too many students." The rest of the conversation stayed with the general topic of the connection to "reality," considering the influence of students' experiences in mathematics classes and whether it is helpful or harmful for them to practice methods of calculation they do not yet understand.

Teacher Perceptions

Again, our conversation about the snippet reflected a variety of interpretations of the students' understanding and participation. To review, these included perceptions of the students'

- **difficulty with the concept of electric potential (voltage).** John opened the conversation with this thought and proceeded to describe a piece of software he has found helpful. Given a simple electrical circuit, this program depicts a mechanical analogy to help students visualize electric potential as analogous to height.
- **treating the mathematics as disconnected from the concepts...** Some students, John noted, were struggling with the conceptual relationship, whereas others were just "looking at it as a mathematical equation," without regard to its meaning.
- **...as well as from the procedure and measurements in lab.** Hilda also felt that the students were "using the equation as though it were pure numbers," rather than as though it involved quantities with physical significance. In particular, Hilda was referring to the lack of correspondence between the students' explanations and the procedure they had followed in the lab.
- **trying to make sense of discrepant data.** Hilda wrote about one case in her snippet, of students who discovered that they had inadvertently used a different type of bulb, and she told us about another in our conversation of students who determined that a resistor was mismarked. Hilda was impressed that "they looked for what could make this happen so that they could talk about it," in contrast to others who simply attributed discrepancies to "experimental error" without looking for any specific cause.
- **ignoring their common sense in thinking about physics.** Toward the end of this conversation we digressed from Hilda's snippet to talk about a general perception of students, that many do not treat physics as connected to reality. Hilda told of her students' saying they expect her to trick them and of her developing the habit of reassuring them that there are no tricks. Ed spoke of Lis's snippet as a means of teaching students to treat physics as real.

It may seem surprising that a group of teachers could find so much to discuss in these snippets, which to the untrained eye are fairly sparse excerpts and observations. The first point in this chapter, however, is that these are not untrained eyes. Working every day with students, teachers become adept at interpreting what they see and hear. Like physicians for whom a handful of symptoms in a patient may indicate a variety of possible conditions and courses of treatment, these teachers have developed a wealth of knowledge and experience, intellectual resources for thinking about students.

It is unusual, however, to understand teaching in this way, including among teachers. Even in these conversations, the teachers often seemed more inclined to talk about instructional materials and techniques than about interpretations of student statements and behavior. If inquiry into student knowledge and reasoning is at the core of teaching and teachers' expertise, as I am suggesting, then why would teachers seem reluctant to have conversations about it? This section concerns the second point of the chapter, that teachers often talk about their interpretations by talking about instructional materials and techniques.

A LANGUAGE OF ACTION

I had tried at the outset of this project to impose a ground rule. In conversations about snippets, we should restrict ourselves to comments that concern students' statements and behavior rather than what the teacher did or should have done. I had two reasons for imposing this rule. The first was to promote a focus on perceptions of students, rather than on the means for addressing them. The second was to encourage an atmosphere of respect for the teacher presenting the snippet. Too often, I had experienced conversations about teaching degenerate into uncomfortable and unproductive criticism of the teacher's actions.

My rule proved difficult to enforce, however, and, in the end, counterproductive. A key example from the first year was John's "turn on the room lights" strategy (which Bruce recalled during our conversation about Robert's snippet). Discussing another teacher's snippet, John had offered this:

> **John:** [I say] things like, "You know, what if the room lights come on and you see the ball already going down the alley. You know somebody pushed it, but you have no idea who pushed it. All you can say is, well, here it is right now. Now tell me what forces act on it." And sometimes they get it when I talk about room lights coming on. You know, I've had some trouble getting them to forget about earlier things.

What we came to recognize was that, while most of John's comment was explicitly to suggest a teaching strategy, it was also implicitly to express his interpretation of what was happening in the snippet, that the students had not distinguished a force *acting* on an object from a force *having acted* on the object a moment ago. Moreover, John's description of his strategy was helpful in communicating his interpretation to the rest of us, including the snippet's author, who as a result came to understand the students' thinking differently: "Yeah...I think maybe [that was the idea] these kids really had. Not so much that they thought it was pushing now. But more that it was pushed then."

This moment led us to the realization that a comment about teaching strategy may also serve to convey interpretation, and we were then able to recognize that this was happening fairly often. In other words, the teachers were often communicating "what to see" in the students' understanding and participation by suggesting ideas of "what to do" to help them. Their suggestions for methods and materials thus often had a dual purpose: explicitly to suggest instructional action, and implicitly to suggest a diagnosis of the situation.

For this reason, to rule out comments about teaching would be to rule out a principal mode by which the teachers discuss their interpretations. From the teachers' perspective, adherence to my rule made our conversations inauthentic, disconnected from their knowledge and experience, and we decided to abandon it. Perhaps it had served a purpose at the beginning, promoting a level of sensitivity and mutual respect in our conversations, but we came to see it as an impediment.

Our conversation about Hilda's snippet contained several examples of comments explicitly concerning ideas for instruction that served as well the role of expressing an interpretation.

The first example was again John's, who identified in Hilda's snippet a pattern he had seen before, of students' difficulty with the concept of voltage: "essentially nobody knows what it is." He went on to describe what he has found to be an effective means of addressing this difficulty, a computer program that displays mechanical analogies of electric circuits, with voltage analogous to height. By describing the computer program, John was not only suggesting it as an effective approach; he was also specifying what he saw as the problem, clarifying considerably what he meant by "nobody knows what it is." In particular, John saw the students as lacking what researchers might call a "mental model."[8] He went on to note that some students "were getting the mathematical equation right, but . . . they have no idea what they're talking about," and Hilda agreed: "That's the thing. . . the inverse proportion is there and the mathematical equation, but it's not there in terms of the concepts." As the conversation continued, however, it became apparent that Hilda and John differed in interpretation of what was the problem, or, at least, they were fixing on different aspects of it.

When Hilda elaborated on her interpretation, she explained that the students had been "using the equation as though it were pure numbers and not [as though it was] a measurement of anything that had significance." She explained how she had tried to address this in class, and this clarified what she meant by "a measurement . . . that had significance:"

> **Hilda:** I'd say to them . . . "You've got things not in sequence. What's controlling what? We often talk about an independent variable, [and] a dependent variable. What's the control here?"

Thus Hilda was primarily concerned that the students did not connect the mathematics with their experience in the lab: They had manipulated voltage by changing the number of batteries, but in their reports they described the voltage changing as a result of changes in the current or resistance. This was a different perception from John's that the students lacked a conceptual understanding of voltage. For example, with different equipment the students could have manipulated current as the independent variable. In that case, a student could appropriately have written, "When the current flowing increases, the circuit voltage increased," and Hilda's concern would not apply. John's still could, however, because that statement, an empirical summary of the experimental findings, does not indicate what the student understands about the concepts.

Hilda agreed with John that the students did not understand the concepts, but she attributed this to a more general problem that they did not expect ideas in physics to make sense. She was saying, in effect, that the students were all capable of keeping track of what quantity they were measuring. That their explanations did not reflect what they had seen suggested they did not expect the relationship they were studying, Ohm's Law, to have tangible meaning. John's interpretation, in contrast, was specific to this content: The students did not understand the concept of voltage. On this interpretation, these students may not have been able to keep track of what they were doing in the lab, because they needed a mental model for reference. To understand that the voltage in the circuit is determined by the number of batteries, for example, requires an understanding of voltage.

It was not our purpose in discussing this snippet, nor is it my purpose here, to decide which of these interpretations is correct. Either could apply for particular students in particular situations. As a matter of principle it is probably best left to the teacher, in this case Hilda, to make that judgment, because, in the end, she has the most information about her students. My purpose here is to suggest that Hilda and John interpreted the students' understanding and participation in different ways, and that we learned this primarily from what they said about instructional action—John describing what he would do and Hilda recounting what she did.

Later in the conversation, I asked why some students do not try to reconcile inconsistencies, as did Hilda's students who had worked hard to understand anomalous data, and whether this was something they could be taught to do. Ed suggested that one answer might be to give students more experiences of the sort we had seen in Lis's snippet, in which a group of students had conducted their own experiments in projectile motion, shooting a BB-gun across a field and tossing pumpkins from a cliff.

Ed's suggestion was one more example of a perception described in a terms of instructional action. He was, in effect, offering another interpretation of why students may not expect physics to make sense. Hilda had told of her students' suspicion that she might do "something to trick them," and how she had responded by saying she would not, an interpretation of the problem as arising out of a specific, articulable belief. Ed was considering the possibility that, for some students, the disconnection between physics and everyday experience lies more deeply, in a more general and less articulate sense of physics as taking place in a different domain of experience from their own, an interpretation similar to perspectives of knowledge and reasoning as "situated."[9] For such students, addressing the problem would not be as simple as telling them common sense applies; it would involve constructing with them a context for physics that directly engages their everyday experience.

In this way, Ed was using the idea of assigning real-world projects to help him express and refine his ideas for why students may not expect physics to make sense. In fact, much of our conversation about Lis's snippet earlier could be seen in this way: Lis had presented us with an example of an assignment designed to address aspects of students' understanding and participation not addressed by more conventional assignments, and our conversation about it drew our attention to those aspects. By referring to Lis's snippet, Ed brought those considerations to bear on the issue we were discussing at this moment, of students not treating physics as meaningful.

Teachers spend much of their time and thought in gathering and interpreting information, trying to gain insight into their students' understanding and participation. In this way, they have much in common with those engaged in formal research on learning. Still, there are important differences. Researchers intend their inquiry to produce explicit, articulate perspectives and claims, supported with arguments and evidence that can withstand peer-review. Teachers inquire toward action, in the contexts of their classes and presumably to the benefit of their students, with little time or opportunity for explicit reflection and awareness, let alone for public articulation. In short, researchers *publish*, whereas teachers *act*, and this difference is reflected in the ways in which they experience and express their respective insights into learning and instruction.

In the following section I present a third and final example from our conversations, illustrating a role for education research. I suggest that the interaction between the practice of teaching and the practice of education research should be understood principally on this common ground of inquiry into student understanding and participation.

INTERPRETING A TEXT ON PLANETARY MOTION: A ROLE FOR EDUCATION RESEARCH

One of the snippets we discussed on December 16, 1996 concerned the students' responses to two questions from a test on planetary motion and gravity, which Bruce had given to his twelfth-grade college-prep students. As part of his snippet, Bruce explained that the class had seen and discussed the PSSC film *Frames of References*. Much of their discussion was of what reasons there were for believing the earth goes around the sun, and what reasons there had been for earlier beliefs that the sun goes around the earth. Bruce noted that they had explicitly addressed whether the apparent motion of the sun across the sky was a reason for believing the earth moves: Airplanes and clouds also move across the sky, but that is obviously not reason to believe the earth is moving.

> **Bruce:** Nevertheless, to the true-false question: "The rising and setting of the sun proves that the earth spins on its axis," eighteen of twenty-five students answered "true." Since we explain this observation today by saying the earth is turning, I can understand such a response from those who forgot the film and our discussion.
>
> However, halfway down the page was this question: "State two reasons why earth-centered models of planetary motion were favored for so long over sun-centered models." Ten of the eighteen who had answered the previously-discussed question "true" nevertheless used the apparent motion of celestial objects as a reason for this too. Typical answers included:
>
> **Student 1:** "...when they saw the sun rising at the east + setting at the west, they concluded that the sun went around the earth."

Student 2: "People believed that the sun traveled around the earth because the sun rose and set every day."
Student 3: "It seemed that the sun rotated around the earth because of the change in day and night."

What surprised Bruce was the number of students who could, on the same test, answer both that people had once thought the sun's apparent motion was actual, and that the sun's apparent motion across the sky "proves" the earth rotates.

> And, although the top two scorers on this test answered these correctly, there was no pattern to who got these right or wrong. This seems to me a perfect (in fact, extreme) example of the "pieces" approach to learning physics—that ideas don't have to fit together or even make logical sense!

Our Conversation

In our conversation, Bruce explained that he sees this behavior often.

> **Bruce:** I see this kind of disconnect a lot. I'm sure we all do. But I was surprised there were so many of them, this time. Particularly when they had seen the film and we had discussed things over. And these two questions were on the same page, about half a sheet apart.

John suggested that some students might have read the question as the rising and the setting of the sun *reflects the fact* that [the earth spins on its axis], rather than *proves*. Bruce agreed that was a good possibility—in fact, one student had told him she answered "true" to the first question "because the rising and setting can be explained by the earth spinning"—but he felt this was consistent with his interpretation: Given the emphasis on this point in the class discussion, it was odd that a student would misinterpret the question in this way. That a student would treat "proves" as equivalent to "can be explained by" suggests the student was not paying attention to the logical connections among ideas.

In a similar vein, Lis noted that both parts of the statement in the first, true/false item are "true": The sun does rise and set, and the earth does spin on its axis. Seeing two true statements joined in a sentence, the students may have been distracted from the logic of their relationship, especially under the duress of a test. Hilda and John talked about these as general liabilities of true/false and multiple-choice questions, that they are open to such misreadings, that they invite test-taking strategies, such as trying to second-guess the test author's intentions, and that it is difficult to know why students answer as they do, even when their answers are correct.

Still, returning to the snippet, Hilda affirmed Bruce's interpretation, in part because of her own similar experience.

> **Hilda:** They don't see that they're answering this one, which contradicts that one. Because I very often have that [happen]. You know, they're doing the exact opposite for those two questions, and they're not seeing the connection when we go over it in class.

Lis's first reaction to the snippet, however, was surprise at a difference from what she had seen in her students. Early in the conversation she remarked that her students seemed to have a head start on this topic, having considered in previous classes the transition from an earth-centered to a sun-centered worldview "at a philosophical level."

> **Lis:** They do a lot in humanities that follows right along with [these ideas in] mechanics. . . . They all do. It's amazing. I mean, they would be using the words "geocentric" [and] "heliocentric." They would be quoting Aristotle. . . .

Lis emphasized that she was not referring to a technical familiarity, the ability to solve physics problems; she was referring to a familiarity with these larger systems of thought and the general shift in popular belief.

Later in the conversation, Lis observed that the students in the snippet had all approached the test question, which asked why people had favored earth-centered models of planetary motion, as a question about physical objects rather than about people and how they form beliefs. Their answers referred to the sun's apparent motion in the sky and the earth's rotation, rather than, for example, to the influence of the Church and popular religious convictions.

As in the previous two examples, the snippet and our conversation about it raised a range of interpretations of the data, in this case student responses to two questions on a test. Research on learning may have contributed to that range.

A Role for Education Research

By and large, the education community tends to think of the connection between research and teaching in terms of instructional methods and materials: Research on learning should have implications for what teachers should do in class, whether to form cooperative groups, adopt microcomputer-based materials, or assess through student portfolios. Research, in short, should establish and prescribe effective methods.

We set out in this project to develop a different understanding of the relationship between teaching and research on learning, at the level of interpretation rather than method. Instead of asking how researchers' findings should inform teachers' techniques, we have been asking how researchers' interpretations may inform teachers' interpretations. To that end, we read articles from the research literature, considered the perspectives they presented, and asked what insights they could provide into the snippets we were discussing. Instead of methods or general principles, we were looking for insights into particular moments of learning and instruction.

I have been especially interested in the possible contributions of my own research on student learning. During the first year of the project, I asked the group to read two of my articles on students' beliefs about knowledge and learning, or student "epistemologies." We read one of my articles in May, 1995 (Hammer, 1995) and another in November, 1995 (Hammer, 1994). Bruce was referring to that work when he called the exam results an "example of the 'pieces' approach to learning physics": "Pieces" was the term I had used to describe the belief that physics knowledge is a collection of independent, disconnected pieces of information, as opposed to a connected, coherent system of ideas.

The important point here is that Bruce's use of the perspective in discussing his snippet reflects an influence at the level of interpretation: He saw his students as not attending to the connections among ideas. In fact, Bruce described this sort of contribution to his thinking on several occasions. In one other case, for example, he was discussing a snippet he had written to recount how three of his better students, working on a problem about light reflection, had concluded that one's image in an ordinary mirror is upside-down which is contrary to everyday experience.

> **Bruce:** He apparently never made the connection even though we'd talked about it, that this is like when you look at yourself in a mirror on the wall. Or else how could he possibly put it upside down? In that sense it seemed to be an example of your [David's] kind of disconnection between reality and physics class.... Prior to reading your article, a couple of years ago I probably wouldn't have thought of it any other way except, well they just confused [ordinary mirrors and curved mirrors] and didn't think what they were doing.

In other words, the perspective gave Bruce a new diagnostic option for understanding his students, one he has applied and found useful in certain circumstances. This is a more modest but, I contend, more appropriate role for research on learning than what is generally assumed in the education community, that research should contribute at the level of instructional method.

If this is the role I as a researcher expect my work to play, then conversations such as these are essential, both in developing the ideas themselves and in understanding how they may or may not contribute to teacher perception and judgment. To be sure, our conversations led me to reconsider both the perspective and how I have presented it. I will not pursue that topic here except as follows, specifically regarding the language of action I discussed in the previous section.

In designing this project, I had assumed a clear distinction between diagnosis and action. That assumption helped shape my thinking about the role of education research. Consistent with the philosophy behind "Cognitively Guided Instruction" (Carpenter, Fennema, & Franke, 1996), I consider teachers to be in a much better position than I to derive methodological implications for their practices. For that reason I had been careful to avoid prescribing methods in writing about student epistemologies.

I was taken aback, therefore, by Bruce's describing what he found useful about my articles: It was not, he explained, the presentation of the theoretical framework but the ideas they contained for what to do in class, which he drew primarily from the classroom episode and discussion of instructional strategy (in Hammer, 1995). On the other hand, it was clear from his comments that he used the perspective as a diagnostic option for understanding his students. That Bruce considered the articles most useful with respect to the ideas they provided for instruction, I contend, is another example of the melding of interpretation and method in the language of action I have described. As we discovered in the failure of the ground rule I tried to impose on our conversations, for teachers diagnoses of student strengths and needs are tightly interwoven with strategies for addressing them.

I maintain that offering insights into student understanding and participation is a more appropriate role for education research than prescribing methods, but it is not inappropriate for education research to *suggest* methods. From our experience in this project, suggestions of method are an important means of communicating those insights.

This project was designed to study how perspectives from research may contribute to teacher perceptions, but there have been signs throughout our conversations of what teacher perceptions may have to offer education research. One example was Lis's observation that the students' had used technical rather than social language to answer the question on the test. This could be the kernel of a doctoral dissertation: What might affect students' choice of a mode of reasoning or discourse? Under what circumstances would they have approached the question as an issue, for example, of how people are swayed by popular opinion? More to the point, Lis's insights in this regard should be of interest to researchers investigating discourse in science teaching (e.g. Lemke, 1990; Roth & Lucas, 1997).

In sum, teacher inquiry overlaps substantially with research on learning. Both involve observing students and examining what they produce, so it is not surprising that they arrive at similar ideas. But there are important differences between the practices of teaching and research: Researchers publish whereas teachers act.[10] Having an insight into student understanding and participation, a researcher asks, in essence, "What can I say about this?" whereas a teacher asks "What can I do about this?"

The differences in practices are reflected in differences of language, as we have found in this project, and these present a challenge to substantive exchange between teacher inquiry and research on learning. At the same time, the differences represent complementary strengths: Researchers can and must focus on developing narrow, articulate views; teachers can and must be more broadly aware and responsive. We have explored the role perspectives from research may play in supporting teacher inquiry, but the benefit should certainly be mutual.[11]

TEACHER INQUIRY AND STUDENT INQUIRY

To return in closing this chapter to the central theme of this book, nowhere is effective exchange of insights among teachers and researchers more important than with respect to student inquiry. There are many calls in state frameworks and national standards for a greater emphasis on student inquiry in science education. As a general nicety, student inquiry seems a simple, desirable goal. In specific contexts of instruction, however, it is not a simple matter at all. No one understands clearly how to discern and assess it, or how to coordinate it with the more traditional but still important agenda of covering the content. This, of course, is not for lack of trying, but attempts by philosophers of science to define what is the scientific method (e.g. Popper, 1992/1968) or by educators to specify "process" skills as appropriate educational objectives (starting with Gagné, 1965) are widely considered unsuccessful. If it is possible to capture the essence of scientific reasoning—and some agree with Feyerabend (1988) that it is not—it has not been done.

The physical sciences have achieved stable, precise, and principled systems of knowledge. Working within these systems there is much that is, at least in practice, objectively true: There are clear, reliable, and reproducible methods, for example, for determining atomic masses or for manufacturing light bulbs. Education research has not achieved this quality of understanding; for good or ill, it is not possible to provide teachers clear, reliable, and reproducible methods for assessment and instruction. Interpreting student understanding and participation remains highly subjective, and this subjective judgment inevitably falls to the teacher, in specific moments of instruction like those recounted in the snippets here.

Moreover, this discrepancy between the quality of knowledge *within* science and the quality of knowledge *about* science and science education has particular significance for teachers trying to coordinate objectives of student inquiry and traditional content. It is, in general, relatively straightforward for a physics teacher to recognize when a student's answer to a question is correct or incorrect, judging it against the established body of knowledge. It is not difficult, for example, to see that the student in Robert's snippet was incorrect, from a Newtonian standpoint, when she said that "the force of gravity is getting bigger." It is not at all straightforward, however, to assess her *understanding*, to determine whether her comment reflects a misconception, which will prevent her from learning Newton's Laws if it is not eliminated, or a valid insight that will help her if she is encouraged to develop it. Nor is it straightforward to assess her reasoning as inquiry, for example to measure the value for her of having contributed an original idea or to weigh that value against the fact that it was incorrect.

Robert often expressed his concern that students learn to engage in scientific reasoning, rather than simply cover the content with a superficial understanding of the ideas. To pursue his objectives, however, to have conversations such as that he presented in his snippet, Robert must compromise the traditional content of the course. He must be able not only to reconcile this for himself, but also to justify it to administrators, parents, and students, all of whom will be aware that his class has not covered as much of the textbook as other classes. What should he tell them? How can he make what they have gained as tangible as what they have lost?

Similar tensions arise in other snippets. To pursue many activities of the sort Lis assigned, which led to the students' videotaped experiments in her snippet, would similarly diminish the traditional content. How should she consider and describe the relative value of those activities, as compared to other more familiar activities, as she plans the distribution of time over her year?

Hilda saw differences among her students, not only in the correctness of their reasoning, but also in the quality of their reasoning. A number of her students had followed the instructions of the lab and arrived at mathematically correct conclusions, but their thinking troubled her. It contrasted with the work of other students who had identified sources of discrepancies in their measurements. Precisely how should she interpret the differences between these students—was it interest, intellectual ability, confidence, all of these?—and what is largely an equivalent question in the practice of teaching, how might she design instruction to promote the more impressive reasoning? And, again, how should she weigh the value of that agenda against the value of covering more material?

Bruce saw, in his students' responses to two test items, an indication that they were approaching physics as a collection of incoherent facts. How should he value that perception against his perceptions of the correctness of the individual responses? Should students who are less consistent in their responses to questions on an exam, but get a greater percentage correct, receive a higher or lower score than students whose answers are more consistent but have a smaller percentage correct? This may be seen as a conflict between valuing "inquiry" (the internal coherence of a student's reasoning) and valuing traditional content (the correctness of a student's individual answers with respect to the intended body of knowledge).

If we are to achieve student inquiry-based science instruction, we must do much more to appreciate and address the intellectual demands that places on teachers. This will require developing conversations among and between teachers and researchers, much more than is currently occurring, and these conversations should begin from specific, authentic episodes of learning and instruction.

ENDNOTES

1. The project described in this chapter was funded by a joint grant from the John D. and Catherine T. MacArthur Foundation and the Spencer Foundation under the Professional Development Research and Documentation Program and by a grant from the DeWitt Wallace-Reader's Digest Fund to the Center for the Development of Teaching at the Education Development Center in Newton, MA. This chapter, however, is solely the responsibility of the author and does not necessarily reflect the views of any of these organizations. I am most grateful to Lis,

Hilda, Ed, Bruce, John, and Robert for participating in this project, for all their help and ideas in designing and redesigning it, for the windows they provided into their practices, as well as for their critical readings of several drafts of this paper. Thanks also to Denise Ciotti, Kass Hogan, June Mark, Jim Minstrell, Peggy Mueller, Barbara Scott Nelson, Mark Rigdon, Ann Rosebery, Annette Sassi, Deborah Schifter, and Emily van Zee for helpful comments, suggestions, and questions.

2. *Exploring the Place of Exemplary Science,* edited by Ann Haley-Oliphant, includes several chapters that discuss teaching as inquiry into student understanding and participation. See especially the chapters by Julia Riley, "Improvisational Teaching"; Robert Yinger and Martha Hendricks-Lee, "Improvising Learning Conversations"; Betty Wright, "How Do I Read my Students"; and Kenneth Tobin, "Learning From the Stories of Science Teachers." Other writings include Duschl and Gitomer (1997) on assessment conversations as tools for teacher inquiry into student understanding, and Hammer (1997) in which I present an account of my inquiry as the teacher in a high school physics class.

3. See Schön (1983) for an account of the nature of expertise in reflective practitioners.

4. All are public secondary schools in Massachusetts with the exception of Dana Hall which is a private school for girls. The teachers were recruited through letters and phone calls and were compensated as consultants. The project began in March 1995 under the auspices of the Teachers' Research Network of the Center for the Development of Teaching at the Education Development Center in Newton, MA. Initial funding came from a grant from the DeWitt Wallace-Reader's Digest Fund. When that ended in June 1996, the MacArthur/Spencer Professional Development Research and Documentation Program provided support for two more years.

5. All of the schools have recognizable distinctions among levels of physics classes. At the top level are the Advanced Placement classes, which are almost always the second year of physics instruction. Among the first-year courses, there are the honors courses, which may be calculus-based; algebra-based college prep courses, typically with two or three sections; and, at some schools, a conceptual level with minimal mathematics.

6. Ellipses indicate where I have omitted portions of the transcript. Square brackets indicate words I have substituted or added to the transcript for clarity.

7. Ohm's Law is a relationship among the electric potential or voltage (V), the current (I), and a resistance (R), usually written "V = IR." It states, in essence, that the voltage across a resistor and the current through the resistor are proportional: The higher the resistance, the greater the ratio of voltage to current.

8. Gentner and Gentner (1983) discussed students' different mental models of electric current and voltage.

9. Brown, Collins, and Duguid (1989) is probably the most well-known reference.

10. It is for this reason that I have mostly referred to teachers' interpretations as perceptions and researchers' as perspectives. This is not to imply that teachers do not have perspectives or that researchers are unperceptive; it is to connote different modes of inquiry, one more characteristic of teaching and one more characteristic of research. The practice of research requires that interpretations be made articulate in presentations, publications, and proposals, whereas the practice of teaching requires action, responding to students during class, choosing or designing materials and assignments. To act responsibly, teachers must perceive more than anyone could articulate; to be articulate, researchers must omit from their perspectives much of what they see.

11. For extended discussions of the value of teacher inquiry for education research, see Cochran-Smith and Lytle (1993) and Schifter (submitted).

REFERENCES

Brown, J. S., A. Collins, and P. Duguid. 1989. Situated cognition and the culture of learning. *Educational Researcher* 18(1): 32-42.

Carpenter, T. P., E. Fennema, and M.L. Franke. 1996. Cognitively guided instruction: A knowledge base for reform in primary mathematics education. *Elementary School Journal* 97(1): 3-20.

Cochran-Smith, M., and S.L. Lytle. 1993. *Inside/outside: Teacher research and knowledge.* New York: Teachers College Press.

Duschl, R. A., and D. H. Gitomer. 1997. Strategies and challenges to changing the focus of assessment and instruction in science classrooms. *Educational Assessment* 4(1): 37-73.

Feyerabend, P. K. 1988. *Against method.* New York: Verso.

Gagné, R. M. 1965. *The psychological bases of science—A process approach.* Washington, DC: American Association for the Advancement of Science.

Gentner, D., and D. R. Gentner. 1983. Flowing waters or teeming crowds: Mental models of electricity. In *Mental models*, edited by D. Gentner and A. Stevens, 75-98. Hillsdale, NJ: Erlbaum.

Haley-Oliphant, A. E. (Ed.) 1994. *This year in school science 1993: Exploring the place of exemplary science teaching.* Washington, DC: American Association for the Advancement of Science.

Hammer, D. 1994. Students' beliefs about conceptual knowledge in introductory physics. *International Journal of Science Education* 16(4): 385-403.

Hammer, D. 1995. Epistemological considerations in teaching introductory physics. *Science Education* 79(4): 393-413.

Hammer, D. 1997. Discovery learning and discovery teaching. *Cognition and Instruction* 15(4): 485-529.

Lemke, J. L. 1990. *Talking science: Language, learning and values.* Norwood NJ: Ablex.

Popper, K. R. 1992/1968. *Conjectures & refutations: The growth of scientific knowledge.* New York: Routledge.

Roth, W.M., and K.B. Lucas. 1997. From "truth" to "invented reality": A discourse analysis of high school physics students' talk about scientific knowledge. *Journal of Research in Science Teaching* 34(2):145-179.

Schifter, D. submitted. Learning geometry in the elementary grades: Some insights drawn from teacher writing.

Schön, D.A. 1983. *The reflective practitioner: How professionals think in action.* New York: Basic Books.

Inquiry in the Informal Learning Environment

Doris Ash and Christine Klein[1]

In informal settings inquiry is a time-honored tradition as a way of learning. In this chapter we begin to explore ways to translate to more formal settings important underlying principles of inquiry in informal settings. Our important objective is to find what they have in common. In short, we look for areas of overlap and potential synergy. We hope to provide readers from each of the two settings with a new perspective on inquiry learning as it takes place in the other.

When asked how they first acquired a love for science, many scientists respond with poignant stories of playing with electronics or engines at museums, taking nature walks with a relative, watching spiders in the window, or making potions (or bombs) in the basement. All these contexts come under the general category of informal science learning environments. It can be argued that real learning occurs most easily in the informal learning environment and that human learning, over the centuries, has been characterized by an intellectual tradition of informal inquiry through observing, posing burning questions, hypothesizing, making predictions, doing research or experimenting to find the answers, interpreting and communicating.

Inquiry in informal learning settings, the basement, the lab, the museum, the zoo, has been defined as free-choice (Falk & Dierking, 1998), learner-driven, open-ended, unhurried, and personal. It is driven by attitudes of curiosity, what

Mikhail Csikszentmihalyi in 1990 identified as intrinsic motivation, and a willingness to be uncertain and to change directions as new evidence dictates. And inquiry is both a methodology and a vehicle for learning content. The processes, the content learned, and the effects are all critical components to inquiry learning. From the time of Socrates through the twentieth century work of John Dewey and Jerome Bruner, inquiry has been a habit of mind limited only by a person's capacity to learn and furthered by selection among the many ways of getting to an answer.

Here, in order to provide a template for translation to the classroom, we begin to examine deep underlying principles of inquiry learning in informal environments. We compare the general characteristics of learning in formal and informal environments, present vignettes from two existing programs that make use of inquiry learning, define common principles of the two settings, provide a template for implementing these principles in other contexts, and offer resources for educators to put them into practice.

FORMAL AND INFORMAL LEARNING ENVIRONMENTS

What do we mean by informal learning environments and how are they different from formal learning situations? The distinction often centers around the <u>environment</u> or context rather than the type of learning involved. Here we define <u>formal</u> environments as based in the classroom and <u>informal</u> environments as museums, science-technology centers, zoos, aquaria, botanical gardens, aboreta, nature centers, and similar settings.[2]

Centering on the <u>inquiry learning</u> involved rather than the setting, we start by looking closely at two successful programs, each concentrating on inquiry learning, each designed by informal learning centers. These are the Institute for Inquiry at The Exploratorium in San Francisco and the Compton-Drew Investigative Learning Center Middle School at the St. Louis Science Center.[3] Between these two programs are some principles common to any inquiry learning setting, and as design features these can be built into many different learning environments.

Though we intend to move away from any stark contrast between informal and formal learning, the distinctions are helpful in planning the future. Table 1 compares the two for the characteristics of inquiry typical of each.[4]

TABLE 1. CHARACTERISTICS OF INFORMAL LEARNING EXPERIENCES

MORE LIKE	LESS LIKE
AFFECT/CHOICE	
fun, enjoyable, playful	repetitive
voluntary	mandatory
MEDIUM	
visually oriented	oriented to text
real objects	secondary sources
authentic task	tasks for others
SOCIAL CONTEXT	
individuals learning together	instruction to a whole group
directed by learning	directed by others
multi-generational experiences	one age group
intrinsic motivation	extrinsic motivation
INTERACTION	
highly interactive, learn by doing	didactic
multi-dimensional interactions	single dimensional
oriented to the process	oriented to topics
COGNITIVELY CHALLENGING	
multiple entry points—various ages	one or two grade levels
TIME	
short term	long term
self-paced	other sets pace
open-ended	limited
single visits	longitudinal

ASSESSMENT

self assessment based on feedback	formal assessment
serendipitous	multiple forms

STRUCTURE

non-structured	highly structured
non-linear	linear, sequential

One striking variable in Table 1 is the locus of control of the activity. In informal settings, the learner is usually in control of the activity and the learning. Another variable is the opportunity to understand what the learner takes away from the interaction. Informal settings are less likely to have strong assessment opportunities and long term observation, strengths of more formal settings.

Synergy Between Informal and Formal Learning Environments

Reviewing the characteristics of learning in the informal setting reveals a powerful and desirable constellation of traits that would be ideal in any classroom. Clearly there is a strong opportunity for reciprocity. Informal learning settings are self-directed, fun, playful, cooperative and highly interactive, traits that appeal to all settings. And indeed, some of the most successful classroom design experiments (Ash, 1995; Brown, 1992; Brown & Campione, 1996) include these elements and are modeled after good informal settings. Similarly some of the more powerful aspects of formal settings are becoming more attractive to informal educators: the prospect for long-term opportunity to understand the kinds of learning, types of reasoning, and change in knowledge that occurs at exhibits is attractive to designers in museums. These two areas, formal and informal, can inform each other, share expertise, and maximize synergy.

Museum schools are one example of a current trend in museum education toward a synergy of formal and informal learning.[5] As classroom teachers and curriculum developers adopt more methods based in inquiry, the classroom takes on more characteristics of informal learning. As informal learning environments like science centers develop partnerships with schools, many programs acquire some of the strengths of formal learning. Here is a hypothetical example.

As students work on their projects, energy is brought to the classroom. The students are taking advantage of an opportunity to design an exhibition on space travel for the science museum. They began the project by generating their own questions about space travel and exhibition design, questions they would need to complete their projects. Now they are exploring those questions in small collaborative groups of their own choosing.

For an exhibit on the race to space, one group of students works at computers gathering additional information from the NASA web site on the history of the space program. One group works on an experiment that demonstrates the effects of zero gravity on various objects and considers how this experiment could be used as a hands-on exhibit based in inquiry. After questioning the one-sided results its members had found, another revisits its research notes on the environmental and economic costs and benefits of space tourism. As part of its background research on the role of science fiction in guiding space programs, still another reads a science fiction short story. A group works on calculations needed to bring a model of the solar system to scale on an exhibit. A sixth group of students is at the science center examining the existing exhibits and conducting evaluations of visitors.

As the extended class period draws to a close, the students gather together in jigsaw groups[6] to share their expertise, teach one another, reflect on their progress, and compare findings. Plans for seeking additional resources and conducting additional experiments are made. All the students are eager to continue work when they return.

In this hypothetical classroom, students are intrinsically motivated and having fun. The exhibition at the science center provides the authentic task that drives their study of science, history, economics, literature, and mathematics. The broad idea for the inquiry is set by the science center and teachers, but students plan the details. Participation is mandatory, but students are eager to do the work. Students work in collaborative, self-directed groups. The teacher provides some structure and helps with project management, setting some of the pace to insure completion of the task on time. The final task acts as a formative assessment for teachers and students. Students receive feedback from their work and their peers as they move through their tasks, from the teacher as individual and group grades are given, and from science center staff as their designs go through the usual review process.

VIGNETTES OF INQUIRY AND RESEARCH FROM TWO PROGRAMS

That hypothetical example provides a picture of the synergy that can occur between formal and informal learning. A look in-depth at the two real examples that follow gives an even clearer image. In The Exploratorium's Institute for Inquiry, teachers, professional development specialists, and administrators develop strategies for bringing inquiry into the classroom and into national professional development efforts. In the museum school at the St. Louis Science Center, the strengths of formal and informal learning are central to the Schools for Thought (SFT) curriculum framework. As you read each vignette, please consider the principles common to both.

Two Inquiring Minds (The Exploratorium)

The Institute for Inquiry at The Exploratorium centers on inquiry learning for elementary school teachers, professional development specialists, and administrators, as a personal core inquiry experience and as basis for translation to the classroom and professional development. In this vignette, we describe the inquiry experience of two elementary school teachers in the Introductory Institute.

The core inquiry experience is designed to include a carefully crafted set of experiences that lead learners towards independently designed, long-term inquiry activities founded on their own questions.

At the beginning, everyone participates in a common set of experiences that give multiple entry points and diverse pathways towards deeper exploration. These experiences include light and color exhibits, paper chromatography, activities designed by artists, mixing color solutions, and many more. Working from their own questions, groups of participants define investigations and pursue them in self-selected small groups. These experiences are accompanied by group and individual "guided model building" initiated by the staff along with mini-lectures, demonstrations, and at appropriate times reference to related exhibitry. A few pivotal, guiding questions are repeatedly posed, for example, "What happens to the path of light as it travels toward your eye?" At the end of the investigations, groups teach one another what they have learned and together make sense of color, light, and pigment.

Light and Color: The Question.[7] Terry and Connie's original experimental question was: "Can we recreate colors of marker pens by separating their pigments and then mix the primary pigments together to get the original colors?" This type of question had been asked in past Institutes and, in and of itself, could not guarantee experimentation beyond the early exploration stage of inquiry. But throughout Terry and Connie's work, there were several clues that something important, beyond simple exploration, was occurring.

Taking Apart: Separating Pigments. Initially, Terry and Connie's focus, by use of a variety of color pens, was on using a diversity of solutions for chromatography color separations; they were careful with their technique; they inspected their results rigorously.

After several hours, Terry and Connie changed directions. "We did three solutions (with many different colors) then stopped." They felt that they were done with this part of their work and that doing chromatography with other solutions would give them no more useful, new, or discrepant information. They redirected their investigation, an act that told us that they were gathering information for a purpose, not just experimenting for the sake of data collection.

Putting Together: Mixing Color Pigments. The next part of their investigation complemented their earlier efforts. They wanted to mix color pigments to match the colors from the markers using the "primary" colors that they had seen separated out in their chromatography experiments. Their efforts to reproduce the original marking pen colors were exhaustive. For them to trust the symmetry of the problem, the process had to work in both directions: to be exactly reversible. They had to be able to take the color apart and then recreate it as well.

A difficult but illuminating instance was the reproduction of color for both light green and dark green pens. Terry and Connie had already determined that mixing cyan and yellow would produce green but no matter what the proportion of each, they could not match the dark green color. When they looked more closely at the results of their chromatography separation, in one sample they found something different. "We found out that to get dark green we needed to add a bit of magenta. It was so exciting...I guess you had to be there." "There" in this case included both the actual color separation and mixing and the excitement of charting new ground in understanding color pigment. By adding magenta to green, the two indicated they knew how to darken green pigment. They had tested complementary by making and breaking apart color pigment but had not understood why magenta darkened green. At this point they had answered their original question and might have felt that they could stop their investigation. Some nagging unanswered questions propelled them forward. Why did they need to add magenta to make dark green?

Making Sense of the Experience: "What Light Gets to Your Eye?" Throughout the color investigation we had advised people to think about a question that would help in making sense of the phenomena being investigated:

"What light gets to your eye?" At this juncture we suggested to Terry and Connie, whose experiments up to then had involved only color pigments, that they begin to explore color light and its interaction with pigments.

In looking back at Terry and Connie's work it is evident that they had begun to use these ideas to anchor their thinking. They spent the bulk of their time working with pigment, while investigations with light by other groups surrounded them. Like the rest of the participants Terry and Connie took part in these investigations, by talking and observing as part of community sharing in formal and informal ways. Investigators from different groups sat side by side talking and sharing information so that Terry and Connie learned about phenomena they hadn't investigated personally. It was the visit to The Exploratorium exhibits that proved to be the key to their understanding of the role of magenta in darkening green, and ultimately, to their creation of a model of light and pigment color in relation to absorption and reflection.

This model was articulated in their final presentation to the group. At that point Terry explained, "After we made the pen colors, we went to Color Removal." Color Removal is an exhibit that uses a projector and prism to produce a bright spectrum. To show what filters do to the color spectrum, you place a color filter into the path of the light. It was in this interaction with light subtraction that the couple made the connection to the mystery of dark green. They saw that a magenta colored filter darkened only the green part of the light spectrum. Terry and Connie were able to use the new information from this exhibit, integrate it with their existing pigment work, and transfer their understanding to light and its relationship to color pigment.

In their final project presentation, according to Terry:

"We came up with the idea that:

<u>for pigment</u>	if you have all pigments / absorb all light / gives black
<u>for light</u>	if you have no light / no light can be reflected / gives black
<u>for pigment</u>	if you have no pigment / reflects all light / gives white
<u>for light</u>	if you have all light / all light can be reflected / gives white

It's all about reflection and absorption."

The analysis was correct. Adding all light colors together gives white light. Adding all pigments together gives black. This is best understood when the question is asked what light gets to your eye and what light has been absorbed or reflected by the pigment? Clearly Terry and Connie were using these ideas to tie together their investigations. The underlying complementary of light and pigment in relation to the process of absorption and reflection in their model seemed a deeply compelling notion for them. They had moved from separating and mixing color pigment to an encompassing model of reflection and absorption in the interplay of light and pigment.

Lessons Learned. A central element to the power of inquiry learning is ownership of the overall process. We had already understood the importance of the question owned by the learner, as well as a personal pathway to inquiry, but along the way Terry and Connie learned some deep content and grew more confident in their abilities to do inquiry. They practiced using the processes of inquiry (observing, questioning, hypothesizing, predicting, investigating, interpreting, and communicating) to learn content. In this case we would not have guessed that extensive work with separation and mixing of pigments would be the foundation for understanding the interaction between pigment and color light. Terry and Connie created a unique path to creating meaning.

St. Louis Meets the Sea (Compton-Drew)

Compton-Drew Investigative Learning Center (ILC) Middle School at the St. Louis Science Center serves approximately 540 students from the St. Louis area in sixth, seventh, and eighth grades. In partnership with the St. Louis Science Center (SLSC), this magnet school is based on the concept of the museum school, in which museums and schools pool their strengths to form a unique educational experience (Klein, 1998).

The curriculum at Compton-Drew is developed by the teachers with support from SLSC staff and faculty from the University of Missouri - St. Louis. Curriculum is based on Schools for Thought, a curriculum framework and philosophy that utilizes the results of cognitive science research, that is, how we learn, think, and remember. This vignette tells the story of one Schools for Thought interdisciplinary unit conducted during the summer of 1997.

As students enter the classroom, they know by the circle of chairs in the room that they will be having a class discussion. This will be their opportunity to talk about what they know and think. Knowing the new unit deals with the sea, they anticipate the topic of the discussion. As they suspected, the teacher introduces the unit's theme, asks them what they already know about oceans and seas, and turns the discussion over to the students. This gives the teacher an opportunity to listen, to assess what the students know, and to identify incomplete understanding that will need to be addressed before or during the unit.

The following day the St. Louis Science Center's ILC Program Liaison introduces students to the unit's challenge.

> The Science Center is showing *The Living Sea* at the OMNIMAX®. We like to make connections between our movies and exhibits and the lives of our visitors. This movie has been a real challenge. Since St. Louis is not on a coast, why should St. Louisians care about the sea? Or should they?
>
> We need your help. We would like you to develop videos or books or give live performances in our galleries to share your ideas with our visitors on whether or not they should care about oceans, and why.

Students now know what they will be examining and the task they will be completing as a result of their research and investigation.

Students walk next door to the St. Louis Science Center to see *The Living Sea* to note any questions raised by the movie and to gather ideas that might help them meet the challenge.

Back at school, the whole team of four classes generates a list of questions it will need to answer to be able to address the dilemma: Should St. Louisians care about the ocean, why or why not? The team puts the questions into categories, and individual students choose a category to work on. Research groups of four to six students are formed by student preferences, with teachers making the final decision from their knowledge of the students.

One research group has come together to study endangered species of animals in the ocean. Its research takes the students to the library to review materials, to the World Wide Web through computers in the classroom to conduct searches for additional materials, to the St. Louis Zoo and the Mid-America Aquatic Center to gather more information on endangered aquatic species, and to their teachers for articles and direct instruction to assist them in their quest for answers and to learn more about how different media—short stories, poetry, movies—convey messages. Students participate in the Big Map program through the SLSC to learn more about the world's ocean and water system. Each resource provides some answers and sparks new questions. The teachers have established benchmarks and timelines for their research, but students want to keep going with their research when it is time to stop.

Once research groups have reached enough of a level of expertise, about halfway through the unit, students share what they have learned by jigsawing. One person from each research group joins representatives from each of the other research groups to form a new jigsaw group. With an expert from each research topic, groups are now ready to work on the consequential task, going back to conduct additional research if needed.

One jigsaw group has decided as its consequential task to make a video for the SLSC. The members visit the traveling exhibition at the SLSC, *Special Effects: A Hands-On Exhibition* to learn more about how messages are conveyed through special effects. They work with the SLSC Producer to learn how to produce videos that demonstrate their understanding of the ocean. To be sure they use the results of their research and convey a clear message through their video, they work with teachers on their message. During the last two weeks of their unit, they produce their video, a newscast with scenes filmed at the SLSC and Compton-Drew.

COMMON PRINCIPLES OF INQUIRY AND RESEARCH

These two previous vignettes illustrate different settings, different audiences, different subjects, different disciplines, and different ages. We also offered two different ways of looking at inquiry: The Exploratorium exercise is based on interactive experiences with phenomena, that of the St. Louis Science Center on research with secondary sources.

We believe that a common thread runs throughout both programs, especially in the fundamental assumptions that each program brings to the design of the inquiry learning experience. While surface features may look different, we suggest that deep underlying structures are very similar and that together the two programs allow us to extrapolate principles that will be appropriate for many different environments, disciplines, and learning settings.

What Is Meant by Inquiry at The Exploratorium?

Fundamental to The Exploratorium's Institute for Inquiry is the notion of inquiry as a way to make sense of the natural world. Its method is to ask questions and answer them with the process skills of science.

Inquiry starts within a defined content rich with real-world phenomena. The learner begins by being curious about those phenomena. This curiosity gives rise to many questions, some of which provide entry points to investigation as well as potential pathways for answering them. Learners group into teams to pursue their questions with investigations that take time. There is subtle facilitation and scaffolding along the way at appropriate moments so that the locus of control is transferred from facilitator to learner through modeling. There are enough materials to support an ongoing investigation and many opportunities to share results by discourse with others. All along the way investigators collaborate, talk, learn how to represent newly learned concepts, and discover how to use the process skills of science: observing, questioning, hypothesizing, predicting, investigating, interpreting, and communicating.

The definition of inquiry on The Exploratorium's World Wide Web site (Table 2) gives a synopsis of this process. This mode of inquiry is active, interactive, self-regulated, and collaborative. It is assumed that different groups will share the fruits of their labor with one another at appropriate moments and that the group builds up an interrelated body of expertise that moves in the direction of deep content principles.

TABLE 2. WHAT DO WE MEAN BY INQUIRY?

Inquiry is an approach to learning that involves a process of exploring the natural or material world, that leads to asking questions and making discoveries in the search for new understandings. Inquiry, as it relates to science education, should mirror as closely as possible the enterprise of doing real science.

The inquiry process is driven by one's own curiosity, wonder, interest or passion to understand an observation or solve a problem.

The process begins by the learner noticing something that intrigues, surprises, or stimulates a question. What is observed often does not make sense in relationship to the learner's previous experience or current understanding.

Action is then taken through continued observing, raising questions, making predictions, testing hypotheses, and creating theories and conceptual models. The learners must find their own idiosyncratic pathway through this process; it is hardly ever a linear progression, but rather more of a back and forth or cyclical series of events.

As the process unfolds more observations and questions emerge, giving occasion for deeper interaction and relationship with the phenomena—and greater potential for further development of understanding.

Along the way, the inquirer is collecting and recording data, making representations of results and explanations, drawing upon other resources such as books, videos, and colleagues.

Making meaning from the experience requires intermittent reflection, conversations and comparison of findings with others, interpretation of data and observations, and applying new conceptions to other contexts as one attempts to construct new mental frameworks of the world.

Teaching science using the inquiry process requires a fundamental reexamination of the relationship between the teacher and the learner whereby the teacher becomes a facilitator or guide for the learner's own process of discovery and creating understanding of the world.

Source: http://www.exploratorium.edu/IFI/about/inquiry.html

What Is Meant by Research in Schools for Thought and Fostering Communities of Learners?

Schools for Thought (SFT) integrates three foundational projects: *The Adventures of Jasper Woodbury* developed at Vanderbilt University, *CSILE (Computer Supported Intentional Learning Environments)* developed at the Ontario Institute for Studies in Education, and *Fostering Communities of Learners* (FCL) developed at the University of California - Berkeley. In addition to researchers at these three institutions, faculty at the University of Missouri - St. Louis, staff at the St. Louis Science Center, and SFT teachers across North America are part of the SFT Collaborative, funded by the James S. McDonnell Foundation.[8]

The SFT research cycle, as used at Compton-Drew ILC Middle School, is based largely on the FCL Principles given in Table 3. As was seen in the vignette, this research process involves generating questions around a rich content, categorizing those questions, looking for answers in self-selected small groups, sharing results of research with classmates, and sharing understanding with the larger community of learners through use of jigsawing and a consequential task, such as the video for the Living Sea unit. The cycle is not as linear as it might first appear. Students find that answers lead to more questions and sharing knowledge with others leads to more research. The emphasis is on dialogue all along the way. The subject is rich enough to support a variety of research groups, yet all research centers around the large topic selected at the beginning of the project and leads toward a consequential task. This task is authentic and allows students of differing expertise to bring together their knowledge and to demonstrate their understanding of the principles of science, or other content, underlying the central idea.

TABLE 3. FOSTERING COMMUNITIES OF LEARNERS PRINCIPLES

ACTIVE, STRATEGIC LEARNING

Systems and cycles
- ▶ repetitive participant structures as part of a research-share-perform activity system

Metacognition
- ▶ awareness, intentional learning
- ▶ reflective practice

Dialogic base
- ▶ shared discourse, negotiation of meaning
- ▶ seeding, migration, and appropriation of ideas

Distributed expertise
- ▶ individual and group expertise
- ▶ diversity, legitimization of differences

Multiple zones of proximal development
- ▶ mutual appropriation
- ▶ guided practice, guided participation

Community of practice
- ▶ shared community values
- ▶ respect, responsibility

Contextualized and situated
- ▶ purpose for activity
- ▶ intellectually honest curriculum
- ▶ transparent and authentic assessment

Source: Ash, 1995 adapted from Brown & Campione, 1994

Inquiry and Research: Equivalent Forms in Different Settings

Inquiry as described by the Institute for Inquiry and research as described in the Compton-Drew example share similar underlying principles. We support this by highlighting commonalties between both programs and by identifying theoretical principles that others might use in their design. These principles allow us to move towards providing a practical template for learning environment design.

Part of the strength of the argument for commonalty between inquiry and research is that even though the programs differ in context and learning setting, they map onto each other in significant ways. FCL/SFT is heavily steeped in L. S. Vygotsky's (1978) concept of zones of proximal development (ZPD)—the area between a child's current abilities and the distance she can traverse with the aid of a more capable peer or teacher—using a variety of participant structures (Brown et al., 1993) designed to create multiple opportunities for classroom learning. At the Institute for Inquiry, inquiry is strongly rooted in curiosity and personal interaction with generative phenomena. And although inquiry can have a variety of interpretations, this cyclic process has a structure and essential elements that characterize it. The learner is engaged with compelling content while using process skills to investigate self-selected questions. Scaffolding of skills and strategies is an essential for translation to classroom settings. Foundational elements—how to use the process skills of science, how to think metacognitively, how to share results with others—need to be modeled. Once laid down as foundation, these skills help build ownership and competence for inquirers and a sense of accomplishment in moving towards independent investigations (Ash, 1999).

We have isolated ten principles common to both programs. They provide a framework beginning to build inquiry programs in many settings.

1. **Delimited content.** Inquirers and researchers move toward a richly generative content that has well-defined principles. Sample conceptual principles might include the independence of ecosystems, as in the Living Sea unit at Compton-Drew, or understanding the interrelationship between color light and color pigment, considered in the Institute for Inquiry.

2. **Research and inquiry are controlled by the learner.** The inquiry question, the nature of the research, the object of inquiry have been determined, at least partially, by the learners, in a group of from two to five. Ideally, the locus of control is gradually given to the learners as they gain competence.

3. **Group meaning making.** As they build an understanding of big ideas, learners engage in group making of meaning. This building of knowledge contributes to the community's knowledge base, which continues to grow as learners share results of their inquiries.

4. **The process of inquiry and research is repetitive, cyclic, and open-ended.** An iterative research or inquiry cycle allows the learner to use the processes of science over and over again, in non-linear fashion. The cycle is somewhat open-ended, allowing many possible direc-

tions and many possible different answers. Research and inquiry often end with more questions. (See Brown & Campione, 1996; White & Frederiksen, in this volume.)

5. **The processes and skills are the same.** The processes of observing, raising questions, making predictions, posing hypotheses, investigating and gathering information, interpreting, and communicating are the same (Harlen & Jelly, 1997).

6. **The cycles are similar.** The research or inquiry cycle begins with curiosity and questioning in the exploration phase. As it continues, there is a deep involvement in a particular set of questions generated by the learners. There can be confusion, redirection, and re-posing of hypotheses and questions. The cycle culminates with teaching and sharing results and interpretations.

7. **Collaborative learning.** Children and adults work together to create a larger understanding. In such a community of learners, everyone knows different parts of the overall topic; creating distributed expertise (Brown et al., 1993; Bereiter & Scardamalia, 1993). Everyone's work is an essential component of the whole group's making of meaning.

8. **Active, strategic learning.** Learners are expected to interact on an ongoing basis with one another, with resources and with phenomena. Learners move beyond hands-on interactions to engagement with one another and the objects of inquiry.

9. **Metacognitive.** At regular checkpoints, learners are asked to reflect upon what they know and what they still need to find out in light of theirs and others' work (Palinscar & Brown, 1984; Bruer, 1993). Learners become aware of their own understanding and know why certain strategies facilitate their learning.

10. **Dialogic.** In order to make sense, learners communicate, have information to share and need the information from others. Thus, learning and teaching become reciprocal activities for everyone involved (Brown et al., 1993; Lemke, 1990).

PUTTING IT ALL TOGETHER

Practical Issues

We next need to consider how these ten principles can help us design an inquiry classroom. We know that inquiry can take on many different looks that depend on the classroom context, the age of students, the disciplinary content, and of course the level of experience of the teacher and the students. A teacher new to inquiry may approach design differently from another more experienced. At the beginning of classroom inquiry, for example, the teacher typically helps her students learn the skills basic to doing independent inquiry. She may take time to develop questioning skills or different ways to interpret evidence. We argue, however, that most inquiry classrooms will eventually embody some aspects of the basic principles we have outlined. We suggest that these principles can be met in any number of practical ways, from simple to more complex and that there are incremental steps toward complex inquiry or research.

Amidst the difficulties of the average classroom, how might a teacher apply the ten principles? This chapter lacks the space to do more than begin to address these issues. Table 4 provides a working model of an inquiry path framed by the practical and social constraints critical to the average classroom. We know full well that this is only the beginning of a model that might be used in a variety of instantiations (Ash, 1999).

Social Context. Of the many aspects of the social context that the ten principles address, two elements deserve special emphasis here: the ethos of questioning, of not knowing the answers; and the notion of scaffolding within the zone of proximal development.

An Ethos of Questioning. The social dynamic is driven by the learner's curiosity: questioning is critical to forming a community of learners. Together learners form a social unit that investigates a series of related questions. Taken together, the results of the inquiry will create a whole that is bigger than any individual piece. Within the community, facilitators are curious learners with questions of their own. Curiosity followed by questioning is the underpinning of a community of inquiry. There needs to be a genuine honoring of "not knowing" and a trust that there are steps toward finding out.

Scaffolding. Scaffolding implies gradually transferring ownership of the inquiry process to the learner. Though facilitators are not willing to give the answers without investigation on the part of the learner, they are ready to model the inquiry path themselves. Facilitation is subtle, but ever-present. Knowing when to intervene and scaffold is a critical factor for those in charge. Asking open-ended questions to guide the learners becomes an important technique in shifting the locus of control toward the learner. In the beginning the facilitator gives more guidance, more modeling. Gradually, as the learner gains competence, the facilitator pulls back increasingly.

TABLE 4. WORKING MODEL OF INQUIRY PATH IN AN AVERAGE CLASSROOM

SOCIAL CONTEXT	CURIOSITY AROUSED	PRACTICAL CONTEXT
Driven by learner	Learners pose questions and choose pathway	Delimited content area
Collaborative learning	Observation	Formative and summative assessment
Ethos: Questioning as a habit of mind	Hypotheses	Assorted, rich materials to invite inquiry
Active, strategic learning	Gathering data	Physical environment
Dialogic	Predicting	Time
Metacognition: Knowing what you know	Interpreting	Multiple entry points
Group making of meaning	Testing ideas	Facilitation
	Communication and reporting to others	Moving toward big ideas

Practical Context. The practical context creates an environment that supports the inquiry path. In designing this environment, the content of the inquiry, assessment, materials, physical space, location, and time must be considered.

Assessment. In research and inquiry, assessment might take the form of examining the students' understanding in an ongoing and formative fashion. Facilitators can assess the students' prior knowledge early on. Student self assessment can occur throughout as groups receive feedback from their peers and teachers. This type of assessment becomes continuous and integral to the process of moving toward deeper understanding of the big ideas of science. Assessing formatively allows the facilitator to change her actions in response to student needs (Vermont Elementary Science Project, 1995; Brown & Shavelson, 1996; Harlen, 1999).

Materials. Materials that are easily accessible and visible to learners invite their curiosity. They should be well planned in advance. There will be enough to get started but also more available to invite further questioning. Materials need not be expensive and many of them can come from the students. Resources and the results of inquiry collected from one class can serve as a foundation for resources for the next semester or year. Materials that are easily accessible and visible to learners invite their curiosity.

Physical Environment. Inquiry can be messy and require a large space. The ideal classroom will be large, easy to clean, and sturdy. Chairs and tables will be arranged in variety of configurations by learners to support their work.

Time. Time to spend in an inquiry process over an extended period is critical. One forty-five minute period is not enough. Block scheduling and stretching the inquiry over weeks provide more time for meaningful inquiry.

GETTING STARTED/RESOURCES[9]

To assist the reader's own inquiry into building the synergy between formal and informal learning and environments, we offer a variety of resources as starting points. (See Table 5.) We hope these will lead to additional resources, additional questions, and new understandings.

TABLE 5. RESOURCES FOR BUILDING SYNERGY BETWEEN FORMAL AND INFORMAL LEARNING AND ENVIRONMENTS

COMMUNITIES OF LEARNERS

Brown & Campione, 1994. Guided discovery in a community of learners.

Brown & Campione, 1996. Psychological theory and the design of learning environments: On procedures, principles and systems.

Matusov & Rogoff, 1995. Evidence of development from people's participation in communities of learners.

DISTRIBUTED EXPERTISE

Brown, Ash, et al., 1993. Distributed expertise in the classroom.

Scardamalia & Bereiter, 1996. Engaging students in a knowledge society.

INTRINSIC MOTIVATION

Csikszentmihalyi & Hermanson, 1995. Intrinsic motivation in museums: Why does one want to learn?

INQUIRY AND LEARNING WEBSITES

Exploratorium Institute for Inquiry—http://www.exploratorium.edu/IFI/

Museum Learning Collaborative annotated bibliographies – http://mlc.lrdc.pitt.edu/mlc/

Center for Museum Studies' database with search capabilities – http://www.si.edu.organiza/offices/musstud/data.htm

MUSEUM EDUCATION

Hein & Alexander, 1998. *Museums: Places of learning.*

Mann, 1997. Extending the curriculum through museums.

MUSEUM SCHOOLS

Klein, 1998. Putting theory into practice: Compton-Drew investigative learning center.

Science Museum of Minnesota, 1995. Case studies of five museum schools.

SCIENCE CENTERS, MUSEUMS, AND OTHER INFORMAL LEARNING ENVIRONMENTS WEBSITES

American Museum of Natural History – http://www.amnh.org

Boston Museum of Science – http://www.mos.org

The Exploratorium – http://www.exploratorium.edu

The Field Museum – http://www.fmnh.org

Franklin Institute – http://www.fi.edu

Informal Learning Environments Research SIG –
http://darwin.sesp.nwu.edu/informal/

Lawrence Hall of Science – http://www.lhs.berkeley.edu

Missouri Botanical Gardens – http://www.mobot.org

New York Hall of Science – http://www.nyhallsci.org

Ontario Science Center – http://www.osc.on.ca

Science Museum of Minnesota – http://www.sci.mus.mn.us

St. Louis Science Center – http://www.slsc.org

St. Louis Zoo – http://www.stlzoo.org

We hope you, the reader, have gained new perspectives on the potential synergy between informal and formal learning environments. We hope this chapter caused you to generate your own questions about how inquiry can build on this synergy in your own practice. Finally, we hope you will continue your inquiry as you put these principles into practice.

ENDNOTES

1. In the vignettes and throughout this chapter, each author wrote from her own experience with formal and informal learning environments and with inquiry and research. Doris Ash is a science educator at The Exploratorium in San Francisco working with the Institute for Inquiry for teachers and administrators new to the inquiry process and the Professional Development Design Workshop for those who will lead others in designing workshops and curriculum based in inquiry. Doris has worked with teachers and students involved in the Fostering Communities of Learners project in the Bay Area since it began in 1989. Christine Klein, Investigative Learning Center (ILC) Program Manager for the St. Louis Science Center, works with students and teachers at the Compton-Drew ILC Middle School, a museum school in partnership with St. Louis Public Schools that uses the Schools for Thought curriculum framework. Both authors are working toward building the synergistic relationship between formal and informal learning discussed in this chapter.

2. Funding agencies like the National Science Foundation began to use the term "informal" to support projects outside "formal" classroom settings. They include public television and community organizations as informal settings. These last two settings are not the focus of this chapter.

3. Schools for Thought, described later in the chapter, provides the theoretical framework and philosophy for the program at Compton-Drew Middle School. The curriculum described in this chapter was designed by Science Center staff within that framework.

4. For further discussion of distinctions and definitions of informal and formal, see "ILER Forum: How Do We Define Ourselves?" in the *Informal Learning Environments Research Newsletter*, February 1998, available from the Informal Learning Environments Research special interest group of the American Educational Research Association or on the group's website, http://darwin.sesp.nwu.edu/informal/.

5. For more information on museum schools, see Science Museum of Minnesota 1995 and the issue of the *Journal of Museum Education* (vol. 23, no. 2, 1998) on museum and school partnerships.

6. A jigsaw group is formed by taking one expert from each research group to form a new group in which each expert has a key piece of the puzzle to share (Aronson, 1978).

7. The longer version of this case study was presented at the Institute of Inquiry Forum in 1996.

8. For more information on these projects, see Bruer, 1993; Lamon et al., 1996; and McGilly 1994.

9. Complete citations for publications are in reference section.

REFERENCES

Aronson, E. 1978. *The jigsaw classroom*. Beverly Hills, CA: Sage.

Ash, D. 1995. From functional reasoning to an adaptionist stance: Children's transition toward deep biology. Unpublished thesis. University of California, Berkeley.

Ash, D. 1999. The process skills of inquiry. In *Inquiry: Thoughts, views and strategies for the K-5 classroom. Foundations*. Vol. 2. Washington, DC: National Science Foundation.

Bereiter, C., and M. Scardamalia. 1993. *Surpassing ourselves: An inquiry into the nature and implications of expertise*. Peru, IL: Open Court.

Brown, A.L. 1992. Design experiments: Theoretical and methodical challenges in creating complex interventions in classroom settings. *The Journal of Learning Sciences* 2(2): 141-178.

Brown, A.L., D. Ash, M. Rutherford, K. Nakagawa, A. Gordon, and J.C. Campione. 1993. Distributed expertise in the classroom. In *Distributed cognitions*, edited by G. Salomon, 188-228. New York: Cambridge University Press.

Brown, A.L., and J.C. Campione. 1994. Guided discovery in a community of learners. In *Classroom lessons: Integrating cognitive theory and classroom practice* edited by K. McGilly, 229-270. Cambridge, MA: MIT Press.

Brown, A.L., and J.C. Campione. 1996. Psychological theory and the design of learning environments: On procedures, principles and systems. In *Innovations in learning*, edited by L. Schauble and R. Glaser, 289-325 Hillsdale, NJ: Lawrence Erlbaum.

Brown, J., and R. Shavelson. 1996. *Assessing hands-on science: A teacher's guide to performance assessment*. Thousand Oaks, CA: Corwin Press.

Bruer, J. T. 1993. *Schools for thought: A science of learning in the classroom*. Cambridge, MA: MIT Press.

Csikszentmihalyi, M. 1990. *Flow: The psychology of optimal experience*. New York: Harper & Row.

Csikszentmihalyi, M., and K. Hermanson. 1995. Intrinsic motivation in museums: Why does one want to learn? In *Public institutions for personnal learning: Establishing a research agenda*, edited by J. H. Falk and L. C. Dierking, 67-77. Washington, DC: American Association of Museums

Falk. J. H., and L. D. Dierking. 1998. Free-choice learning: An alternative term to informal learning? *Informal Learning Environments Research Newsletter* 2(1): 3.

Hein, G. E., and M. Alexander. 1998. *Museums: Places of learning.* Washington, DC: American Association of Museums.

Harlen, W., and S. Jelly. 1999. How assessment can help learning through inquiry. Paper presented at Supporting Inquiry Through Assessment Forum, The Exploratorium, San Francisco.

Harlen, W., and S. Jelly. 1997. *Developing primary science.* 2nd rev. ed. London: Longman.

Klein, C. 1998. Putting theory into practice: Compton-Drew investigative learning center. *Journal of Museum Education* 23 (2):8-10.

Lamon, M., T. Secules, A. J. Petrosino, R. Hackett, J. D. Bransford, and S. R. Goldman. 1996. Schools for Thought: Overview of the project and lessons learned from one of the sites. In *Innovations in Learning,* edited by L. Schauble and R. Glaser, 243-288 Hillsdale, NJ: Lawrence Erlbaum.

Lemke, J. L. 1990. Talking science: *Language, learning and values.* Norwood, NJ: Ablex.

Mann, L. 1997. Extending the curriculum through museums. *Curriculum Update.* Newsletter of the Association for Supervision and Curriculum Development.

Matusov, E., and B. Rogoff. 1995. Evidence of development from people's participation in communities of learners. In *Public institutions for personal learning: Establishing a research agenda,* edited by J.H. Falk and L.D. Dierking, 97-104. Washington, DC: American Association of Museums.

McGilly, K. (Ed.) 1994. *Classroom lessons: Integrating cognitive theory and classroom practice.* Cambridge, MA: MIT Press.

Palincsar, A.S., and A.L. Brown. 1984. Reciprocal teaching of comprehension-fostering and monitoring activities. *Cognition and
Instruction* 1(2): 117-175.

Scardamalia, M. and C. Bereiter. 1996. Engaging students in a knowledge society. *Educational Leadership* 54(3).

Science Museum of Minnesota. 1995. *Museum schools symposium
1995: Beginning the conversation.* St. Paul, MN: Science Museum of
Minnesota.

Vermont Elementary Science Project. 1995. Inquiry based science:
What does it look like? *Connect* 8(4): 13.

Vygotsky, L.S. 1978. *Mind in society: The development of higher psychological processes.* Cambridge, MA: Harvard University Press.

The Need for Special Science Courses for Teachers: Two Perspectives

Lillian C. McDermott and Lezlie S. DeWater*

A physics professor and a classroom teacher present their perspectives on the type of preparation that K-12 teachers need in order to be able to teach science as a process of inquiry. Both have had more than twenty-five years of experience in teaching at their respective levels and in working with precollege teachers. Although the context for much of the discussion is physics, analogies to other sciences can readily be made.

THE PERSPECTIVE OF A PHYSICS PROFESSOR

In the United States, precollege teachers are educated in the same universities and colleges as the general population. In most institutions, two independent administrative units are involved: a college or school of education that offers courses on the psychological, social, and cultural aspects of teaching, and a college of arts and sciences (or equivalent) that provides instruction in various disciplines. Whereas the preparation of K-12 teachers may be central to faculty in education, such a function is often considered peripheral to the mission of a science department. Most faculty in the sciences take the position that responsibility for the professional development of teachers resides solely within colleges of education.

* Lezlie S. DeWater is currently on leave from the Seattle Public Schools as a Visiting Lecturer in the Department of Physics at the University of Washington.

This point of view ignores the fact that almost all the instruction that precollege teachers receive in the sciences takes place in science departments. If the current national effort toward reform in K-12 science education is to succeed, science faculty must take an active role in the preparation of teachers in their disciplines.

The perspective of the physics professor that is presented here is based on the cumulative experience of the Physics Education Group at the University of Washington.[1] For many years, the group has been conducting research on the learning and teaching of physics and using the results to guide the development of curriculum for various student populations at the introductory level and beyond.[2] In addition to participating in the regular instructional program in the Department of Physics, the group has been conducting intensive programs for in-service teachers during the summer and for both preservice and in-service teachers during the academic year. This experience has provided a structure for the ongoing development and assessment of *Physics by Inquiry*, a laboratory-based curriculum designed to prepare K-12 teachers to teach physics and physical science as a process of inquiry.[3] The group is also producing a research-based supplementary curriculum to help improve student learning in introductory physics.[4]

Inadequacy of Traditional Approach

Science departments offer a number of courses that can be taken by prospective teachers. Some of these courses may be required for certification to teach a particular science in high school. Others may be taken by future elementary and middle school teachers to satisfy a general science requirement for graduation. Some departments also offer short survey courses that they recommend for teachers. As the discussion below in the context of physics illustrates, this traditional approach does not work well for preparing teachers at the elementary, middle, or high school levels.

Many physics faculty seem to believe that the effectiveness of a high school teacher will be determined by the number and rigor of physics courses taken. Accordingly, the usual practice is to offer the same courses to future high school teachers as to students who expect to work in industry or enter graduate school. That overlooks the character of K-12 teaching.

The content of the high school curriculum in physics is closely matched to the first-year in college. However, the first-year college course is not adequate preparation for teaching the same material in high school. The breadth of topics covered allows little time for acquiring a sound grasp of the underlying concepts. The routine problem solving that characterizes most introductory courses does not help teachers develop the reasoning ability necessary for handling the unanticipated questions that are likely to arise in a classroom. The laboratory courses offered by most physics departments do not address the needs of teachers. Often the equipment is not available in the teachers' schools, and no provision is made for showing them how to plan laboratory experiences that make use of simple apparatus. A more serious shortcoming is that experiments are mostly limited to the verification of known principles. Students have little opportunity to start from their observations and go through the reasoning involved in formulating principles. As a result, it is possible to complete the laboratory course without confronting conceptual issues or understanding the scientific process.

For those students who progress beyond the first year of university physics, advanced courses are of little direct help in teaching. The abstract formalism that characterizes upper division courses is not of immediate use in the precollege classroom. Sometimes in the belief that teachers need to update their knowledge, university faculty may offer courses on contemporary physics for preservice or in-service teachers. Such courses are of limited utility. The information may be motivational but does not help the teachers recognize the distinction between a memorized description and substantive understanding of a topic. Although work beyond the introductory level may help teachers deepen their understanding of physics, no guidance is provided about how to make appropriate use of this knowledge in teaching high school students.

Elementary and middle school teachers often lack the prerequisites for even the standard introductory courses, especially in the physical sciences. They are unlikely to pursue the study of any science in depth because the vertical structure of the subject matter requires progression through a prescribed sequence of courses. In physics, in particular, the need for mathematical facility in the standard courses effectively excludes those planning to teach below the high school level. The only courses generally available are almost entirely descriptive. A great deal of material is presented, for which most preservice and in-service teachers (as well as other students) have neither the background nor the time to absorb. Such courses often reinforce the tendency of teachers to perceive physics as an inert body of information to be memorized, not as an active process of inquiry in which they and their students can participate.

The total separation of instruction in science (which takes place in science courses) from instruction in methodology (which takes place in education courses) decreases the value of both for teachers. Effective use of a particular instructional strategy is often specific to the content. If teaching methods are not studied in the context in which they are to be implemented, teachers may be unable to identify the critical elements. Thus, they may not be able to adapt a strategy that has been presented in general terms to specific subject matter or to new situations. Detailed directions in teacher's guides are of little use when teachers do not understand either the content or the intended method of presentation.

The traditional approach to teacher preparation in science departments has another major shortcoming. Teachers tend to teach as they have been taught. If they were taught through lecture, they are likely to lecture, even if this type of instruction may be inappropriate for their students. Many teachers cannot, on their own, separate the physics they have learned from the way in which it was presented to them.

Development of Special Physics Courses for Teachers

To counter the public perception that physics is extremely hard, a teacher must be able to teach in a way that allows students to achieve adequate mastery of a topic and to develop the confidence necessary to apply this knowledge in daily life. Since neither traditional physics courses nor professional education courses provide this type of preparation, there is a need for special physics courses for teachers.[1,5]

An effort to meet this need led to the establishment by A.B. Arons in 1968 of a course for preservice elementary school teachers in the Department of Physics at the University of Washington.[6] Shortly after, preservice courses for middle and high school teachers were added.[7,8] In-service versions soon followed. Modifications of the original courses constitute the core of the present teacher preparation program of the Physics Education Group. The preservice courses, which are supported by the department, are taught during the academic year. The inservice program consists of an intensive, six-week National Science Foundation summer institute for K-12 teachers and a continuation course that meets weekly during the academic year. Most teachers enroll for more than one year.

In addition to their instructional function, the preservice and in-service courses have provided a context for research on the learning and teaching of physics, as well as a setting for the development of *Physics by Inquiry*. Presented here is a distillation of what we have learned about the preparation of teachers and what we have tried to incorporate in our curriculum. The discussion is not an exhaustive summary of all that should be done. For example, important practical matters, such as laboratory logistics and classroom management, are not addressed. The focus is on intellectual aspects.

Intellectual Objectives

The curriculum used in courses for teachers should be in accord with the instructional objectives. The emphasis should be on the content that the teachers are expected to teach. They need the time and guidance to learn basic physics in depth, beyond what is possible in standard courses. Teachers should be given the opportunity to examine the nature of the subject matter, to understand not only what we know, but on what evidence and through what lines of reasoning we have come to this knowledge. Conceptual understanding and capability in scientific reasoning provide a firmer foundation for effective teaching than superficial learning of more advanced material.

A primary intellectual objective of a course for teachers should be a sound understanding of important concepts. Equally critical is the ability to do the qualitative and quantitative reasoning that underlie the development and application of concepts. Instruction for teachers should cultivate scientific reasoning skills, which tend to be overlooked in traditional courses. It has been demonstrated, for example, that university students enrolled in standard physics courses often cannot reason with ratios and proportions.[9] Proportional reasoning is obviously a critically important skill for high school science teachers, but it is also essential for elementary and middle school teachers who are expected to teach science units that involve concepts such as density and speed.

The emphasis in a course for teachers should not be on mathematical manipulation. Of course, high school teachers must be able to solve textbook problems. As necessary as quantitative skills are, however, ability in qualitative reasoning is even more critical. Courses for teachers should avoid algorithmic problem solving. Questions should be posed that require careful reasoning and explanations. Teachers need to recognize that success on numerical problems is not a reliable measure of conceptual development. It is also necessary for them to develop skill in using and interpreting formal representations such as graphs, diagrams, and equations. To be able to make the formalism of physics meaningful to students, teachers must be adept at relating different representations to one another, to physical concepts, and to objects and events in the real world.

An understanding of the nature of science should be an important objective in a course for teachers. Teachers at all grade levels must be able to distinguish observations from inferences and to do the reasoning necessary to proceed from observations and assumptions to logically valid conclusions. They must understand what is considered evidence in science, what is meant by an explanation, and what the difference is between naming and explaining. The scientific process can only be taught through direct experience. An effective way of providing such experience is to give teachers the opportunity to construct a conceptual model from their own observations. They should go step by step through the process of making observations, drawing inferences, identifying assumptions, formulating, testing, and modifying hypotheses. The intellectual challenge of applying a model that they themselves have built (albeit with guidance) to predict and explain progressively complex phenomena can help teachers deepen their own understanding of the evolving nature, use, and limitations of a scientific model. We have also found that successfully constructing a model through their own efforts helps convince teachers (and other university students) that reasoning based on a coherent conceptual framework is a far more powerful approach to problem solving than rote substitution of numbers in memorized formulas.

The instructional objectives discussed above are, in principle, equally appropriate for the general student population. However, teachers have other requirements that special physics courses should address. For example, teachers need practice in formulating and using operational definitions. To be able to help students distinguish between related but different concepts such as between velocity and acceleration, teachers must be able to describe precisely and unambiguously how the concepts differ and how they are related. It is important that teachers be able to express their thoughts clearly. Discussions and writing assignments that require them to reflect on the development of their own conceptual understanding of a particular topic can enhance both their knowledge of physics and their ability to communicate.

In addition to having a strong command of the subject matter, teachers need to be aware of difficulties that students encounter in studying specific topics. There has been a considerable amount of research on difficulties common to students at all levels (K-20) of physics education.[10] Instructors in courses for teachers should be thoroughly familiar with this resource and, when appropriate, refer them to the literature.

Courses for teachers should also help develop the critical judgment necessary for making sound choices on issues that can indirectly affect the quality of instruction in the schools. Teachers must learn, for example, to discriminate between meaningful and trivial learning objectives. When instruction is driven by a list of objectives that are easy to achieve and measure, there is danger that only shallow learning such as memorization of factual information will take place. Teachers need to develop criteria for evaluating instructional materials, such as science kits, textbooks, laboratory equipment, and computer software. They should be able to identify strengths and weaknesses in school science programs. Aggressive advertising and an attractive presentation often interfere with objective appraisal of the intellectual content of printed materials or computer software. We have seen teachers react with enthusiasm to an appealing program format, while they ignore serious flaws in physics. Through service on district committees, individual teachers can often have an impact that extends beyond their own classrooms. A poor curriculum decision can easily deplete the small budget most school districts have for science without resulting in an improvement in instruction.

Instructional Approach

If the ability to teach by inquiry is a goal of instruction, then teachers need to work through a substantial amount of content in a way that reflects this spirit. Teachers should be prepared to teach in a manner that is appropriate for the K-12 grades. Science instruction for young students is known to be more effective when concrete experience establishes the basis for the construction of scientific concepts.[11] We and others have found that the same is true for adults, especially when they encounter a new topic or a different treatment of a familiar topic. Therefore, instruction for prospective and practicing teachers should be laboratory-based. However, "hands-on" is not enough. Unstructured activities do not help students construct a coherent conceptual framework. Carefully sequenced questions are needed to help them think critically about what they observe and what they can infer. When students work together in small groups, guided by well-organized instructional materials, they can also learn from one another.

Whether intended or not, teaching methods are learned by example. The common tendency to teach physics from the top down, and to teach by telling, runs counter to the way precollege (and many university) students learn best. The instructor in a course for teachers should not transmit information by lecturing, but neither should he or she take a passive role. The instructor should assume responsibility for student learning at a level that exceeds delivery of content and evaluation of performance. Active leadership is essential, but in ways that differ markedly from the traditional mode. This approach, which can be greatly facilitated by a well-designed curriculum, is characterized below in general terms and illustrated in the next section in the context of specific subject matter. Other examples are given in published articles.[12]

The instructional materials used in a course for teachers should be similar to those used in K-12 science programs, but the curriculum should not be identical. Teachers must have a deeper conceptual understanding than their students are expected to achieve. They need to be able to set learning objectives that are both intellectually meaningful for the topic under study and developmentally appropriate for the students.

The study of a new topic should begin with open-ended investigation in the laboratory, through which students can become familiar with the phenomena of interest. Instead of introducing new concepts or principles by definitions and assertions, the instructor should set up situations that suggest the need for new concepts or the utility of new principles. By providing such motivation, the instructor can begin to demonstrate that formation of concepts requires students to become mentally engaged. Generalization and abstraction should follow, not precede, specific instances in which the concept or principle may apply. Once a concept has been developed, the instructor should present new situations in which the concept is applicable but may need to be modified. This process of gradually refining a concept can help develop an appreciation of the successive stages that are involved in developing a sound conceptual understanding.

As students work through the curriculum, the instructor should pose questions designed to help them to think critically about the subject matter and to ask questions on their own. The appropriate response of the instructor to most questions is not a direct answer but another question that can help guide the students through the reasoning necessary to arrive at their own answers. Questions and comments by the instructor should be followed by long pauses in which the temptation for additional remarks is consciously resisted. Findings from research indicate that the quality of student response to questions increases significantly with an increase in "wait time," the time the instructor waits without comment after asking a question.[13]

As mentioned earlier, a course for teachers should develop an awareness of common student difficulties. Some are at such a fundamental level that, unless they are effectively addressed, meaningful learning of related content is not possible. Serious difficulties cannot be overcome through listening to lectures, reading textbooks, participating in class discussions, or consulting references (including teacher's guides). Like all students, teachers need to work through the material and have the opportunity to make their own mistakes. When difficulties are described in words, teachers may perceive them as trivial. Yet we know that often these same teachers, when confronted with unanticipated situations, will make the same errors as students. As the opportunity arises during the course, the instructor should illustrate instructional strategies that have proved effective in addressing specific difficulties. If possible, the discussion of a specific strategy should occur only after it has been used in response to an error. Teachers are much more likely to appreciate important nuances through an actual example than through a hypothetical discussion. Without specific illustrations in the context of familiar subject matter, it is difficult for teachers to envision how to translate a general pedagogical approach into a specific strategy that they can use in the classroom. The experience of working through the material themselves can help teachers identify the difficulties their students may have. Those who understand both the subject matter and the difficulties it poses for students are likely to be more effective than those who know only the content.

Illustrative Example

To illustrate the type of instruction discussed above, we present a specific example based on a topic included in many precollege programs: batteries and bulbs. Below we describe how students, including preservice and in-service teachers, are guided to develop a conceptual model for a simple dc circuit.[14] Mathematics is not necessary; qualitative reasoning is sufficient.

The students begin the process of model-building by trying to light a small bulb with a battery and a single wire. They develop an operational definition for the concept of a complete circuit. Exploring the effect of adding additional bulbs and wires to the circuit, they find that their observations are consistent with the assumptions that a current exists in a complete circuit and the relative brightness of identical bulbs indicates the relative magnitude of the current. As the students conduct further experiments—some suggested, some of their own devising—they find that the brightness of individual bulbs depends both on how many are in the circuit and on how they are connected to the battery and to one another. The students are led to construct the concept of electrical resistance and find that they can predict the behavior of many, but not all, circuits of identical bulbs. They recognize the need to extend their model beyond the concepts of current and resistance to include the concept of voltage (which will later be refined to potential difference). As bulbs of different resistance and additional batteries are added, the students find that they need additional concepts to account for the behavior of more complicated circuits. They are guided in developing more complex concepts, such as electrical power and energy. Proceeding step by step through deductive and inductive reasoning, the students construct a conceptual model that they can apply to predict relative brightness in any circuit consisting of batteries and bulbs.

We have used this guided inquiry approach with teachers at all educational levels from elementary through high school. The process of hypothesizing, testing, extending, and refining a conceptual model to the point that it can be used to predict and explain a range of phenomena is the heart of the scientific method. It is a process that must be experienced to be understood.

It is important that teachers be asked to synthesize what they have learned, to reflect on how their understanding of a particular topic has evolved, and to try to identify the critical issues that need to be addressed for meaningful learning to occur. They also need to examine the interrelationship of topics in the curriculum in order to be able to teach science in a coherent manner. Through direct experience with the intellectual demands of learning through inquiry, teachers can become better equipped to meet the challenge of matching their instruction to the developmental level of their students. We have found that the sense of empowerment that results from this type of preparation helps teachers develop confidence in their ability to deal with unexpected situations in the classroom.

THE PERSPECTIVE OF A CLASSROOM TEACHER

In this section, an elementary school teacher describes her early days as a teacher and the impact of the type of instruction described above on her professional development. Today, more than 25 years later, she reflects on how this experience has affected the way in which she teaches science in her classroom.

I earned my B.A. with a major in French and a minor in elementary education. I had initially intended to be a high school teacher but at the eleventh hour changed my mind, hoping that my 5' 3" stature might put me above the eye level of most elementary students. With my diploma and certificate in hand, I was confident that I knew at least as much English and mathematics as a sixth grader, certain that I could learn the rest from the teacher's guides for the student textbooks. I applied for only one position and was promptly hired by a large urban school district as a second-grade teacher. All went as I had expected. I knew more than the second graders and was quick to employ all that I had learned in my first eight years of education—teaching, of course, in the same manner in which I had been taught in a crowded Catholic school classroom. I did, however, make some attempts to form small groups for reading. I felt very fortunate that my class of twenty-eight students was twenty fewer than the classes that I had experienced as a child.

In late spring of my first year of teaching, I was informed that a drop in enrollment would result in the elimination of the position that I held. The good news was that I was welcome to take a newly created position as the science specialist for grades K-4. Not wanting to relocate and not stopping to consider that my major in French might not have appropriately prepared me for this new position, I quickly agreed to take it for the following year. Shortly after I accepted the job, the district science supervisor contacted me, suggesting we start with a couple of *Elementary Science Study* units, "Clay Boats" and "Primary Balancing."[15] The unit guides and equipment were ordered. I was all set to begin my new teaching role.

Never having had a science lesson in elementary school, I was not predisposed, as I had been with the other subjects, to teach it as I had been taught. In fact, without any real textbook to guide the students, I was left with the materials and a few rather general instructions in the teacher's guide. And so it was that we, my students and I, became explorers of materials. We had a great time. The students were engaged; they talked a lot about what they were doing and we all asked a lot of questions. But I wanted to do more than just explore and ask questions. I wanted to learn some basic principles and have a clear vision of where we were going. I wanted to lead my students to discover and understand something as well. But what was it that we should be understanding? I hadn't a clue. This is when I first came to recognize that to become a truly effective teacher, I would need scientific skills and understandings that I had not developed, nor been required to develop, during my undergraduate years.

Not long after I recognized my deficiencies, I happened to glance through the school district's newsletter and came across a notice for a National Science Foundation Summer Institute in Physics and Physical Science for Elementary Teachers. I had been turned down the previous year for lacking sufficient teaching experience. I was certain that they would now recognize that I did not know enough science to teach. I was right; this time I was accepted.

I walked away from that summer feeling that my brain had been to boot camp. No course of study, no teacher, had ever demanded so much of me. I had never before been asked to explain my reasoning. A simple answer was no longer sufficient. I had been expected to think about how I came to that answer and what that answer meant. It had been excruciating at times, extricating the complicated and detailed thought processes that brought me to a conclusion, but I found it became easier to do as the summer progressed. I realized that it was not only *what* I learned but also *how* I learned that had provided me with new-found self esteem and confidence. The carefully sequenced questions had helped me come to an understanding of science that I had always felt was beyond me. I wanted to be able to lead my students to that same kind of understanding. It became clear to me that the key to teaching by inquiry was first understanding the content myself.

As a result of the Summer Institute, I had developed a sound understanding of several basic science concepts including balance, mass, and volume. Along with these concepts I discovered an appreciation for the need to control variables in an experiment. I was now better equipped to take a more critical look at the science units I had used the previous year. I recognized that "Clay Boats" had probably not been the best choice for a teacher with only a budding understanding of sinking and floating, but "Primary Balancing" seemed to be an appropriate choice. I had worked with very similar materials in the Summer Institute and had some ideas about how I could lead students to discover, through experiments in which they would come to understand the need to control variables, which factors seem to influence balance and which do not.

Unlike many of the professional development courses that I have taken since that time, the Summer Institute was certainly not in the category *"been there, done that."* I had been there, but there was still so much I hadn't done, so much I felt I needed and wanted both to do and to learn. Science, with all the skills and concepts that term connotes, is an overwhelming body of knowledge. You just scratch the surface and you find that what lies underneath extends much deeper than you had ever anticipated. It is for this reason that I felt compelled to return. And so I did, again and again and again. I participated in several Summer Institutes and academic year continuation courses, both as a student and as a member of the instructional staff. Assimilation is not a process to be rushed, nor is application. It must be thoughtful. It must be deliberate. It must be evaluated. The process takes time, lots of time. Time that we as teachers have difficulty finding. Time that administrators are reluctant to relinquish.

After many years of professional development in science education, I feel comfortable teaching most, if not all, of the science concepts covered in elementary and middle school. An understanding of the content allows me to teach with confidence units such as electric circuits, magnetism, heat and temperature, and sinking and floating. But simply understanding the content did not assure that I could bring my students to an understanding appropriate for them.

How does one begin to develop some expertise in these strategies we call inquiry? I can only suppose that for me it began by reflecting upon my personal experience. I don't believe that this was a deliberate exercise on my part. In subtle ways, over many years, I began to teach in the manner in which I had been taught in the Summer Institutes.

I know that early on I began to pay attention to the questions that I asked, for the questions stood out in my mind as the tools that, when deftly wielded, resulted in the desired state of understanding. I knew, too, that questions would help me to discover the intellectual status of my students: to tell me where they were with the necessary conceptual understanding. Aware of several "pitfalls" (misconceptions) that I had personally encountered, I was prepared to think about questions that would help me find out where I needed to start. I envisioned the terrain between the students and their conceptual understanding. I liken the terrain to an aerial photograph that clearly details the various roads that lead to the designated destination, along with dead ends and the hazards. I am well acquainted with this terrain, because I have traversed it on more than one occasion myself, and have conversed with others who have, perhaps, taken a different path to the same destination. I want my students to encounter some difficulties and to resolve conflicts and inconsistencies and to grow intellectually from these experiences. But I do not want them to wander aimlessly or to plunge over a cliff. For this reason it is crucial that, like a vigilant parent, I continue to offer support in their intellectual insecurity. I question and listen carefully. I scan the territory to find where the explanations and responses to my questions place them, and then plan my next strategy to keep them moving ahead. I recall from being a learner that sometimes this next strategy is a question such as, "What would you need to do to find out?" Sometimes it is a suggestion of some experiment to try. Sometimes it is a comment such as, "Why don't you think about that for a bit." It has taken many years of trying out these strategies to learn how to gauge which tactic is appropriate at what time and with which student.

There are, of course, other considerations in the teaching of inquiry-based science that must accompany all that I have said. It is necessary to think about the engagement of students and developmental appropriateness. For the elementary school students with whom I have worked, engagement has never been a problem. Science is naturally engaging if the teacher shows the least bit of enthusiasm. Students are intrigued by the world around them and have already begun to develop their own explanations for how and why Mother Nature operates in the way she does. The trick is to capitalize on this curiosity and channel it so that students develop better explanations for basic phenomena.

The question of developmental appropriateness is another matter. I have come to a much clearer recognition of what will "fly" and what will not, as a result of working through *Physics by Inquiry* in the Summer Institutes.[3] These materials were carefully designed to build conceptual understanding in logical, sequential steps. You do not, for instance, begin to think about why things sink or float without first understanding by concrete operational definitions what is meant by mass and what is meant by volume. Only then can you begin to think about how these two variables may influence sinking and floating. I have also come to appreciate the difficulty of these concepts for the adult learner. I think long and hard about the research that gives us some notion of what children of a particular age are capable of doing. Although I may explore the concepts of mass and volume with eight- and nine-year olds, I certainly would not expect most of them to come to an understanding of density. I will sometimes stretch the limits slightly, knowing that we may not really know how far each student can go. Elementary school teachers have not had a great track record for teaching science to their students, but I believe that if we structure what we teach more carefully, we can do better. After all, who would have thought that a high-school-bound French teacher would come to understand a relatively large body of physical science, including physics?

In summary, I would like to repeat that what seems to have been most important for me in becoming a more effective teacher of science at the elementary level was both to gain a sound understanding of the content and to learn it through inquiry-based instruction. It was then necessary to reflect explicitly on my experience as a learner so that I could put into practice what had been modeled for me.

CONCLUSION

Significant improvement in the learning of science by elementary, middle, and high school students can take place only when the problem of inadequate teacher preparation is successfully addressed. Although not sufficient, this is a necessary condition for effective reform. Since the type of preparation that addresses the needs of teachers is not available through the standard university science curriculum, a practical alternative is to offer special courses for teachers in science departments. The instructors in such courses must have a sound understanding of the subject matter, of the difficulties that it presents to students, and of effective instructional strategies for addressing these difficulties. It is important for science faculty to recognize that the teachers completing these courses must be prepared to teach the material at an appropriate level in K-12 classrooms. The choice of an appropriate curriculum is critical. We have found that teachers often try to implement instructional materials in their classrooms that are very similar to those they have used in their college courses. Therefore, even though it has not been our intent to have young students work directly with *Physics by Inquiry*, we have designed the curriculum so that it can be used in this way by experienced high school teachers.

Our experience indicates that it is not easy to develop good inquiry-oriented instructional materials. Therefore, unless faculty are prepared to devote a great deal of effort over an extended period to the development of a course for teachers, they should take advantage of already existing instructional materials that have been carefully designed and thoroughly tested. The development of any new curriculum should be based on research, with rigorous assessment an integral part of the process. In this way, cumulative progress at all levels of science education can become possible.

ACKNOWLEDGEMENT

The authors deeply appreciate the many contributions over the years to the teacher preparation program by all members of the Physics Education Group (past and present, resident and visiting). We recognize our debt to Arnold B. Arons for his insights into the needs of teachers and for his initiative in introducing the first course for teachers in the Department of Physics at the University of Washington. Special acknowledgment is due to the National Science Foundation for the continuing support that has enabled the group to conduct a coordinated program of research, curriculum development, and instruction and to provide an environment in which the teacher preparation program has thrived.

ENDNOTES

1. This section draws on L.C. McDermott, "A perspective on teacher preparation in physics and other sciences: The need for special science courses for teachers," *American Journal of Physics* 58, 734–42 (1990).
2. The process of research and curriculum development as implemented by the Physics Education Group is illustrated in many papers that appear in the *American Journal of Physics*. For an overview and more general discussion, see L.C. McDermott, "Millikan Lecture 1990: What we teach and what is learned—Closing the gap," *American Journal of Physics* 59, 301–315 (1991).
3. L.C. McDermott, P.S. Shaffer, and the Physics Education Group at the University of Washington, *Physics by Inquiry* (Wiley, New York, 1996), Vols. I and II. A third volume is under development.
4. L.C. McDermott, P.S. Shaffer, and the Physics Education Group at the University of Washington, *Tutorials in Introductory Physics*, Preliminary Edition (Prentice Hall, Upper Saddle River, NJ, 1998).

5. For another discussion of the need for special physics courses for teachers, see K.G. Wilson, "Introductory physics for teachers," *Physics Today* 44 (9), 71–73 (1991).

6. A.B. Arons, *The Various Language* (Oxford U.P., New York, 1977). This book was developed for the original version of the course for elementary school teachers.

7. L.C. McDermott, "Combined physics course for future elementary and secondary school teachers," *American Journal of Physics* 42, 668–676 (1974).

8. L.C. McDermott, "Improving high school physics teacher preparation," *Physics Teacher* 13, 523–529 (1975).

9. A.B. Arons, *A Guide to Introductory Physics Teaching* (Wiley, New York, 1990), pp. 3–6.

10. A selection of articles can be found in L.C. McDermott and E.F. Redish, "Resource Letter PER–1: Physics Education Research," *American Journal of Physics* 67, 755–767 (1999). Although most of the studies cited in this resource letter refer to students at the university level, similar difficulties have been identified among younger students.

11. See, for example, J. Griffith and P. Morrison, "Reflections on a decade of grade-school science," *Physics Today* 25 (6), 29–34 (1972); R. Karplus, Physics for beginners," *Physics Today* 25 (6), 36–47 (1972); and J.W. Renner, D.G. Stafford, W.J. Coffia, D.H. Kellogg, and M.C. Weber, "An evaluation of the Science Curriculum Improvement Study," *School Science and Mathematics*, 73 (4), 291–319 (1973).

12. See, for example, K. Wosilait, P.R.L. Heron, P.S. Shaffer, and L.C. McDermott, "Development and assessment of a research-based tutorial on light and shadow," *American Journal of Physics* 66, 906–913 (1998); L.C. McDermott and P.S. Shaffer, "Research as a guide for curriculum development: An example from introductory electricity. Part I: Investigation of student understanding," *American Journal of Physics* 60, 994–1003 (1992); P.S. Shaffer and L.C. McDermott, "Research as a guide for curriculum development: An example from introductory electricity, Part II: Design of instructional strategies," *American Journal of Physics* 60, 1003–1013 (1992); and M.L. Rosenquist and L.C. McDermott, "A conceptual approach to teaching kinematics," *American Journal of Physics* 55, 407–415 (1987).

13. M.B. Rowe, "Wait time and rewards as instructional variables, their influence on language, logic, and fate control: Part one—wait time," *Journal of Research in Science Teaching* 11, 81–94 (1974).

14. See the third article in Note 12.

15. *Elementary Science Study* (Educational Development Center, Newton, MA); *Science Curriculum Improvement Study* (University of California, Berkeley, CA).

Inquiry Teaching in Biology

Kathleen M. Fisher

INTRODUCTION

M any organizations have recently called for reform of science education in the United States, among them the Carnegie Commission on Science, Technology, and Government in 1991, the National Commission on Excellence in Education and the National Science Board on Precollege Education in Mathematics, Science and Technology in 1983, and in 1978 the National Science Foundation (NSF) Directorate for Science Education. In 1989, 1993, and 1998 the American Association for the Advancement of Science established standards for science learning, as did the National Research Council in 1996. Many states are in the process of establishing their own standards or have just finished doing so. At this point, a very large number of efforts are under way to improve American science teaching so as to produce both a scientifically literate public and better prepared science majors.

For example, I am currently involved in a project sponsored by the NSF that is national in scope and aims to promote college and university learning based in inquiry (Ebert-May & Hodder, 1995). Working in this project gives me an opportunity to hear what science faculty want to know about such learning. "Is it effective?" "How do you know?" and "What is the evidence?" are understandably some of their most common questions.

These questions are hard to answer, in part because there are so many different varieties of learning based in inquiry and partly because we are building on a research foundation that has spanned a quarter of a century and involves numerous fields of study, including but not limited to cognitive science, linguistics, psychology, sociology, and science education. The evidence has grown so gradually and is now so pervasive that, to my knowledge, there isn't a single definitive paper that sums it all up.

I have tried to organize this chapter to provide brief answers to some of the questions raised by my biology colleagues and to provide some useful references for further reading. I consider what inquiry-based learning is, why it is useful, when to avoid it, some strategies for its employment, lecturing in active learning, some features of inquiry-based and active learning, a thinking tool, learning communities, and the evidence for the feasibility of learning by inquiry.

WHAT IS INQUIRY-BASED LEARNING?

Different types of learning by inquiry include guided inquiry, open-ended inquiry, project-based learning, inquiry in collaboration with teachers, and learning by problems or cases (see McNeal & D'Avanzo, 1997 and "Some Inquiry-Based Strategies" below). The term "inquiry-based" is sometimes interpreted to mean that students are engaged in studying a phenomenon and at other times to mean they are engaged in questioning. Often both features are included in courses based in inquiry.

There is also a growing collection of strategies to bring active learning into classes large and small that were previously taught by traditional lecture methods (see Ebert-May, Brewer, & Allred, 1997 and "Lecturing with Active Learning" below). I describe my version of an active learning lecture course as well as reciprocal teaching, dyads, and jig-saws. These would fit under the broader definition of inquiry learning, but I have chosen to separate them because I tend to fix on the narrower definition of inquiry as study of a phenomenon.

WHY USE INQUIRY-BASED TEACHING?

Inquiry Mimics Everyday Learning

Consider how a child learns. A parent and toddler encounter a neighborhood dog. The toddler squeals and claps. The dog prances around, wags her tail, licks the child's hands. Soon the mother and child see another dog, and then another. Each day that they go out, the child sees a few more dogs. The dogs come in a wide variety of shapes, sizes, and colors. They may have long hair or short, long tails or short, ears that stand up or ears that hang down. But there are some constants as well: The shape of the dogs' heads, the shape of their paws, the way they bark and wag their tails, the way they move. When the toddler first begins seeing dogs, the mother often says the name: dog, puppy dog, Collie dog, big dog. Eventually the child says the names, too.

From these specific encounters with specific dogs, a child's mind automatically and subconsciously forms the general idea of dog. In the same way, a child acquires many, many other general concepts—cat, house, living room, run, walk—gradually building from specific experiences general knowledge of the world. Much of this knowledge of the world is intuitive and implicit rather than conscious and explicit. We are not very good at describing how we recognize a dog or how we distinguish a dog from a cat, even though our minds become very skilled at doing it.

There is much to be learned about how the mind automatically generates and uses general categories. But several observations about the process have strong implications for teaching and learning. Formation of categories occurs spontaneously upon exposure to multiple examples. In many learning events, experiences with specific instantiations of an idea come first; extraction of the more general idea follows, often subconsciously. And opportunities to interact with an object help learners to deepen their understanding and make finer discriminations.

Escaping the Deadly Dull

Almost any professor or teacher today will tell you that students are often absent, often uninterested, and often inclined to favor memorization over making sense. Why is that? We seldom stop to think how deadly dull higher education has become. The great God Coverage drives teachers to pack ever more facts into each lecture and to use labs as just another means of transmitting information. Multiple-choice tests reward simple low level learning: recognition of key phrases and ideas. The student's responsibility is to accept and absorb volumes of content provided by the experts. Where is the creativity? The engagement? The challenge? The self-expression? The opportunity to pursue an interest? And where do students learn about the other half of science—the *processes* involved in creating the content they are so dutifully expected to learn?

Restoring the Balance

Among many forms of learning by inquiry are some common features. Students typically are engaged in active learning. They are expected to take a greater share of the responsibility for their own learning. Most important is that they are rewarded for higher levels of thinking. There is also greater emphasis on conveying the ways in which new knowledge is generated and old knowledge is evaluated and applied in new contexts. Inquiry learning is therefore intrinsically more interesting and rewarding for both students and teachers than our established patterns of teaching and learning.

Building Bridges Between Concrete and Abstract

Concrete ideas provide important tools for thinking (Lakoff, 1987). Through a process of metaphorical mapping, George Lakoff and Mark Johnson observed in 1980, familiar concrete experiences help us think and talk about abstract things. Love, for instance, a most abstract concept, is understood as a journey. "We're flying high." "We've taken a wrong turn." "We're spinning our wheels." "We're off track." We're not usually conscious of our cognitive mappings but our expressions give them away. The mappings are consistent across people, across time, and often across cultures.

Yet in spite of the recognized value of concrete referents, academic presentations often skip the practical and jump straight into the abstract. It is one of the many ways in which "fat is trimmed" and "corners are cut." Read about Gregor Mendel, for instance, in most introductory biology or genetics textbooks. You'll get numbers of peas, lists of traits, and important ratios. But will you learn how to grow a pea plant? How to prevent it from self-fertilizing? What is involved in crossing one pea plant with another? What each pea represents? All of these practical aspects of Mendel's work—the things he spent most of his life doing—are deemed too trivial to describe (there are some exceptions as J.H. Postlethwait and J.L. Hopson demonstrated in 1995). Yet these concrete beginnings provide the visual images that allow us to comprehend what it means to cross two pea plants and they help us to make sense of the rest of it. Learning founded in inquiry provides motivation for us to get our feet back on the ground.

WHEN TO AVOID INQUIRY-BASED TEACHING

If you believe it is your responsibility to convey as many facts as possible to your students in the shortest possible time, then inquiry teaching is probably not for you. The lecture is a paradigm of efficiency for transfer of information. Or you may be intrigued but not quite yet ready to take the leap. I thought for several years about introducing active learning methods into my lecture course on Human Heredity before I actually took the plunge. Trying something new can be scary. Reading a pioneering paper on the topic, that of D. Ebert-May et al. in 1997, gave me both specific strategies and courage. Participating in the FIRST project sponsored by the NSF (Ebert-May & Hodder, 1995), which aims to promote inquiry learning through the use of biological field stations, pushed me over the edge. It helps when teachers experience active learning strategies themselves (in the student role) and when they acquire a handful of strategies or more for their toolbox. In that sense, perhaps the brief descriptions here and the accompanying references will be useful.

SOME INQUIRY-BASED STRATEGIES

This section provides very brief descriptions of some common strategies for teaching by inquiry.

Guided Inquiry

Guided inquiry, so termed by Ann McNeal and Charlene D'Avanzo in 1997, is also called privileging, as C.W. Keys named it in 1997. Its goal is to provide access to difficult scientific ideas. I use guided inquiry in my biology course for prospective elementary school teachers.

Before each lesson, I elicit the students' prior knowledge about the topic, usually through a class discussion. The goal is to bring to the surface their underlying assumptions. Some assumptions serve as foundations for learning, but others actually interfere with comprehension. Ponder questions—questions asking "why"—at the beginning of each lesson help in eliciting student understandings of everyday events. Students work in groups to perform activities, such as observing osmosis or modeling mitosis, and make observations, such as how a bean grows. The activities are quite simple but they illustrate important biological phenomena and many are designed to challenge naïve conceptions. Before each activity, students predict the outcome. This is another valuable means for eliciting underlying assumptions. In order to predict an outcome, students must make sense of the activity, construct a mental model of it, and run a mental simulation. In doing so, they draw upon their assumptions and beliefs, which are often different from those of scientists. The students discuss their predictions within their groups and attempt to come to a consensus, although this is not required. Once they have made their predictions in writing, their interest in the outcome of the experiment is heightened. The student groups work through the lesson, collecting data and responding to the questions in it. Questions embedded in the lesson aim to prompt their higher order scientific thinking about the phenomenon being examined.

A very interesting and consistent outcome is that, when the consequence of an activity differs from the students' predictions, they usually assume that they did the experiment wrong. It doesn't even occur to them that their mental model might need rethinking. To persuade students to reconsider their predictions, we compare results of all student groups. If most or all groups observed the same results, the outcome is validated. At that point, students are often reluctantly willing to reconsider their expectations.

These biology lessons can be viewed and downloaded from the World Wide Web (Fisher, 1996). There have been more then forty-five thousand visitors to the Biology Lessons web site in the past two years, and the site has been mirrored in both South Africa and Hong Kong. There is some evidence that this approach succeeds in promoting higher order thinking (Christianson & Fisher, submitted; Gorodetsky & Fisher, 1996).

Open-Ended Inquiry

Open-ended inquiry engages students even earlier in the scientific process and they take on more responsibility for the lesson. The teacher creates the context and establishes the constraints within which the students work. It is up to the students to identify the question to be asked and to design an experiment to provide the answer. These are, of course, two of the most challenging and engrossing steps in the entire scientific process. Students usually start out with very general questions that the teacher helps them trim down to something testable. The teacher also provides guidance and resources. Open-ended inquiry into the processes of decomposition by fourth graders working with an insightful teacher, Terez Waldoch, has been captured on a videotape available from Annenberg/CPB (Schnepps, 1994b). Open-ended inquiry at the university level is described in Ann McNeal and Charlene D'Avanzo's publication of 1997.

Project-Based Learning

In learning based on a project, a context is created for open-ended inquiry. Often the context is both real and current, and a news film or newspaper article may be used to describe the situation. An ecology course, for example, might draw upon a quarrel between a farming conglomerate and a group of environmentalists over a piece of land. A chemistry course might examine a conflict between towns-people and a paper plant about effluent contaminating a local stream. Whatever the situation, students are asked to identify the pertinent scientific questions and to determine what data can be collected to help resolve the dispute.

This has the advantage of showing students that science has practical application outside the classroom and can have a significant role in social and political conflicts. Such exercises can help students to learn to think objectively in emotional situations. And they often prompt original research into situations where several parties beyond the classroom are interested in the outcome. This raises the status of the students and increases the pressure on them to do a completely professional job in their research.

Like all original research, these projects are uncertain in outcome and there is potential for frustration. It may not be possible for the same groups to repeat the experiment as researchers often do in real life. When several student groups are working on the same question, the project may result in mutual validation or in magnifying confusion. One thing for sure: it is an easier situation for the teacher to manage when all groups are pursuing the same question.

Teacher-Collaborative Inquiry

Collaboration between teacher and students works best in small classes. Student groups may examine several different facets of the same problem. The problem is authentic, the research is original, and often several parties are interested in the outcome. This approach can be both satisfying and challenging. It is probably difficult at times for the teacher to treat the students as equal co-investigators, but then skillfully managed collaboration can create growing and learning situations for all participants.

Problem-Based or Case-Based Learning

Methods addressing specific cases have been used extensively for years in the professions, especially medicine, law, and business. They are not yet common in science. In these cases, students analyze existing situations rather than develop their own projects. The usual approach is to give students a real-life problem, ask them to identify the essential medical or legal or scientific or business issues, and to do the research necessary to produce a resolution. Students generally work in groups. The method lies at the intersection of the society and the discipline. Proponents find it is a way of humanizing science, bringing together scientific methodology and social values. For more information about case-based learning in science see C.F. Herreid's web site developed in 1998.

LECTURING WITH ACTIVE LEARNING

With some trepidation, this semester I changed into an active learning course my traditional lecture course in human heredity. The class meets for three fifty-minute lecture periods a week in a small room with fixed chairs and raised platforms. The room holds sixty-eight students and is filled to capacity. Human Heredity is a general education course for students not majoring in biology and so is relatively easy to modify.

On the first day I introduced students to the format of the course using a colorful poster to display our intended schedule for each fifty-minute period. I subsequently modified the schedule slightly:

11:00 – 11:10 Small Group Discussion
11:10 – 11:20 Class Discussion
11:20 – 11:35 Presentation
11:35 – 11:45 Quiz
11:45 – 11:50 Journal

We are working through a chapter a week in a human heredity text and students turn in assigned problems weekly. I encouraged the students to move around and form new groups daily for the first week, and at the fourth class meeting we formed fixed groups of three or four students. We are three weeks into the semester and so far, my students and I are very happy with the situation. It is fun! I follow the schedule loosely rather than rigorously, extending the presentation or the discussion when it seems useful. I generally use the discussion time for eliciting prior knowledge and the subsequent class discussions to build bridges from students' prior knowledge to the material to be presented. The quiz—at least one a week—centers on material from the previous class meeting, sometimes taken by individual students and sometimes by groups. Students have developed an ease and willingness to speak out in the large class discussions, which is very refreshing. Interactive lecture classes use little of the study of phenomena that is one component of learning by inquiry.

This approach gives me an entirely new way of thinking about the content. Instead of trying to cover everything in each chapter in reasonable detail, I think about what the students most need to know to function well in society, what they are already likely to know, and what I know about their naïve conceptions. This leads me to be quite selective and discriminating in my choice of topics. It is clear that students will be exposed to fewer topics, but I believe they are more likely to comprehend and remember them.

The students report that they like the group work best of all. They are less fearful of learning science and actually enjoy coming to class. Their attendance, which remains very high, verify this. Some initially worried about what the Right Answer is, but I turn that around and say, "If you have any questions please, please ask them" or, "Please feel free to consult your book for the correct answers." They have a much greater responsibility for their own learning and for determining what are believed to be the right answers.

Reciprocal Teaching

Anne Marie Palinscar and Ann Brown, as they report in an article published in 1984, introduced reciprocal teaching into the middle schools of a large school system. The process aims to foster monitoring and fostering of comprehension. It produces active rather than passive reading. The two were working with students reading science texts. In small groups with an assistant or aid, the students read one paragraph at a time and then summarize what it said, identify a question it raises, identify a point in the paragraph that needs clarification, and then predict what is likely to be in the next paragraph. The impressive thing about this active reading method is that the researchers began with the poorest twenty percent of the readers in each class, yet at the end of six months the participants were the best readers. This is the most effective educational intervention I have ever seen and you would think it would be in every school system in American by now—but it isn't.

Dyads

In a lecture course, it can help to stop after describing a difficult topic and ask students to discuss the material with a neighbor, to determine what makes sense to them and what their questions are. Tell the students they will have to stop talking when you send the signal, which will be in X minutes. A bell or chime can then restore order very quickly. The student pairs can then ask their questions and get the clarifications they need for comprehension. Or they can write a one-minute paper about the muddiest point, and you can address their comments in the next class. You can learn something about what erroneous assumptions you are making with respect to students' background knowledge and how to improve your presentation next time. Some people use this approach even with very large lecture classes.

Jig Saws

A jig saw is generally a variation on group work. For simplicity, let's suppose there are four groups of four students apiece. Each group may concentrate on obtaining information about a different topic, possibly through laboratory observations or literature review. When the students feel they have mastered their topic, new groups are formed, each containing a student from each of the four previous groups. Every student is then responsible for teaching the other three in the group about the topic the student has previously worked on.

SOME IMPORTANT FEATURES OF INQUIRY-BASED AND ACTIVE LEARNING

Eliciting Prior Knowledge

Eliciting prior knowledge can reveal solid starting points for a new lesson, critical gaps in the students' knowledge, misunderstandings, and important alternative conceptions. An *alternative conception* (a term with many synonyms, including *misconception* and *naïve conception*—I will use these three terms interchangeably) is an idea held by a learner that differs significantly from the scientific conception. A misconception is an error or misunderstanding to which the learner has a strong commitment. Naïve conceptions are *persistent*, well embedded in an individual's cognitive ecology, and difficult to correct, especially by didactic methods. Misconceptions are also widely shared, often by from twenty to sixty percent of students in a given class. For a compendium of misconceptions that have been studied see the articles by J.H. Wandersee, J.J. Mintzes, and J.D. Novak along with that by H. Pfundt and R. Duit, both published in 1994.

It is astounding, for example, how few students actually understand the basic idea of photosynthesis even though the topic tends to be "covered" in middle school, in high school, and again in college. The misconception that completely blocks comprehension appears to be that gas has no weight. If gas has no weight, then how could carbon dioxide be used to construct trees? Students dutifully memorize the formula for photosynthesis but look for more logical explanations: "The weight of a tree comes from the soil, the nutrients, the water," according to M. Schnepp and the Science Media Group in a video of 1994. The same misconception probably contributes to the belief that the bubbles in boiling water consist of hydrogen and oxygen for unlike water vapor, which many students think is visible and only temporarily present in air, these are known gases that also are assumed to have no weight.

Instructors in most lecture courses today have little knowledge about what their students are thinking. Without free discussion, without an awareness of some of the most persistent misconceptions that researchers have discovered, without any existing feedback mechanisms other than multiple-choice tests in which the instructor sets the questions, the answers, and the distractors, the disjunction between what students "know" and what teachers "know" grows ever wider.

Prediction

Dr. Roger Christianson, Chair of the Biology Department at the University of Oregon, notes that engaging students in making predictions is probably the single biggest difference between inquiry and standard biology labs. Predicting requires students to construct a mental model of an event, run a simulation, and commit themselves to an anticipated outcome. Like betting on a race, prediction heightens their interest in observing the actual outcome. Videotapes of student groups demonstrate that the most interesting discussions occur when students are making predictions and again when they are comparing their predictions with their results. Prediction and the resolution of differences between predictions and observations is the heart of the learning process in these labs, the place where conceptual change is most likely to occur. Among facets of inquiry teaching I would put prediction and explanation on the absolutely essential list.

When observations differ from predictions, students typically make the assumption that they must have done the experiment wrong. It doesn't occur to them to question their assumptions. It usually isn't until they see that all or nearly all the groups observed the same outcome that they begin to question their mental model and predictions. Many teachers deliberately use anomalous, unexpected events to challenge known misconceptions and to help students see in a new way.

Engagement with a Phenomenon

Each lesson here has students observe a phenomenon, event or simulation that illustrates the scientific principle being studied. The activity draws them into the problem, generating interest and motivation. It also provides the platform for making predictions. Guided interaction with the phenomenon gives students an opportunity to understand it better. It also serves as an anchor in memory for the related abstract ideas, since we tend to remember concrete experiences more easily than other kinds of knowledge. If the activity has been designed to challenge a misconception, the surprise associated with the outcome can increase its memorability.

One problem in biology is that so many events are invisible to the naked eye. We can't observe respiration or photosynthesis. But if we seal a plant in soil in a bottle for a semester and see that it survives so long as it gets plenty of light, we may be a little more convinced that plants both respire and photosynthesize. (Many students believe that photosynthesis is respiration in plants, or that animals respire while plants photosynthesize.) They seem to be able to imagine that in this situation, respiration produces enough carbon dioxide to support photosynthesis, and photosynthesis produces enough oxygen to support respiration. If Americans are ever to understand the issues of global warming, they need to grasp some of these basic ideas.

Group Work

When young children are learning a new skill, you often see them talking out loud but to themselves. Adults do the same thing but they tend to do their self-talk silently. Since talking is vital to understanding, students need to find their voice in the classroom. Through conversation, students gradually move from perceptual knowledge founded in images and other sensations to conceptual knowledge, based in the words that are inseparable from thought. This transition is the key to learning an academic subject such as biology.

Peer evaluations are useful for monitoring progress and evaluating each individual's contribution. In my class, students evaluate themselves as well as others in their group. The consensus on who contributed what is usually typically very high. This allows me to give a grade on a group project and adjust it up or down for individual students.

Higher Order Thinking

One of the teacher's roles is to prompt students to seize on central issues and think about deeper levels of interpretation (Resnick, 1983, 1987). In studying boiling water, for example, it is always useful to ask questions such as: Why do the bubbles form at the bottom of the container? (It is closest to the heat source.) What is in the bubbles? (On average, perhaps one student in the whole class will know the answer is water vapor; most think the bubbles contain hydrogen and oxygen.) Why do the bubbles rise to the top? (Most will know it goes into the air.) How does it change? (Most do not understand that the water molecule remains intact and simply separates from other water molecules.) In general, questions that are so basic that they are almost never asked in standard science classes provoke the greatest amount of rethinking on the part of the students.

It takes time and patience to ask the students to think about questions such as these and to guide their discussions toward appropriate answers. At the same time, the pressure to move on leads to the constant temptation to slip into teaching by telling. It is OK, even desirable and necessary, to slip into the telling mode sometimes. The important thing is to keep raising good questions. To sail through a lesson without asking deep questions is as serious an oversight as failing to prompt students to make predictions and generate explanations.

A good rule of thumb is that students spend at least as much time making sense of a lesson as they do in performing an experiment or activity, and often more. This is where the science illustrated by the activity is developed or lost. Reflection occurs as students talk to their peers within their groups, in the class discussions before and after lessons, as student groups work together to represent their knowledge with a thinking tool, which I'll describe shortly, and in the assessments.

In an inquiry class, assessment is an ongoing process. A teacher can learn a lot from the level of engagement and understanding of each student during the class discussions, by talking with student groups individually, and by listening in on conversations. It also helps to ask students to give presentations, write essays, search the web for related information, and perform other skills of a higher order.

Frequent opportunities for assessment and feedback are preferred. The important thing in testing is that it is not business as usual. Multiple-choice tests are not the preferred method. A teacher needs to concentrate on higher order thinking skills: synthesis, analysis, evaluation, application, performance.

Student-Centered Classes

For a teacher to share with students the responsibility for learning means an important shift of power. Students turn from recipients to actors. They ask questions that teachers had never before thought to answer. They bring in learning from other classes, from the TV and other media, and from their parents, and try to fit it all together.

Such an enterprise can seem like risk-taking for a teacher, in part because it is impossible to know all the answers when students are asking the questions. But the important thing to realize is that it is OK not to know all the answers. The teacher can prompt students to seek answers to their questions and share them with the class. The camaraderie and respect that grows between teacher and students in a class centered in students is a wonderful reward.

A THINKING TOOL TO SUPPORT CONSTRUCTION OF KNOWLEDGE

Many different tools are available to help us record our knowledge, analyze it, organize it, understand it, see it in new ways, and share it with others. Using such tools can make us smarter, more capable learners and thinkers. We are able to do things with such tools that we couldn't accomplish as easily or at all with our minds alone. This phenomenon has been described as distributed intelligence (Saloman, 1993; Pea, 1985; Pea & Gomez, 1992).

I will describe one such tool here: a thinking and knowledge analysis tool, the SemNet® software (Fisher et al., 1990; Fisher, 1991; Fisher & Kibby, 1996; Fisher, Wandersee, & Moody, in press). SemNet®, which has been used by third graders up through graduate students as well as by professionals, is an application for Macintosh computers that can create a map of ideas having many complex interconnections. It can serve as a tangible representation of the user's thinking that then supports the user's reflection, revision, and polishing of ideas. SemNet® provides a space in which groups of students can think together about their experiences and how to make sense of them. It provides a forum for interpretation and negotiation of ideas.

The *process* of constructing knowledge (net-building, mapping) generates much thought and discussion among students. This is where the major pay-off occurs (see, for example, Christianson & Fisher, submitted; Gorodetsky & Fisher, 1996). Students who construct semantic networks spend a lot more time on task thinking about biology than students who don't. The product (the semantic network) is also useful as a reference, a resource, and a record of what was thought in a given context at a given point in time.

Among many things that studying semantic networks constructed by students, teachers, and scientists has taught about how people think is that while there is no constrained set of relations that is useful for representing all biology knowledge, just three relations are used half the time. There is no consensus on the particular words used to describe the relations—people describe them in different ways (some alternatives are shown below), but the meaning remains similar. They are:

▶ Has part / is a part of (has component, contains)
▶ Has type / is a type of (has example, has class, set has member)
▶ Has characteristic / is a characteristic of (has attribute, has trait)

We have also become aware that students for whom English is a second language (ESL) have greater difficulty understanding the relations in biology than the concepts. Most relations are verb phrases, and these little words are simply more difficult to master than nouns, in part because their meanings vary with the context. This follows the pattern seen as children learn their first language: nouns before verbs. Being able to help ESL students to master essential relations is a significant benefit of using SemNet®, since the relations are used over and over again throughout biology. Figure 1 shows the corresponding frames of an English/Spanish SemNet®. SemNet® simplified the representation of complexly interlinked, ill-structured knowledge by showing one central concept at a time with all of its links to related concepts. Seeing words embedded within robust descriptions may be helpful to new second language learners, especially when the representations are accompanied by pictures.

**FIGURE 1. ENGLISH/SPANISH REPRESENTATION OF THE CENTRAL
CONCEPT, CONEJO (RABBIT), IN A FOOD WEB NET**

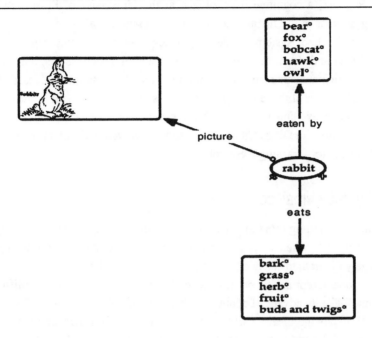

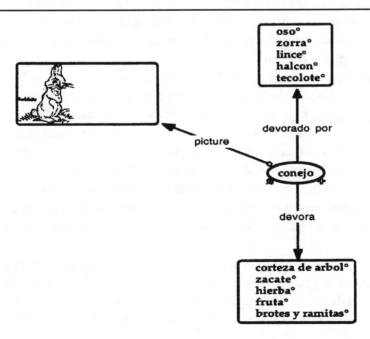

NESTED LEARNING COMMUNITIES

Lauren Resnick, a cognitive psychologist at the University of Pittsburgh and founding director of the Institute for Learning Research, has been working with several entire school systems to develop a new theory of school organization, everybody up and down the hierarchy of the school system is responsible for learning. Participants are responsible for continuous learning themselves, and for promoting learning of those in their care (that is, those below them in the hierarchy). This is an exciting vision, and we are delighted that Dr. Resnick is now working with the San Diego City Schools and is involved in the creation of a San Diego Institute for Learning Research.

WHAT IS THE EVIDENCE?

Motivation for finding more effective teaching and learning strategies in science began about 1980, when it was becoming strikingly clear that the American public is largely illiterate in science (Fisher & Lipson, 1980; National Science Board, 1983; National Commission on Excellence in Education, 1983). A scientifically and technologically advanced country with a scientifically illiterate public is at a serious disadvantage in the world. To this day, many of our decision makers in Washington, D.C. and in the states lack more than the most rudimentary understanding of the issues. The evidence for the need for change in the ways we teach and learn, as well as for the benefits of change, is so extensive that I can only pick a few examples to cite here.

Among the most interesting and persuasive comparative studies are those that A. Van Heuvelen published in 1991. He examined many dozens of physics courses being taught by high school and college teachers across the country. He found that students in courses using active instructional strategies based in cases or other research significantly and consistently outperformed students in traditional courses. D. Hestenes and I. Halloun, publishing in 1995, have developed the force concept inventory, a test being used by physics teachers across the country to determine how well their students are understanding mechanics. The general finding is that in traditional lecture courses, student understanding does not go very deep.

Another interesting comparative study comes from William Schmidt and the International Education Association's Third International Mathematics and Science Study (TIMSS, 1998; Schmidt, McKnight, & Raizen, 1996). It shows that while the United States has improved in some areas during the past twenty years, it is still far from being a top-performing country. Even more interesting are the comparisons among curricula. In top performing countries such as Japan and what was West Germany, a small number of science topics is taught each year and each topic is taught only once throughout the student's career. In contrast, in the American curriculum, sixty-five topics or so are taught a year per science course. And each topic is repeated again and again throughout the curriculum. Schmidt describes our curriculum as a mile wide and an inch deep. In other countries science books are small and focused, not intimidating. Students can carry the books in their pockets and know they are responsible for everything in them. In the United States, science books are like encyclopedias—too heavy to carry to school, too overwhelming to take seriously, too superficial to make a lot of sense, and so full of topics that any given teacher can only "cover" a fraction of them.

In 1983 and 1987 Lauren Resnick published interesting reviews that provide a cognitive science perspective on the evidence to support higher order thinking. Anna Sfard (in preparation) offers a review from the mathematics education perspective on the benefits of constructivist teaching. J.L. Lemke, whose study came out in 1993, as well as E.H. van Zee and J. Minstrell published in 1997, are among the researchers who are looking at the impact of language and reflective discourse on science learning. There are many wonderful qualitative studies on the benefits of the new teaching strategies such as that published by R. Driver, H. Asoko, J. Leach, E. Mortimer, and P. Scott in 1994; Driver, A. Squires, P. Rushworth, and V. Wood-Robinson the same year; J.R. Baird and I.J. Mitchell in 1986; and D. Brown and J. Clement in 1989. This is only a sliver of the research supporting the reform movement, but should be enough to get an interested reader started.

The evidence, in sum, is compelling both for the need and for the benefits of change in the ways we teach and learn science in the United States. Introducing significant change into our mammoth educational system is a major challenge that has engaged many branches of the government for the last quarter century. Inertia is the biggest problem, coupled with major pockets of resistance such as the conservative movement in California. We are still searching for an ideal instructional method that fits within the budget we are willing and able to allocate for education. In the meantime, an array of methods based in inquiry is being implemented and tested, and this seems like the most promising approach. So long as we remain focused on the effective features of instruction by inquiry such as *eliciting prior knowledge, prediction, engagement with a phenomenon, group work, higher order thinking and classes centered in students,* I believe, we will remain on the right track. To the extent that we are able to reorganize our classrooms, school systems, and textbooks to emphasize higher order learning, and we are able to effectively use computers and other tools to help our learners think smarter, we will succeed in advancing science learning.

REFERENCES

American Association for the Advancement of Science. 1993. *Benchmarks for science literacy*. New York: Oxford University Press.

American Association for the Advancement of Science. 1998. *Blueprints for reform*. New York: Oxford University Press.

Baird, J.R., and I.J. Mitchell. (Eds.) 1986. *Improving the quality of teaching and learning: An Australian case study—The Peel project*. Melbourne, Victoria: Monash University Printery.

Brown, D., and J. Clement. 1989. Overcoming misconceptions via analogical reasoning: Abstract transfer versus explanatory model construction. *Instructional Science* 18: 237-61.

Carnegie Commission on Science, Technology, and Government. 1991. *In the national interest: The federal government in the reform of K–12 math and science education*. New York: Carnegie Commission on Science, Technology, and Government.

Christianson, R.G., and K. M. Fisher. Submitted. Comparison of student learning about diffusion and osmosis in constructivist and traditional classrooms.

Driver, R., H. Asoko, J. Leach, E. Mortimer, and P. Scott. 1994. Constructing scientific knowledge in the classroom. *Educational Researcher* 23(7): 5-12.

Driver, R., A. Squires, P. Rushworth, and V. Wood-Robinson. 1994. *Making sense of secondary science: Research into children's ideas*. New York: Routledge.

Ebert-May, D., C. Brewer, and S. Allred. 1997 Innovation in large lectures—teaching for active learning. *BioScience* 47(9): 601-7

Ebert-May D., and J. Hodder. 1995. Faculty institutes for reforming science teaching through field stations. Final report for a project support by the National Science Foundation, Division of Undergraduate Education.

Fisher, K.M. 1991. SemNet: A tool for personal knowledge construction. In *Mindtools*, edited by D. Jonassen and P. Kommers. Heidelberg: Springer-Verlag.

Fisher, K.M. 1996. Biology lessons for prospective and practicing teachers. http://www.BiologyLessons.sdsu.edu/.

Fisher, K.M., J. Faletti, H.A. Patterson, R. Thornton, J. Lipson, and C. Spring. 1990. Computer-based concept mapping: SemNet software—a tool for describing knowledge networks. *Journal of College Science Teaching* 19(6): 347-52.

Fisher, K.M., and J. Kibby. (Eds.) 1996. *Knowledge acquisition, Organization and use in biology*. Heidelberg: Springer-Verlag.

Fisher, K.M., and J. Lipson. 1980. Crisis in education. Unpublished manuscript, Center for Research in Mathematics and Science Education, San Diego State University, San Diego, CA.

Fisher, K.M., J.H. Wandersee, and D. E. Moody. In preparation. *Mapping biology knowledge*. The Netherlands: Kluwer Academic Publishers.

Gorodetsky, J., and K.M. Fisher. 1996. Generating connections and learning in biology. In *Knowledge acquisition, organization and use in biology*, edited by K.M. Fisher and M. Kibby, 135-54. Heidelberg: Springer-Verlag.

Herreid, C.F. 1998. Case studies in science. http://ublib.buffalo.edu/libraries/projects/cases/case.html.

Hestenes, D., and I. Halloun. 1995. Interpreting the force concept inventory. *The Physics Teacher* 22: 502-506.

Keys, C.W. April, 1997. Perspectives on inquiry-oriented teaching practice: Clarification and conflict. Symposium presented at the Annual Meeting of the National Association for Research in Science Teaching, Chicago.

Lakoff, G. 1987. *Women, fire and dangerous things; What categories reveal about the mind.* Chicago: University of Chicago Press.

Lakoff, G., and M. Johnson. 1980. *Metaphors we live by.* Chicago: University of Chicago Press.

Lemke, J.L. 1993. *Talking science: Language, learning, and values.* Norwood, NJ: Ablex.

McNeal, A.P., and C. D'Avanzo. (Eds.) 1997. *Student active science: Models of innovation in college science teaching.* San Diego: Saunders College Publishing (Harcourt Brace).

National Commission on Excellence in Education. 1983. *A nation at risk: The imperative for educational reform.* Washington, DC: U.S. Government Printing Office.

National Research Council. 1996. *National science education standards.* Washington, DC: National Academy Press.

National Science Board on Pre-College Education in Mathematics, Science and Technology. 1983. *Educating Americans for the 21st century: A plan of action for improving mathematics, science, and technology education for all American elementary and secondary students so that their achievement is the best in the world by 1995.* Washington, DC: National Science Foundation. Publication No. CPCE-NSF-03

National Science Foundation. 1978. *The status of pre-college science, mathematics, and social studies educational practices in U.S. schools.*

Washington, DC: National Science Foundation. Publication #038-000-00383-6.

Palincsar, A.S., and A.L. Brown. 1984. Reciprocal teaching of comprehension-fostering and comprehension-monitoring activities. *Cognition and Instruction* 1(2): 117-75.

Pea, R.D. 1985. Beyond amplification: Using the computer to reorganize mental functioning. *Educational Psychologist* 20(4): 167-82.

Pea, R.D., and L. J. Gomez. 1992. Distributed multimedia learning environments: Why and how? *Interactive Learning Environments* 2(2): 73-109.

Pfundt, H., and R. Duit. 1994. *Bibliography: Students' alternative frameworks and science education*. Fourth Edition. Kiel, Germany: Institute fur die Padagogik der Naturwissenschaften (Institute for Science Education) at Kiel University

Postlethwait, J.H., and J. L. Hopson. 1995. *The nature of life*. Third Edition. San Francisco: McGraw-Hill.

Resnick, L.B. 1983. Mathematics and science learning: A new conception. *Science* 220: 477-78.

Resnick, L.B. 1987. *Education and learning to think*. Washington, DC: National Academy Press.

Rutherford, F. J., and A. Ahlgren. 1989. *Science for all Americans*. New York: Oxford University Press.

Saloman, G. 1993. *Distributed cognitions: Psychological and educational considerations*. New York: Cambridge University Press.

Schmidt, W.H., C.C. McKnight, and S.A. Raizen, with P.M. Jakwerth, G.A. Valverde, R.G. Wolfe, E.D. Britton, L.J. Bianchi, and R.T. Houang. 1996. *A splintered vision: An investigation of U.S. science and mathematics education*. Executive Summary. Michigan State

University: U.S. National Research Center for the Third International Mathematics and Science Study.

Schnepp, J., and the Science Media Group. 1994a. *Into thin air*. Video program produced at the Harvard/Smithsonian Center for Astrophysics. New York: Annenberg Foundation and the Corporation for Public Broadcasting.

Schnepp, J., and the Science Media Group. 1994b. *Decomposition*. Video program produced at the Harvard/Smithsonian Center for Astrophysics. New York: Annenberg Foundation and the Corporation for Public Broadcasting.

Sfard, A. March, 1998. Balancing the unbalanceable: The NCTM standards in the light of theories of learning mathematics. Paper presented at the Conference on the Foundations to NCTM Standards, Atlanta.

Third International Mathematics and Science Study. 1998. http://timss.enc.org/TIMSS/timss/achieve/index.htm. Washington, DC: U.S. Department of Education.

Van Heuvelen, A. 1991a. Learning to think like a physicist: A review of research-based instructional strategies. *American Journal of Physics* 59(10): 891-97.

Van Heuvelen, A. 1991b. Overview: Case study physics. *American Journal of Physics* 59(10): 898-907.

van Zee, E.H., and J. Minstrell, 1997. Reflective discourse: Developing shared understandings in a physics classroom. *International Journal of Science Education* 19(2): 209-28.

Wandersee, J.H., J.J. Mintzes, and J.D. Novak. 1994. Research on alternative conceptions in science. In *Handbook of Research in Science Teaching and Learning*, edited by D. Gabel, 177-210. Hillside, NJ: Earlbaum.

Part 3

What Issues Arise with Inquiry Learning and Teaching?

Instructional, Curricular, and Technological Supports for Inquiry in Science Classrooms[1]

Joseph Krajcik, Phyllis Blumenfeld,
Ron Marx, and Elliot Soloway

New approaches to science instruction feature inquiry on the part of students as essential for student learning (Lunetta, 1998; Roth, 1995). The assumption is that students need opportunities to find solutions to real problems by asking and refining questions, designing and conducting investigations, gathering and analyzing information and data, making interpretations, drawing conclusions, and reporting findings. In the spirit with recommendations by the American Association for the Advancement of Science (AAAS) (1993), the National Research Council (NRC) (1996) argues that "there needs to be a de-emphasis on didactic instruction focusing on memorizing decontextualized scientific facts, and there needs to be new emphasis placed on inquiry-based learning focusing on having students develop a deep understanding of science embedded in the everyday world."

Evidence indicates that engagement in inquiry can bring students to deeper understanding of science content and processes (e.g., Brown & Campione, 1994; Cognition and Technology Group at Vanderbilt, 1992; Metz, 1995). But our work (Krajcik et al., 1998), along with that of others (Brown & Campione, 1994; Linn, 1998; Roth, 1995), has demonstrated that the cognitive demands that inquiry places on learners require considerable support. Students need help to become knowledgeable about content, skilled in using inquiry strategies, proficient at using technological tools, productive in collaborating with others, competent in exercising self-regulation, and motivated to sustain careful and thoughtful work over time. Describing problems students encounter as they engage in inquiry and finding ways to ameliorate those problems have received considerable attention recently (Hmelo & Williams, 1998; McGilly, 1994; Blumenfeld et al., 1998). In this paper, we describe inquiry in more detail, discuss ways to aid students via instructional, curriculum, and technological supports, and then illustrate how these have been applied to specific phases on inquiry where students encounter difficulties.

WHAT IS INQUIRY AND WHY USE IT?

Broadly conceived, inquiry refers to the diverse ways in which scientists study the natural world and propose explanations based on the evidence derived from their work (NRC, 1996). Inquiry is not a linear process. Phases interact; preliminary findings, for instance, may result in a decision to revise the original question or to alter data collection procedures. Figure 1 shows a model of inquiry (Krajcik et al., 1998).

Renewed interest in inquiry comes from research that shows that students' understanding of scientific ideas and scientific process is limited, so that many who do well on tests cannot apply their knowledge outside the classroom. New approaches to instruction assign primary importance to the way in which students make sense of what they are learning, rather than to how teachers deliver information. The assumption of such constructivist programs (Fensham, Gunstone, & White, 1994) is that integrated and usable knowledge develops when learners create multiple representations of ideas and are engaged in activities that require them to use that knowledge. Inquiry promotes development, transformation, and representation of ideas and helps learners understand how knowledge is generated in different disciplines. The emphasis is on depth, not breadth. In addition, conversation with others is an important way for students to exchange information, explain, and clarify their ideas, consider others' ideas, and expand their understanding.

FIGURE 1: THE INVESTIGATIVE WEB

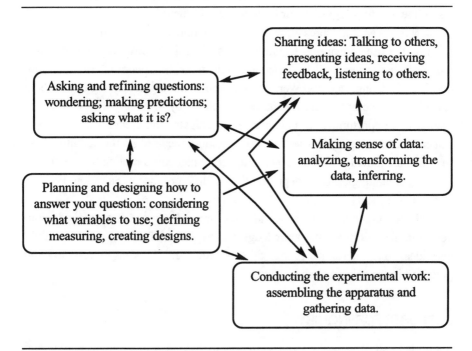

Among approaches that use inquiry, which include authentic tasks, artifacts, alternative assessments, technological tools, and collaboration, our work has students pursue investigations to answer a driving question related to their everyday experience. In finding answers to the question students learn scientific concepts, engage in scientific processes, and gain a better understanding of the discipline (Blumenfeld et al., 1991; Krajcik et al., 1994). Others rely on anchored experiences created on videotape (Sherwood, Petrosino, & Lin, 1998). Constructivist approaches often emphasize the production of artifacts such as multimedia documents, models, presentations, or demonstrations. Producing such artifacts allows students to apply information and represent knowledge in a variety of ways. Artifacts serve as a way for teachers to see how students are thinking, and for students to share their ideas and receive feedback which can be incorporated into revisions. Rather than rely on standardized tests, which have been criticized for concentrating on knowledge of isolated facts, the use of alternative assessments that have some value beyond the classroom is encouraged (for example, Newmann & Archibald, 1992; Perkins, 1992). These assessments, such as public performances, creation of museum exhibits or reports to local groups, require students to exhibit mastery of the discipline and to integrate prior knowledge with new.

Tool use is another core element of these approaches. Recent interest has centered on the use of learning technologies, such as interactive video technology (compact discs or videodisks), telecommunications, microcomputer-based laboratories, modeling, and the World Wide Web. Learning technologies can help learners solve complex and ambiguous problems by providing access to information and opportunities to collaborate, investigate, and create artifacts. Tools can extend and amplify learners' thinking because they reduce the cognitive load for students, moving from students to the computer some routine tasks like calculating, creating graphs, or depicting data in different forms (Salomon, Perkins, & Globerson, 1991).

Collaboration and conversation also are stressed. As students engage in conversation, they draw on the knowledge and expertise of others, reflect on their own ideas, and internalize modes of knowledge and thinking represented and practiced in the subject (Bruer, 1994).

SUPPORTS FOR INQUIRY

Supports to students in the inquiry process include instructional, curricular, and technological aids. These can work independently as well as in conjunction.

Instructional Supports

During inquiry the teacher serves as a learner as well as a guide or facilitator. Benchmark lessons (Krajcik, Czerniak, & Berger, 1999) introduce students to relevant content and skills before and during inquiry. The teacher helps students develop the thinking strategies used by experts, like heuristics for generating questions or interpreting data. They also help students become more metacognitive, attentive to planning, monitoring work, and evaluating their progress.

Scaffolding. For the teaching and learning situation, Collins, Brown, and Newman (1989) use the analogy of a cognitive apprenticeship. The teacher scaffolds instruction by breaking down tasks, using modeling and coaching to teach strategies for thinking, provides feedback that helps students diagnose their problems, and gradually releases responsibility to learners to perform these functions on their own. The emphasis is on helping students to become more like experts in their thinking about generating questions, using strategies to design inquiries to find solutions to questions, and evaluating the results of their efforts by mirroring heuristics and stratagems that experts have been found to use. These types of scaffolds can be used during each phase of inquiry.

Krajcik, Czerniak, and Berger (1999) offer these definitions and examples of scaffolds.

- Modeling is the process by which a more knowledgeable individual illustrates to the learner how to do or think about a task. For example, a teacher could demonstrate how to use the concept of "mean" to analyze data or how to read a pH meter. Many science processes can be modeled for students. Some of these include illustrating for students how to ask questions, plan and design investigations, or form conclusions.
- Coaching involves providing suggestions and asking questions to help the student improve knowledge or skills. For example, a teacher could make suggestions to a student about how to make more precise measurements when reading a spring scale. Other forms of coaching can include asking thought provoking questions (such as "How do your data support your conclusion?"), giving students sentence stems (for example, "My data supports my conclusion because…"), and supplying intellectual or cognitive prompts (such as asking students to write down predictions, give reasons, and elaborate answers).
- Sequencing is breaking down a larger task into step-by-step sub-tasks so a learner can focus on completing just one sub-task at a time rather than the entire task at once. For instance, the teacher might break down the process of investigations into various components and not allow the learner to proceed to the next step until completing the previous step. For example, the teacher could require the learner to complete a plan before moving on to building an apparatus.
- Reducing complexity involves hiding complex understandings or tasks until the learner has mastered simpler understandings or sub-tasks. The classical example here is helping a child learn to ride a bicycle by using training wheels. In science classrooms, this might mean a teacher uses an analogy to reduce the complexity of a concept. For instance, the teacher could compare DNA to the instructions for building a model airplane.
- Highlighting the critical features of a concept or task is another way a knowledgeable other can support the learning of another person. A teacher could point out to young students that animals called mammals

all have hair—hair is a key feature. As another example, in teaching a student how to focus a microscope, the teacher might point out that a basic step in focusing the object on the slide is to start always with the lowest powered lens first.

▶ Using visual tools can help students understand a concept or task (Hyerle, 1996; Parks & Black, 1992). Visual tools are pictorial prompts that help students understand their own thinking process. Visual tools also help make abstract ideas more concrete by organizing ideas or illustrating relationships. To develop understandings of kinetic molecular theory, students could use a computer simulation that represents the particle nature of matter.

Prediction, Observation, and Explanation Cycles. Another frequently used technique that promotes linking prior and new knowledge is the cycle of prediction, observation, and explanation (POE) (White & Gunstone, 1992). Students are asked to draw on prior knowledge to make predictions about what will occur during a demonstration, what they might find when searching for information, or what the results of an experiment might be. They can make individual predictions, share them with a group, discuss reasons for their predictions and come to some consensus about what might occur as they exchange ideas. Teachers can also make a prediction, thinking aloud to model how they draw on what they know to determine what might happen, or coach students as they consider possibilities, pointing out things to consider. Next students observe the phenomena and record their observations. As in the prediction phase, students can make individual observations, share them with a group, and come to some consensus about what they observed. Teachers provide scaffolds by demonstrating how they record and organize data, how students may reduce complexity of complicated observations by creating charts, or how teachers insure that the data are complete and correct. Finally, students compare their predictions with the observations and develop explanations about inconsistencies. Here again teachers can model how they generate explanations and consider whether the explanation is adequate, coach students as they develop explanations, and highlight various essential features to consider. The point is to emphasize justification and exchange of ideas.

Concept Maps. Concept maps are visual representations of the relationship among ideas. The maps are organized hierarchically with most important and inclusive concepts at the top. Related ideas are clustered around the overarching concepts and are linked. Maps are judged on the accuracy of the hierarchy and linking of ideas. Such mapping helps students organize, structure, and connect information and results in more meaningful understanding of ideas. Mapping also aids in the retrieval and facilitates the transfer of ideas.

Novak and Gowin in 1984 created the concept map as a tool to assess changes in learning. But concept mapping can be used as well to elicit student understandings prior to exploring a question. They are also an excellent way for students to track concepts that are being explored during inquiry. As the investigation continues, students make new concept maps that integrate new information with previous understandings. Comparing earlier and later versions of their maps can show students how their conceptual understanding is developing. Another useful approach is to have students compare their concepts maps with those of other students; to discuss and resolve differences. Conversational aids developed by Coleman (1998) to improve discussion in small groups of ideas and quality of explanations during the construction of concept maps are described in the section on collaboration.

Writing. Writing is another way to enhance student understanding. As students write they must retrieve, synthesize, and organize information. Production of a written document requires learners to clarify their thoughts (Santa & Havens, 1991) and also provides teachers with a window on student thinking. Keys (1994) shows that using collaborative writing guided by a series of prompts during the POE cycle improved the student's ability to draw conclusions, formulate models, and compose explanations that synthesized prior knowledge, observations, and other sources of information.

As in other fields, journals are one form of writing that is receiving considerable attention in science education (e. g., Audet, Hickman, & Dobrynina, 1996; Britsch & Shepardson, 1997). Bass and Baxter in 1998 studied how fifth-grade teachers made use of such notebooks, such as to write and draw, as a way to take notes, record observations, practice science skills, and summarize information and as a resource for teachers to monitor completion of tasks, assess understanding of specific concepts, and provide feedback. They found that rather than use the writing as a way for students to work through and demonstrate their understanding, teachers often controlled the writing, dictating what should be included and how. For instance, the authors determined that the notebooks were used to record procedure and findings rather than to explain conclusions and reasons for them and underrepresented the types of conversations students had about strengths and weaknesses of different methods and about the meaning of their data. The authors argue that for notebooks to fulfill their potential students should be asked to record their decisions and explain their thinking, keep track of their ideas and points made in conversations, and in general, center on substance rather than procedures.

Design of Curriculum Materials to Support Inquiry

In response to educational recommendations, many new curricular packages have been designed to promote inquiry, which also incorporate technology. Examples include Scientists in Action developed by the Cognition and Technology Group at Vanderbilt (Sherwood et al., 1998), Linn's (1998) Computers as Learning Partners, Songer's (1993) Kids as Global Scientists, and Edelson's (1998) WorldWatcher. Evidence indicates that these approaches help students achieve deeper understanding. Under the auspices of the Center for Learning Technologies in Urban Schools, The University of Michigan in collaboration with Detroit Public Schools is developing year-long curriculum materials for middle school students. The design principles[2] underlying these materials incorporate learning theory, our own experience, and the experience and suggestions of teachers and professional educators. The curriculum materials are organized into projects that promote understanding of science concepts via inquiry, are predicated on constructivist principles (Marx et al., 1998), and address the needs of diverse students (Atwater, 1994; Ladson-Billings, 1995). These design principles are the basis for curriculum materials.

1. **Standards.** Materials are designed to meet school district curriculum guidelines, which are congruent with AAAS' Benchmarks (1993) and the NRC's *Standards* (1996).
2. **Contextualization.** A "driving question" that draws on students' experiences gives context to scientific ideas and makes inquiry authentic. In the process of exploring answers to the question, students encounter and come to understand these scientific ideas. The question must therefore encompass rich scientific content so that it is intellectually worthwhile. It is chosen with the advice of teachers, parents, and content experts. For instance, students study chemistry by investigating the question, "Why does our air smell bad and is it bad for us?"
3. **Anchoring.** Students begin exploring the question via a common experience they can refer back to during the course of the project. These experiences, such as collecting and analyzing samples of water from the local river, help to anchor the question (CTGV, 1992).
4. **Inquiry.** In exploring the driving question, students raise questions, design investigations, apparatus, and procedures for collecting data, gather and analyze data, and present results (Krajcik et al., 1998).

5. **Technology tools.** Each project is designed to incorporate technology tools that are most appropriate for finding solutions to the question. In the project "Why do I need to wear a bicycle helmet?" students use motion probes to explore distance-time graphs and velocity-time graphs. In "What is the quality of water in our river?" students use Model-It to create relations among various factors affecting water quality and use probes to monitor the water.

6. **Collaboration.** Students work with peers and with others outside the classroom: community members, university students, and students in other schools.

7. **Community involvement.** The questions on which students work mesh science with issues such as environmental quality and disease that are likely to be of interest to community organizations and to the family. Community organizations serve as sources of information about local problems and local expertise with respect to the question under study, as sites where students access technology after school, and as audiences for student work.

8. **Scaffolding.** The curriculum materials are scaffolded within projects so that students are introduced to concepts and to science processes in a manner that guides their learning. The emphasis is on modeling of skills and heuristics, such as how to evaluate the quality of a question, how to create charts to keep track of data collection or how to represent data in different ways. The teacher, the structure of the tasks, and the technology provide scaffolds within a project. Teachers are given suggestions about when to model, coach, give feedback, and present benchmark lessons. Tasks are structured to reduce complexity so that certain concepts or inquiry strategies are highlighted and questions that foster thoughtfulness provided. Technology scaffolds students by providing multiple representations, hiding complexity, and ordering and guiding processes such as planning, building, and evaluating.

9. **Sequencing.** The curriculum materials also provide support for students by sequencing inquiry processes and scientific concepts. Early in the middle school years, projects are structured tightly to minimize complexity. Tasks are chosen to illustrate particular inquiry strategies and the enabling power of technologies. This tight structuring affords students the opportunity to experience all phases of the inquiry process and to build an understanding of how all the phases fit

together. Later students are given more responsibilities for designing and conducting investigations on their own. Projects also are sequenced so that throughout the middle school years concepts are revisited. As a result, students develop rich understandings of how ideas are related to one another and to different scientific phenomena.

10. **Development of artifacts.** Throughout the projects students create a variety of artifacts such as investigative designs, plans for data collection, laboratory notebooks and models that both represent and help build understandings. These artifacts serve as embedded assessments by the teacher. Also, they can be shared, critiqued, and revised to enhance understanding. Students also create final artifacts such as oral or written presentations or multimedia documents that are exchanged with classmates, and with others in the school and the community. Having students demonstrate their learning in ways that go beyond the classroom is one feature of authentic instruction (Newman & Welage, 1993). Detailed rubrics assist teachers in evaluating artifacts to gauge student understanding.

An Example Project

In a project on motion and force students explore the driving question "Why do I have to wear a bike helmet when I ride my bike?" During this eight-week unit, designed for eighth graders, students inquire into the physics of collision. It begins with a dramatic short videotape illustrating how bike accidents can result in brain injury. Then comes a series of demonstrations using an unprotected egg riding a cart, representing a student riding a bicycle, to illustrate the possible results of a collision. This demonstration is revisited periodically throughout the project and serves as the anchoring experience that students return to as they explore concepts of inertia, velocity, acceleration, force, and the relationships among them. It is also the focus of the final artifact; students design a helmet to protect the egg during a collision.

While exploring aspects of the driving question, students participate in several investigations supported by technology. They design experiments to examine the relationship between mass and inertia. Students study velocity and acceleration by collecting real time data using motion probes, which allows them to see these data immediately on the computer screen. They also learn how to read and interpret motion graphs. An investigation of gravity and mass involves collecting and interpreting information with the use of photogates to determine velocity. Students use motion probes again in designing and testing their egg helmets. These designs and the results of the testing are presented and discussed with the class. We also encourage the teacher to invite visitors from local safety and community organizations who attend the presentations.

Technology Design to Support Inquiry

Although inquiry can be done in classrooms without the aid of technology, learning technologies expand the range of questions that can be investigated, the types of information that can be collected, the kinds of data representations that can be displayed, and the products that that students can create to demonstrate their understandings. Such tools enable students to gather information about their questions on the World Wide Web, collect real time data using probes and other portable technologies, make models, graphs, and tables as a means of visually displaying data and quickly comparing different results, and illustrate their understandings in a variety of ways (for example, multimedia presentations). Students can work collaboratively with others in and outside the classroom. Examples of these tools are Knowledge Integration Environment developed at the University of California, Berkeley (Linn, 1998), and Worldwatcher, developed at Northwestern University (Edelson, 1998). The systems are integrated, designed to promote different fields of inquiry and allow for sharing. The tools are not specific to any particular content and they can be used to solve a range of problems and concepts. Because they can be used in different science classes across different grades, students can become proficient users of the tools and knowledgeable about the process of inquiry they support. The Investigators' Workshop is an example of learning technologies developed at the University of Michigan.

The Investigators' Workshop. The Investigators' Workshop,[3] is a suite of computational tools, based on learner-centered design (see next section), developed to enable sustained inquiry (Soloway & Krajcik, 1996). As described in Table 1, the tools support data collection, data visualization and analysis, dynamic modeling, planning, information gathering from the University of Michigan digital library and the Internet and web publishing (Jackson et al., 1996; Soloway, 1997; Soloway & Krajcik, 1996; Spitulnik et al., 1997). These tools have been revised several times in response to how students use them, the supports needed, and the types of artifacts produced by students.

TABLE 1. THE INVESTIGATORS' WORKSHOP: SCAFFOLDED TOOLS FOR LEARNERS ENGAGED IN SUSTAINED SCIENCE INQUIRY

NAME	FUNCTION	INQUIRY SUPPORT
Artemis	Supports on-line search and information gathering and evaluation using the UM Digital Library and Internet	Information gathering
Middle Years Digital Library Website	Provides support to students and teachers for carrying out on-line search activities to support inquiry	Information gathering and evaluating
Portable Computers and Probes	Microcomputer-based laboratories for portable computers; allows students to collect experimental data outside classroom by connecting various probes to the serial port	Data gathering
RiverBank	Water quality database tied to GREEN's field guide to water monitoring	Data sharing and storage
DataViz	Data visualization tool; supports students as they strive to see relationships and patterns in data both self-collected and gathered from on-line sources using visualization and analysis techniques	Data visualization and analysis

EChem	3-D molecular visualization tool; allows students to build three dimensional representations of complex molecules. Future developments will allow students to link structure of the molecules to physical and chemical properties.	Molecular visualization
Model-It	Modeling tool for dynamic systems; allows students to build, test, and evaluate dynamic qualitative models	Dynamic modeling
Web-It	HTML conversion tool to enable students to publish on the World Wide Web	Web publishing

The tools work together to support each phase of the inquiry process. When students are exploring the quality of a local stream, river, or lake, for instance, they can use probes (pH, temperature, dissolved oxygen, pressure) attached to portable technology and accompanying software to carry out collection of real-time data. The data can be uploaded to DataViz, where students can determine relationships and patterns using statistical analysis tools. By a variety of techniques such as digital photographs, graphs, and text, students visualize multiple types of data. In addition, student can link to representations and available animations to view the dynamic changes in different types of data.

Students can also use Artemis, an interface to the University of Michigan Digital Library. The digital library contains selected materials that are at appropriate levels of difficulty for middle and high school students. Supports in Artemis allow learners to sort, select, and organize documents and then easily return to them for further work.

Students can then use Model-It to build, test, and evaluate qualitative, dynamic models. For this, they import the functional relationships they developed in DataViz. They plan their models and create objects and factors. Using qualitative and quantitative representations, they then build relationship links among the factors. A graphical view of each relationship is also provided. For visualization of the values of factors, Model-It provides meters and graphs. As students test their models they can change the values and immediately see the effects. Finally, students can use Web-It to publish their results on the Web.

Learner-Centered Design. We have learned a considerable amount about creating technology tools to support learners at different levels of expertise. The principles of learner-centered design (LCD) (Soloway, Guzdial, & Hay, 1994) recognize that students differ in a number of ways from professionals who use computational software. Students do not initially know the content they are exploring and must be supported as they engage in inquiry. They also differ from one another in technological expertise and in how they prefer to learn; the tools must therefore be adapted to different levels of complexity and represent information and data in multiple ways. And professionals in a field are more likely than students to be committed to their work. Technology must so design computational activities as to entice students to concentrate on substantive cognitive issues and problem solving.

The incorporation of learning supports or scaffolding that addresses the differences between learners and professionals is central to LCD. Scaffolding software enables the learner to achieve goals or accomplish processes that would otherwise be too difficult to obtain. Our software guides learners through steps within phases of inquiry; when constructing a model, for example, students are reminded to make a plan of variables to include before building and testing. Scaffolds support learners' metacognitive activities, prompting them to test individual relationships or a sequence of relationships before evaluating the entire model. The software supports testing and debugging, allowing students to determine which relationships work and which may need revision. Intrinsic scaffolding software supports different levels of expertise; it makes the simplest function available to the novice learner, but allows learners to gain access to advanced features as their capability grows. At first, for example, students build qualitative models; but as they gain experience they select a weighting tool to make their relationships quantitatively more precise.

SUPPORTING PHASES OF INQUIRY

Asking Questions

Good questions are both feasible to investigate and scientifically worthwhile, so that in exploring the answers students learn important science concepts. For instance, one class was working on a project "Where does all our garbage go?" and conducting experiments on the effect of worms on decomposition. Groups of students were asked to generate sub-questions and to design their own investigations. One group of students asked the question: "Which types of material decompose and which don't in light or dark, with worms?" This question allows students to explore important content addressed by the project and design an experiment to answer their question. To answer this question students needed to set up an experimental situation to test the impact of light on a decomposition environment. Several studies, however, have shown that initially student questions do not reflect these criteria (Erickson & Lehrer, 1998; King, 1990; Scardamalia & Bereiter, 1992).

One temptation among students, which Krajcik et al. (1998) found in seventh graders, is to choose their topics out of personal interest and preference. That is legitimate, even desirable, except that the interest may remain unrelated to the scientific merit of the questions. In the class working on the project "Where does all our garbage go?" one group of students asked: "When there is water in one [decomposition] bottle and apple juice in another which decomposed faster?" because one of the students liked apple juice. Only one student raised concerns about the merit of the question. It was not until the teacher conversed with the students about what apple juice might represent in nature that they realized that the experiment was about acid rain.

Students also fell back on personal experience. In a project on water quality one group asked the question: "Does the water in various places around Ann Arbor have fecal coliform? If so, then to what degree?" because one of the students in the group observed professionals testing his pond for fecal coliform. The investigation allowed students both to answer the question itself and to explore important content related to the project. Initially, however, the students did not seize on the scientific merit of the issue and the teachers needed to help them see it. The challenge we face as educators is how to capitalize on the personal experience of students as well as their interest, yet at the same time help the learners explore powerful scientific ideas.

There are several possible explanations for why students generate these types of questions. Students may not have enough experience with inquiry to fashion meaningful questions that are also feasible to carry out. In fact, students may view their task as generating a question that is acceptable to the teacher and capable of accomplishment in the classroom rather than as a task of building knowledge.

Deliberately allocating time and effort to identifying problems or questions, actively seeking explanations for them, and trying to build knowledge are behaviors that take time to develop in students. Bereiter (1990) has pointed up the difficulty. Some educators will claim that the way to promote the asking of good questions lies in traditional didactic instruction. But several studies show that asking good questions comes from experience at it and from learning how fruitful a well-thought question can be. Content and inquiry should be closely intertwined: a student's ability to generate questions is fostered through active engagement. In fact, Roth and Roychoudhury (1993) have reported that over time, as students engage in experimentation, the questions they use to guide inquiry become more specific and include particular variables and relationships. Scaradamalia and Bereiter (1992) have found that the level of questions students pose improve as they explore a topic and gain more background knowledge. Krajcik et al. (1998) show that even early in their introduction to inquiry students are able to profit from suggestions of the teacher and also use what they have already learned.

Among strategies that have been shown to help students ask more productive questions is that proposed by Erikson and Lehrer (1998) of having the class as a whole develop critical standards for generating questions. Rather than imposing definitions of good questions, the teacher helps students to see that certain questions are less effective than others for building knowledge. For example, teachers might suggest that students should consider the breadth of the question, the ideas that need to be explored, the accuracy with which it reflects what students want to know, and the feasibility of finding information. Erikson and Lehrer demonstrate that over time students may develop critical standards that include interest along with potential for learning and for generating explanations and complex searches. The types of questions posed by the students the two investigated changed in accordance with these standards and came to require integration of multiple sources of information, selection of areas in which to search, and the generation of new questions. In attempting to answer these questions, moreover, students showed greater understanding and involvement.

To generate worthwhile questions, students must receive timely, informative, and critical feedback from teachers, peers, and others. They must also have opportunities to revise their questions and generate new ones. Whole class sessions can be committed to evaluating questions. To support such discussion and student self-assessment, teachers can provide skill templates as scaffolds. But in their concern to complete assigned work, students may be hesitant about devoting time to revising their work even when suggestions are offered. They may also fail to understand how feedback can be used for improving their questions. Teachers must therefore emphasize the importance of revision and allocate time for it.

Gathering Information: Inquiry on the World Wide Web

For seeking information to refine or answer a question, to design investigations, or to interpret findings, one increasingly popular source is the World Wide Web. Soloway, Krajcik, and their colleagues (Wallace et al., 1998; Hoffman & Wallace, 1997) have been exploring how students seek out and make use of information on the Web. Like others (Bereiter, 1990), they find that many students do not behave as intentional learners who aim to increase or build knowledge as they search. The task of seeking information about a question students often interpret as a matter of getting the right answer or good hits. Their background knowledge about their question, moreover, may be too limited to permit any keywords other than whatever is in their question. Failure to create synonyms may also be due to lack of appreciation for the significance of keywords or of understanding about how the technology works. Wallace et al. report that students do not have efficient ways of monitoring what they have accomplished; if the search continues over a period of time they lose their place, often repeating what they have done before or not making use of the information they have already gathered. Nor do they have sufficient strategies for reading or evaluating material online. Perhaps because students are used to looking up brief answers in textbooks or other reference sources, they may have neither skills nor inclination to criticize what they find (Mergendoller, 1996).

These findings point to some of the assistance students need if they are to conduct effective searches, and suggest that a major challenge to using digital information resources is to provide tools that enable the students to embed information seeking in a sustained process. Such tools must support both searching for simple facts and complex exploration of information when learners are trying to understand a multifaceted problem.

Classroom observations of students led to the creation of Artemis as an aid to students as they access and use digital information over the World Wide Web (Wallace et al., 1998). Artemis allows students to accomplish multiple tasks within a single computer environment. This keeps work from becoming fragmented and permits users to return to where they left off in prior sessions. The workspace provides for recording of searches and includes links to actual documents, helping students sustain the search process over time.

One feature, the question folders, supports students in thinking about and organizing the information they find in a way that most effectively addresses their query. They also help students to note what other questions or information they might usefully pursue. Students can store in question folders links to items they find interesting and can create multiple folders that reflect different components of the search or the refinements of an initial question. The folders allow flexibility in storing links and are available across many work sessions so that students can draw on what they have done before. Students can add or delete items or evaluate what they have found to date. Windows of results keep a live list of student searches so that they can see how they searched previously and what they have found. Observations indicate that students forget which queries they have submitted, and consequently repeat the same questions.

A broad feature includes a list of topics organized by domain. The topics present a hierarchy of terms that can be browsed or searched as the first step in creating a query. The feature is intended to help students generate keywords and draw upon prior knowledge as well as giving them a view of the structure of the content area they are exploring and providing them with alternative and productive ways to search.

Artemis is connected to the University of Michigan Digital Library, which contains a collection of relevant sites for middle grade students (see http://umdl.soe.umich.edu). The objective is to alleviate the frustrating problem students often have of getting numerous irrelevant hits in a Web search. Teachers and students also have the ability to criticize and recommend sites. Reading others' recommendations and their accompanying rationales, and contributing their own critiques can help students learn to evaluate information and sites, besides increasing motivation.

Designing and Planning Investigations

Krajcik et al. (1998) found that during their initial experience with designing investigations, middle school students created experimental and descriptive designs differing in complexity from using only one variable to comparing several levels of different variables. Small group discussions about designs primarily centered on feasibility and procedures. Many students, for instance, considered the types of samples to use, ways to create or obtain the samples, and the amount of material needed. They also discussed the need for controls. Some groups, however, had difficulty grasping how to create controlled environments, confounded variables, and misjudged the feasibility of what they were trying to do. The students' planning for data collection ranged from thoughtful to haphazard. Good plans included measurements related to the question, specifying what students were looking for as they measured or observed, and indicating the number of times measurements would be taken. They also detailed procedures to follow and included a way that data would be tracked and organized. Some planning problems students had were with qualitative techniques that involved drawing or writing a brief statement of what they observed. Generally, groups specified neither what they were looking for nor how the observations would help in answering their questions. Those who planned to use quantitative data often included measures with which they were familiar, like pH, but were not always appropriate for their purposes.

Students who have had little experience at gathering or interpreting data are not proficient at eliminating uninformative measures and do not realize the importance of being clear about their purpose. Thus students would benefit from having to explain how the measures selected relate to their questions, and be specific about what particular observations will indicate about the problem under study. Students need help in creating realistic plans. They sometimes overestimate how much they can accomplish within the time allocated and run out of time to complete the complicated set of things they have decided to do.

Although the students observed by Krajcik et al. presented their designs and plans to the class for suggestions, the presentations as well as many of the comments concerned specifics of the procedures rather than their purpose. Templates, which include questions about how the design and measures answer the question, could be used to guide the content of presentations and of the questioning that might accompany them. Such templates will also make it possible for peers to plan or students to engage in self-assessment of their plans. Allowing time for students to incorporate feedback, revise their plans, and emphasize the scientific merit of the inquiry is crucial in helping students create better design and plans.

Carrying Out Investigations

It is important for students to be thorough, systematic, and precise in collecting and describing data. Krajcik et al. (1998) report that many students were careful in setting up experimental procedures and constructing apparatus, following directions precisely. But though many were quite careful to create charts to help them track and organize data, they varied considerably in how systematic they were in following through on their plans. Some groups ran out of time because they did not share responsibility for data collection and consequently failed to complete necessary measurements. Others did not collect the measures they had planned but fixed on phenomena that attracted their attention, like bad smells or strange looking molds. They did not indicate how these phenomena were related to the scientific issues under study.

These problems illustrate students' need for help in managing complexity and time, and centering their attention on both the inquiry question and on the immediate needs entailed in collecting data. Students especially had trouble when plans called for numerous observations or complex procedures for data collection. Often students did not specify what they should record when collecting qualitative data or the reason for recording it. Doing both probably would help students focus on what data are important to gather. When first introduced to inquiry, moreover, students may not appreciate the need for consistency in measurement, following through on procedures, or maintaining experimental controls.

One solution for helping students handle complexity is to simplify and specify procedures so that learners can think about content. But even in more structured laboratory experiments, students tend to concentrate more on coping with procedures than on what they are supposed to learn (Hofstein & Lunetta, 1982). Students also need opportunities to think about how to make procedures precise and complete. Perhaps most challenging is that when students generate questions that are multifaceted, educators need to determine ways to help them reduce the complexity of the phenomena under investigation so that they can manage the work without compromising the integrity of the science and the authenticity of the problem.

Student interest in incidental observations must be for teachers an occasion for explaining how they bear on the larger scientific concepts under study. Krajcik et al. (1998) report that students were excited about what they were building and frequently asked about one another's work. They occasionally had animated conversations about unexpected changes that attracted their attention and tried to find out more about what they saw. But they rarely pursued the scientific implications of the observations or considered what they might suggest about other related questions or investigations. Surprise and curiosity can be an initial step in heightening interest in the work (Renninger, Hidi, & Krapp, 1992). The teacher should be ready to turn such moments into sustained cognitive engagement.

Microcomputer-based laboratories (MBL) can reduce complexity of data collection and representation that interfere with students' thinking about conceptual aspects of the inquiry. Students can use probes to monitor the temperature of a pond, to measure the pH of the pond, or to determine how dissolved oxygen varies at different locations in it. Although many of these measurements can be done with traditional laboratory equipment, using MBL has a number of advantages. Probes can save time. They are also more reliable instruments. They can display the results both graphically and numerically so that children can more easily interpret the findings. A major advantage is the simultaneous collection and graphing of data visually and numerically, which contributes to the students' understanding (Brasell, 1987; Mokros & Tinker, 1987). Another advantage is that probes allow students to do explorations not typically possible in the science classroom. For instance, using a temperature probe, students can continuously track the temperature of a decomposition column. They can answer questions like, "Does the temperature of the column change at night?" or set up an experiment in an aquarium to see whether dissolved oxygen changes with amount of light.

Analyzing and Interpreting Data

Krajcik et al. (1998) note that though the students had prepared charts and tables to record and organize their observations, they did not make graphs of quantitative data or create summary columns of qualitative data to facilitate comparisons across time and conditions even when teachers suggested that they do so. Perhaps because students did not look for patterns, their reports provided little interpretation. Instead, they tended to list findings with minimal elaboration, and failed to articulate how they had arrived at conclusions or to create logical arguments in which data were used to justify conclusions. Nor did they consistently draw upon background information to help interpret their findings. Penner et al. (1998) also tell of students who in creating models tended to describe data rather than identify principles that had produced them. Linn (1992) notes similar omissions; students using Computers as Learners Partners experienced difficulties using the results of laboratory experiments to explain everyday experience, and relied instead on intuitive ideas rather than the ideas under study.

One reason for this problem may be that students have had limited experience with these tasks and also may not know how to develop logical arguments to support their claims. Coleman (1998) reports on students who judged explanations as scientific if they included information that not everyone knew or could see with their own eyes, or information that needed to be discovered rather than looked up in a book. Palinscar, Anderson, and David (1993) have shown that students need considerable assistance in the process of argumentation, and have developed a program to help them systematically consider alternative explanations for phenomena and to provide justification for their reasoning.

Teachers, it is clear need to model how students might go about the process of data analysis and interpretation. But many teachers may not have experience with this phase of inquiry; they are more likely to have dealt with data from highly structured laboratory experiments where the findings are known ahead of time. Exciting new software tools are now available to support students in interpreting data. Model-It allows them to build models that illustrate qualitative and quantitative relationships among data. In developing models, students specify objects and articulate relationships. As they construct and revise models students examine patterns and trends in data and consider the match between the phenomenon under study and the model they have created. DataViz enables students to link various data types together; for instance, pictorial data can be connected to numeric data. Viewing data in these new ways may help enhance student understanding. Evidence from several studies indicates that students can build fairly complicated and accurate models that illustrate deep understanding of science concepts and their relationships (Stratford, Krajcik, & Soloway, 1998; Spitulnik et al., 1997).

ROLE OF METACOGNITION IN INQUIRY

Metacognition or self-regulation involves planning a course of action, monitoring progress to determine whether goals are being reached efficiently and effectively, and evaluating whether a change in plans or approaches is warranted. To stay organized students must track progress and stay focused on their problem, rather than getting confused or sidetracked by its elements. Doing so requires tactical and strategic metacognition. The tactical need is for regulation of cognition so that students can monitor their thinking as they work through details of tasks, such as who will be responsible for collecting data or using all the data collected in creating models. More strategically, students must think through what might seem to be disconnected elements to organize their efforts in service of the large purpose of the inquiry, such as how the data collection relates to the driving question, what data might be omitted if time runs out, or in what ways the model generated represents an answer to the driving question rather than just a representation of the data. Both types of thinking are needed for students to be systematic, accurate, and thorough and to make appropriate modifications or to adjust their strategies during inquiry; otherwise investigation runs the danger of becoming more like activity-based science where connections among activities and links to the overall issue or question often are not evident.

White and Frederiksen (1995) have explored ways to promote metacognition during inquiry. (See also their chapter in this collection.) They argue that metacognitive competence requires students to acquire the language to recognize and report on cognitive activities. ThinkerTools Curriculum provides this language through seeding conversations with categories chosen to represent metacognitive functions such as reflection on goals and on process. Examples of language for goals include formulating hypotheses and designing investigations. Labels are designed to help students recognize, monitor, and communicate about cognitive activities like generating multiple options or employing systematic strategies. Each of these can be further broken down into particular strategies and methods that are employed in each stage of the research process. For instance, being systematic means being careful, organized, and logical in planning and conducting work. When problems come up, helping students focus on their thought processes promotes effective decision-making. Students also use these criteria to do reflective assessments in which they evaluate their own and their classmates' research. White and Frederiksen (1995) have shown that engaging in reflective assessment enhances students' understanding of content and of science inquiry, and is especially beneficial for low achievers.

THE ROLE OF COLLABORATION IN INQUIRY

The aim of collaboration is to build communal knowledge through conversation. It can occur within a whole class, among groups in a class, and with people and groups outside the classroom. Collaboration helps students construct knowledge and introduces them to disciplinary language, values, and ways of knowing. As students converse, they must articulate their ideas clearly, and consider and draw on the expertise of others (Bruer, 1994). In collaborations of this sort, groups are not as highly structured as are small cooperative groups. The aim is to share ideas with the whole class or community in order to enhance knowledge of all individuals. In contrast to cooperative learning programs, there is little emphasis on assigning roles, group rewards, or group competitions (see Slavin, 1990; Blumenfeld et al., 1996).

Effective collaboration requires students to share ideas, take risks, disagree with others and listen to them, and generate and reconcile points of view. As they work together, they must manage substantive, procedural, and affective matters. Often they direct attention to the latter two concerns rather than to substantive issues (Anderson, Holland, & Palincsar, 1997). Students do not spontaneously or naturally generate highly efficient questions or explanations on their own and do not productively evaluate or respond to the explanations of others. Attempts to promote interactions include instructing in listening, resolving conflicts, and appreciating the skills and abilities of others (Webb & Palincsar, 1996).

One popular approach to facilitating student discussion and comprehension is reciprocal teaching (Palincsar & Brown, 1984; Rosenshine, Meister, & Chapman, 1996). As students read, teachers raise aloud a series of questions such as "What is likely to happen next? What do we know already?" Teachers model how more expert readers deal with text, eventually releasing responsibility to learners. Coleman (1998) has used conversational aids to improve the discussion of ideas and quality of explanations. During small group sessions, students used these prompts as they constructed concept maps. For instance the prompt, "Can you explain this in your own words?", encouraged students to construct explanations. Another prompt, "Can you explain why you think this answer is correct?", brought them to justify their responses. "Can you explain this using scientific information learned in class?" induced students to draw on background knowledge. Although they clearly benefited from such supports, the prompts did not always engender productive discussion; at times no one responded to the prompt, or the discussion went off track, or the discussion did not result in an explanation. Even when such conversational aids are employed, teachers need to monitor groups carefully.

Several tools are also available to promote collaboration and improve the quality of discourse (Pea & Gomez, 1992; Songer, 1998). The Computer Supported Intentional Learning Environment (CSILE), developed by Scardamalia and Bereiter (1991), advances understanding of subjects through electronic conversations centered on building a common database. CSILE has been used to support student investigations of topics such as endangered species, fossil fuels, evolution, and human biology. At the beginning of the year CSILE is empty; throughout the year it is populated by students' contributions of text and graphical notes. The electronic database includes four categories of notes or thinking types. These categories correspond to stages in the investigation process. The first two, "what I know" and "high-level questions," are used at the beginning of an investigation. They then use "plans" to generate a strategy for proceeding, and "new learning" to build a knowledge base. Their notes as they proceed in gathering information can be in text or graphical form. They can be commented on or added to by other students. The notes are structured to aid student conversation. They include opening phrases like, "One thing I don't understand is...." or "A reference I thought you might find useful is…." The purpose is to assist students in asking further questions, raising counter arguments, suggesting additional sources of information, or offering feedback. Ultimately students write reports that synthesize the results of the class's investigation.

CONCLUSION

Achieving understanding of science concepts and processes requires supports through each phase of inquiry. These can come from a variety of sources—from the teacher, from curriculum materials, from technology, and from peers within and outside the classroom.

In almost all cases, the supports described here are designed to encourage students to be thoughtful as they explore ideas through investigation. Inquiry demands tactical self-regulation with respect to particulars, such as generating a question or designing an investigation, examining whether the question will actually allow for exploration of the problem at hand or whether the design is adequate for generating useful information. It demands as well long-term self-discipline in setting goals and making modifications according to constraints like time and resources, or to discoveries that might result in revision of designs or procedures. Initially, students will lack experience at being such intentional learners and are likely to need a great deal of assistance.

Inquiry also poses challenges for teachers. It requires different types of instruction that give equal weight to promoting thinking and teaching content, different management routines as groups of students work on various aspects of phases of inquiry, different ways of promoting student interaction and conversation, and different ways of monitoring student progress and understanding. It also necessitates different uses of time—for scaffolding, feedback, discussion and sharing, revision, and reflection. Allocating time can be disconcerting for teachers who worry about curriculum coverage. Nevertheless, although teachers at first find difficult a pedagogy based on inquiry, they report considerable satisfaction in seeing students motivated to learn, becoming proficient at asking questions and devising ways of answering them, and demonstrating deep understanding of scientific concepts. Sharing experiences and continued exploration of techniques is essential to meeting the recommendation of the National Research Council (1996) that inquiry become a predominant mode of instruction.

ENDNOTES

1. The authors would like to thank Ann Rivet from the University of Michigan for her helpful editorial comments.
2. These principles are being instantiated into the curriculum by a team including Margaret Roy, Jon Singer, and Becky Schneider from the University of Michigan and Karen Amati from the Detroit Public Schools.
3. The development of the Investigators' Workshop has been supported by grants from the National Science Foundation (REC 9554205 and 955719).

REFERENCES

American Association for the Advancement of Science. 1993. *Benchmarks for science literacy*. New York: Oxford University Press.

Anderson, C.W., J.D. Holland, and A.S. Palincsar. 1997. Canonical and sociocultural approaches to research and reform in science education: The story of Juan and his group. *Elementary School Journal* 97: 360-383.

Atwater, M.M. 1994. Research on cultural diversity in the classroom. In *Handbook of research on science teaching and learning*, edited by D.L. Gabel, 558-576. New York: MacMillan.

Audet, R.H., P. Hickman, and G. Dobrynina. 1996. Learning logs: Classroom practice enhancing student sense making. *Journal of Research in Science Teaching* 33: 205-222.

Bass, K.J., and G.P. Baxter. 1998. Writing in science...for what purpose? An analysis of notebooks in hands-on science classrooms. Paper presented at the annual meeting of the American Education Research Association, San Diego.

Bereiter, C. 1990. Aspects of an educational learning theory. *Review of Educational Research* 60: 603-624.

Blumenfeld, P.C., E. Soloway, R. Mark, J. Krajcik, M. Guzidal, and A. Palincsar. 1991. Motivating project-based learning: Sustaining the doing, supporting the learning. *Educational Psychologist* 26: 369-398.

Blumenfeld, P.C., R.W. Marx, J.S. Krajcik, and E. Soloway. 1996. Learning with peers: From small group cooperation to collaborative communities. *Educational Researcher* 25(8): 37-40.

Blumenfeld, P.C., R.W. Marx, H. Patrick, and J.S. Krajcik. 1998. Teaching for understanding. In *International handbook of teachers and teaching,* Vol. 2, edited by B.J. Biddle, T.L. Good, and I.F. Goodson, 819-878. Dordrecht, The Netherlands: Kluwer.

Brown, A.L., and J.C. Campione. 1994. Guided discovery in a community of learners. In *Classroom lessons: Integrating cognitive theory and classroom practice*, edited by K. McGilly, 229-270. Cambridge, MA: MIT Press/Bradford Books.

Britsch, S.J., and D.P. Shepardson. 1997. Children's science journals: Tools for teaching, learning and assessing. *Science and Children* 34: 12-17.

Brasell, H. 1987. The effect of real time laboratory graphing on learning graphic representation of distance and velocity. *Journal of Research and Science Teaching* 24(4): 385-395.

Bruer, J. 1994. Classroom problems, school culture, and cognitive research. In *Classroom lessons: Integrating cognitive theory and classroom practice*, edited by K. McGilly, 273-290. Cambridge, MA: MIT Press.

Cognition and Technology Group at Vanderbilt. 1992. The Jasper series as an example of anchored instruction: Theory, program description, and assessment data. *Educational Psychologist* 27: 291-315.

Coleman, E.B. 1998. Using explanatory knowledge during collaborative problem solving in science. *The Journal of the Learning Sciences* 7 (3&4): 387-428.

Collins, A., J.S. Brown, and S.G. Newman. 1989. Cognitive apprenticeship: Teaching the craft of reading, writing, and mathematics. In *Knowing, learning, and instruction: Essays in honor of Robert Glaser*, edited by L.B. Resnick, 453-494. Hillsdale, NJ: Erlbaum.

Edelson, D. 1998. Realizing authentic science learning through the adaptation of scientific practice. In *International handbook of science education*, edited by B.J. Fraser and K.G. Tobin, 317-331. Dordrecht, The Netherlands: Kluwer.

Erickson, J., and R. Lehrer. 1998. The evolution of critical standards as students design hypermedia documents. *The Journal of Learning Science* 7 (3&4): 351-389.

Fensham, P., R. Gunstone, and R. White. 1994. *The content of science: A constructivist approach to its teaching and learning*. London: Falmer Press.

Hmelo, C.E., and S.M. Williams. 1998. Learning through problem solving. Special Issue. *The Journal of Learning Science* 7(3&4).

Hoffman, J., and R. Wallace. 1997. Structuring on-line curriculum for the science classroom: Design, methodology, and implementation. Paper presented at symposium, Using online digital resources to support sustained inquiry learing in K-12 science. Annual meeting of the American Educational Research Association, Chicago.

Hofstein, A., and V.N. Lunetta. 1982. The role of the laboratory in science teaching: Neglected aspects of research. *Review of Educational Research* 52: 201-217.

Hyerle, D. 1996. *Visual tools for constructive knowledge*. Alexandria, VA: Association for Supervision and Curriculum Development.

Jackson, S., S. Stratford, J. Krajcik, and E. Soloway. 1996. Making system dynamics modeling accessible to pre-college science students. *Interactive Learning Environments* 4: 233-257.

Jackson, S., J. Krajcik, and E. Soloway. 1999. Model-It: A design retrospective. In *Advanced designs for the technologies of learning: Innovations in science and mathematics education,* edited by M. Jacobson and R. Kozma, pgs. Hillsdale, NJ: Erlbaum.

Keys, C.W. 1994. The development of scientific reasoning skills in conjunction with collaborative writing assignments: An interpretive study of six ninth-grade students. *Journal of Research in Science Teaching* 31: 1003-1022.

King, A. 1990. Enhancing peer interaction and learning in the classroom through reciprocal questioning. *American Educational Research Journal* 27: 664-687.

Krajcik, J.S., C.M. Czerniak, and C. Berger. 1999. *Teaching children science: A project-based approach.* Boston: McGraw-Hill.

Krajcik, J., P. Blumenfeld, R.W. Marx, K.M. Bass, J. Fredericks, and E. Soloway. 1998. Middle school students initial attempts at inquiry in project-based science classrooms. *The Journal of Learning Sciences* 7 (3&4): 313-350.

Krajcik, J.S., P.C. Blumenfeld, R.W. Marx, and E. Soloway. 1994. A collaborative model for helping teachers learning project-based instruction. *Elementary School Journal* 94: 483-497.

Ladson-Billings, G. 1995. Toward a theory of culturally relevant pedagogy. *American Educational Research Journal* 32: 483-491.

Linn, M. 1992. The computer as learning partner: Can computer tools teach science? In *This year in school science 1991: Technology for teaching and learning,* edited by K. Sheingold, L.G. Roberts, and S.M. Malcom, 31-69. Washington, DC: American Association for the Advancement of Science.

Linn, M.C. 1998. The impact of technology on science instruction: Historical trends and current opportunities. In *International handbook of science education,* edited by K. Tobin and B.J. Fraser, 265-294. Dordrecht, The Netherlands: Kluwer.

Lunetta, V.N. 1998. The school science laboratory: Historical perspectives and contexts for contemporary teaching. In *International handbook of science education*, edited by K. Tobin and B.J. Fraser, 249-264. Dordrecht, The Netherlands: Kluwer.

Marx, R.W., P.C. Blumenfeld, J.S. Krajcik, and E. Soloway. 1998. New technologies for teacher professional development. *Teaching and Teacher Education* 14: 33-52.

McGilly, L. 1994. *Classroom lessons: Integrating cognitive theory and classroom practice*. Cambridge, MA: MIT Press.

Mergendoller, J.R. 1996. Moving from technological possibility to richer student learning: Revitalized infrastructure and reconstructed pedagogy. *Educational Researcher* 25: 43-46.

Metz, K.E. 1995. Reassessment of developmental constraints on children's science instruction. *Review of Educational Research* 65: 93-128.

Mokros, J.R., and R.F. Tinker. 1987. The impact of microcomputer-based labs on children's ability to interpret graphs. *Journal of Research in Science Teaching* 24: 369-383.

National Research Council. 1996. *National science education standards*. Washington, DC: National Academy Press.

Newmann, F.M., and G.G. Wehlage. 1993. Standards for authentic instruction. *Educational Leadership* 50:8-12.

Newmann, F.M., and D.A. Archibald. 1992. Approaches to assessing academic achievement. In *Toward a new science of educational testing and assessment*, edited by H. Berlak, F.M. Newmann, E. Adams, D.A. Archibald, T. Burgess, J. Raven, and T.A. Romberg, 71-83. Albany, NY: SUNY Press.

Novak, J.D., and D.B. Gowin. 1984. *Learning how to learn*. Cambridge, England: Cambridge University Press.

Palincsar, A., and A. Brown. 1984. Reciprocal teaching of comprehension fostering and comprehension monitoring activities. *Cognition and Instruction* 1: 117-175.

Palincsar, A.S., C. Anderson, and Y.M. David. 1993. Pursuing scientific literacy in the middle grades through collaborative problem solving. *The Elementary School Journal* 93: 643-685.

Parks, S., and H. Black. 1992. *Organizing thinking: Graphic organizers*. Pacific Grove, CA: Critical Thinking Press and Software.

Pea, R., and L. Gomez. 1992. Distributed multimedia learning environments: Why and how? *Interactive Learning Environments* 2 (2): 73-109.

Penner, D., R. Lehrer, and L. Schauble. 1998. From physical models to biomechanics: A design-based modeling approach. *The Journal of Learning Sciences* 7 (3&4): 429-450.

Perkins, D. 1992. *Smart schools: From training memories to educating minds*. New York: Free Press.

Renninger, K.A., S. Hidi, and A. Krapp. 1992. *The role of interest in learning and developments*. Hillsdale, NJ: Erlbaum.

Rosenshine, B., C. Meister, and S. Chapman. 1996. Teaching students to generate questions: A review of the intervention studies. *Review of Educational Research* 66: 181-221.

Roth, W.M. 1995. *Authentic school science*. Dordrecht, The Netherlands: Kluwer.

Roth, W.M., and A. Roychoudhury. 1993. The development of science process skills in authentic contexts. *Journal of Research in Science Teaching* 30: 127-152.

Salomon, G., D.M. Perkins, T. Globerson. 1991. Partners in cognition: Extending human intelligence with intelligent technologies. *Educational Researcher* 20: 2-9.

Santa, C.M., and L.T. Havens. 1991. Learning through writing. In *Science learning, processes and applications*, edited by C.M. Santa and D.E. Alvermann, 122-133. Newark, DE: International Reading Association.

Scardamalia, M., and C. Bereiter. 1991. Higher levels of agency for children in knowledge building: A challenge for the design of new knowledge media. *The Journal of the Learning Sciences* 1: 37-68.

Scardamalia, M., and C. Bereiter. 1992. Text-based and knowledge-based questioning by children. *Cognition and Instruction* 9: 177-199.

Sherwood, R., A. Petrosino, X.D. Lin, and the Cognition and Technology Group at Vanderbilt. 1998. Problem based macro contexts in science instruction: Design issues and applications. In *International handbook of science education*, edited by B.J. Fraser and K. Tobin, 349-362. Dordrecht, The Netherlands: Kluwer.

Singer, J., J. Krajcik, and R. Mark. 1998. Development of extended inquiry projects: A collaborative partnership with practitioners. Paper presented at the annual meeting of the American Educational Research Association, San Diego.

Slavin, R.E. 1990. *Cooperative learning: Theory, research, and practice*. Englewood Cliffs, NJ: Prentice Hall.

Soloway, E. 1997. Using on-line digital resources to support sustained inquiry learning in K-12 science. Symposium presented at the annual meeting of the American Educational Research Association, Chicago.

Soloway, E., M. Guzdial, and K.E. Hay. 1994. Learner-centered design: The challenge for human computer interaction in the 21st century. *Interactions* 1: 36-48.

Soloway, E., and J.S. Krajcik. 1996. *The investigator's workshop: Supporting authentic science inquiry activities*. Washington, DC: National Science Foundation.

Songer, N.B. 1993. Learning science with a child-focused resource: A case study of kids as global scientists. In *Proceedings of the Fifteenth Annual Meeting of the Cognitive Science Society*. Hillsdale, NJ: Erlbaum.

Songer, N.B. 1998. Can technology bring students closer to science? In *International handbook of science education*, edited by B.J. Fraser and K.G. Tobin, 333-347. Dordrecht, The Netherlands: Kluwer.

Spitulnik, M.W., S. Stratford, J. Karjcik, and E. Soloway. 1998. Using technology to support student's artifact construction in science. In *International handbook of science education*, edited by B.J. Fraser and K.G. Tobin, 363-381. Dordrecht, The Netherlands: Kluwer.

Stratford, S.J., J. Krajcik, and E. Soloway. 1998. Secondary students' dynamic modeling processes: Analyzing, reasoning about, synthesizing, and testing models of stream ecosystems. *Journal of Science Education and Technology* 7(3): 215-234.

Wallace, R., E. Soloway, J. Krajcik, N. Bos, J. Hoffman, H.E. Hunter, D. Kiskis, E. Klann, G. Peters, D. Richardson, and O. Ronen. 1998. ARTEMIS: learner-centered design of an information seeking environment for K-12 education. In *Human Factors in Computing Systems, CHI-98 Conference Proceedings,* 195-202.

Webb, N.M., and A.S. Palincsar. 1996. Group processes in the classroom. In *Handbook of research in educational psychology*, edited by D. Berliner and R. Calfee, 841-873. New York: Simon & Schuster.

White, B.Y., and J.R. Frederiksen. 1995. *The ThinkerTools inquiry project: Making scientific inquiry accessible to students and teachers*. Berkeley: University of California, School of Education. Causal Models Research Group Report No. 95-02.

White, R., and R. Gunstone. 1992. *Probing understanding*. London: Falmer Press.

Constructing Scientific Models in Middle School

Karen Amati

Located in Detroit, Michigan, Lessenger Middle School is a Title 1 school with about 850 students, most of them African American and from single parent homes. When Dr. Joseph Krajcik, a professor from the University of Michigan, suggested we pilot a scientific modeling program, my initial response was "No way!" Scientific models are built upon statistics, and our students have a difficult time with fundamental word problems. How could children model their understanding of science? How could constructing a model help them better understand science? Isn't it necessary to understand thoroughly a topic before constructing a model? When we examined models constructed by children, the program spoke for itself. Because the models used verbal quantifiers such as "some," "a lot" and "none," I realized that our children could do these.

Our first scientific use of Model-It was as a culminating assessment of the understanding of weather our students had gained in Kids as Global Scientists. Dr. Nancy Songer, the project investigator, and the co-director Dr. Perry Samson along with their staff analyzed our students' pretests and posttests and found that they excelled over participants from schools across the country. The curriculum materials were essentially the same for all of the groups. The difference was the use of Model-It. This was impressive! We are not accustomed to seeing our students excel over children from school systems that experience much higher test scores than we.

Students and teachers view Model-It as a tool to make sense of the many pieces of information acquired in the process of exploring a topic. Although concept maps also help students make meaning, Model-It takes it one step further allowing students to test their ideas. The *National Science Education Standards* (National Research Council, 1996) stress that students need to make meaning rather than just memorize information. As the science and technology resource teacher, I have helped teachers successfully use Model-It in grades six through eight, and I was thrilled to find that students remembered how to use the program the following year.

WHAT IS MODEL-IT?

Model-It is a computer program designed and developed by Shari Jackson when she was a graduate student in Computer Science at the University of Michigan and by two professors from the University of Michigan, Elliot Soloway, in Educational Technology and Joseph Krajcik in Science Education. The program allows students to construct computer models of complex, dynamic science systems using graphics and verbal descriptions. Mathematical quantification is possible. At first glance, the scheme seems to be an elaborate concept map. Concept maps, however, are static; students have no way to verify relationships. Model-It is dynamic. Students can easily test, analyze, and revise their models, and that helps students make meaning of the phenomena they are modeling. Another advantage of the program is that detailed descriptions must be written for each relationship. The descriptions are much more effective than the word or two used in a concept map. In the paragraphs here, each section of Model-It will be described and illustrated by screen shots of student models.

WHAT DOES A MODEL LOOK LIKE?

This is the beginning of a model constructed by a sixth-grade group called "Clouds 'R Us." Each group of students became topic specialists. Prior to beginning the weather unit, they studied the phases of matter and the effect of heat on evaporation. During the first part of the weather unit, they became experts on clouds. Following the arrows, this model shows that as the amount of heat from the sun increases, the amount of water vapor increases. Another arrow indicates that as the amount of water vapor increases, the amount of cloud cover increases. The bottom arrow shows that as the amount of dust in the air increases, so too does the amount of cloud cover. The model demonstrates that students were able to use various concepts to construct complex relationships of cause and effect concerning one facet of weather. The model graphic does not reveal the depth of understanding required to construct a model. Some students, for example, may consider the model to be complete without the dust factor. The teacher may ask them whether they think clouds form whenever there is water vapor. Through continued questioning the students will recall that dust particles are needed. When they define the relationship, they cannot simply state that as dust particles increase the amount of cloud cover increases. They must explain that the dust particles act as a nucleus for gathering moisture to form a droplet of water.

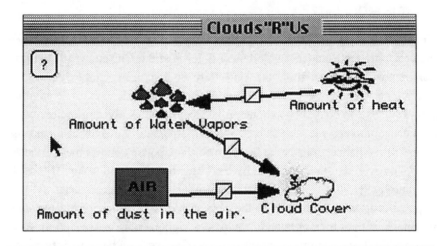

WHAT IS THE PROCESS OF CREATING A MODEL?

Defining Objects and Factors

First the students, either individually or in groups, draw a picture of those things that affect weather. Next, the teacher helps them consolidate into a manageable number of objects the things in their drawings. In the model above, the objects are air, weather, sun, and water. After the objects are identified, the students choose an object and a measurable factor. These students chose the sun as the object and the "Amount of heat" as a factor of the effect of the sun on weather. Notice that text rather than numbers quantifies the factor. The students are free to be creative when they discuss the science concepts and choose the best words for describing the maximum, medium, and lowest amount of the factor. The descriptors are entered in the three text range boxes. In this case, students chose A lot, Some and None. The dialog box calls for a description of the factor. We feel that class time should go to discussing the concepts rather than typing, which most of our students do only slowly. The descriptions are homework on an object/ factor worksheet that must be completed before the students go to the computer. The two students working on a model compare their papers and choose the best parts of each one. Each student is required to enter factors into the model. The students will describe two or three factors and then move on to relationships or create many factors before linking them into relationships.

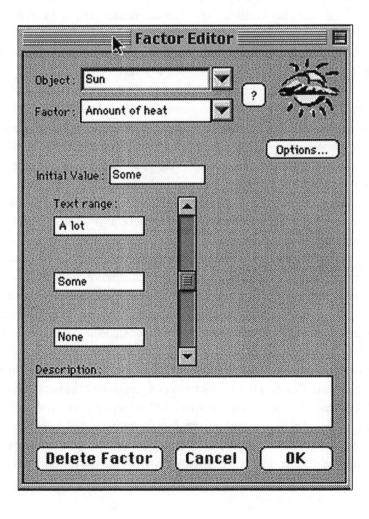

Building Relationships

The students return to their drawing of the things that affect weather. They are asked to look for pairs of objects that form relationships of cause and effect. This is easier than it might seem because it is a common task in our Language Arts classes. The students draw an arrow from the cause to the effect and explain why the relationship exists. Each student constructs some relationships on paper as homework. The computer partners discuss them and choose some from each paper. The dialog between the students, as they decide which are the best relationships, is a valuable opportunity for them to internalize the concepts. They must evaluate each others' information and offer evidence for their choices.

Anyone familiar with the operation of the program can click on the appropriate boxes and construct a relationship. The assessment criteria for an acceptable model require a detailed explanation for each. It is supposed to begin with a statement of the relationship followed by a "because" statement. This beginning model has only one acceptable explanation of a relationship.

The first relationship should explain why heat energy increases evaporation. This group would be encouraged to review its notes or one of its information sources to complete the explanation. A better relationship description would be that as the amount of heat increases the amount of evaporation increases because heat causes the liquid molecules to move faster and farther apart to become a gas.

The second relationship, that between water vapor and cloud formation, has a good explanation for sixth-grade students. The explanation would have been better if they had indicated that increasing moisture was responsible for the increasing amount of cloud cover.

The last explanation demonstrates a common error. Many of our students feel that a restatement of the relationship is an explanation or a justification for the relationship. Personal conferences with the students reveal that they do not understand the difference between a restatement of a question and an explanation. One of the greatest strengths of Model-It is that it forces students to complete a detailed explanation statement. Most of the students find this extremely difficult. Students who have succeeded in school because of their ability to memorize become especially frustrated. Conscientious students will agonize over the explanations, and pester their teachers with frequent requests for help with a relationship. One group nagged us for more than a week concerning the formation of hail. The students constantly quoted the memorized definition, but could not fit it into the relationship using the weather objects. They wanted a special object for hail. In our use of Model-It, students are required to state the relationship, add the word "because," and end with an explanation. Usually, they cannot use memorized definitions or bits of information. For example, the last relationship could say, "As the amount of dust in the air increases, the amount of clouds increase because dust particles make it easier for water vapor to condense into water droplets to make clouds." If students did not learn anything else during the unit, they learned that there is a world of difference between a memorized definition and understanding.

As teachers move from computer to computer, they observe only a sample of each group's relationships. How many are incorrect? Are we allowing students to learn incorrect information? Many teachers express these concerns. Requiring as homework a paper version of the relationships allows the teacher to assess the quality of the contributions by each student and to target groups for increased individual attention. There are instructional advantages in that every student is personally engaged in relationship construction because students are not allowed to work at the computer until they do the paper version. Then in their workgroups they compare their relationships and either choose the best one or combine their ideas. This further compels them to think critically. The paper version is supportive of the teachers who are concerned that students will build incorrect relationships. Errors are valuable opportunities for reflection and learning. When groups having difficulties are targeted, teachers can have conferences with the students, and help them develop true relationships. For the teacher to mark the relationship wrong with the traditional red pen and lower the grade accordingly would be counterproductive. That is not the role of the facilitator. The evaluation of the content of the relationships should occur when the model is completed and the students have had adequate time to test their relationships.

Testing

The final phase of the modeling process is testing the relationships. Each relationship may be tested after it is constructed, or a number of relationships can be constructed and then tested.

In the example, the test meters show the initial midway settings using the quantifiers "Some" and "Medium" as chosen by the students. Note that the meters differ; one meter has a slider. The meter with the slider is the independent variable, the cause of change. The other meter is the dependent variable, the effect. Students are encouraged to test one relationship at a time to simplify troubleshooting should the relationship not work properly. Once the "Run" button is activated, students move the slider up and down changing the amount of the sun's heat and noting the effect on the amount of water vapor.

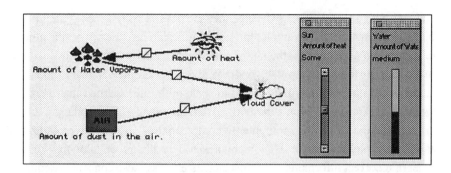

The slider first moved to the "A lot" position, causing the other meter to move to the "high" position. This indicates that if the sun's heat is intense, the amount of water vapor will be high. The underlying assumption is that there is an ample supply of water.

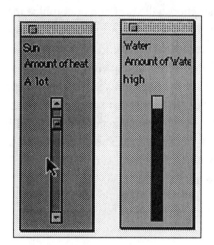

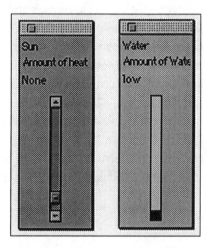

Next, the slider moved toward the bottom of the meter, causing the water vapor meter to decrease.

There are at least three different approaches to teaching the testing of relationships, and each requires its own level of sophistication. The simplest is to have the students open the meters, move them around and see whether the meters behave as they expected. A second level emphasizes the experimental nature of testing by having the students form hypotheses and test them with the meters. The most difficult approach identifies the independent and dependent variables. Middle school science standards expect students to be able to differentiate between variable types, and all but the brightest students find it very

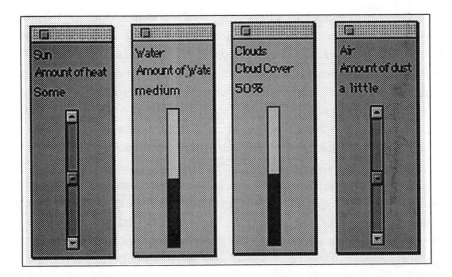

difficult. Model-It is an excellent tool for teaching the difference. The independent variable is the meter that can be changed freely, whereas the dependent variable does not have a slider and is moved by another meter. Repeated sessions of model building and testing allow multiple opportunities for students to practice identifying and explaining the difference between dependent and independent variables.

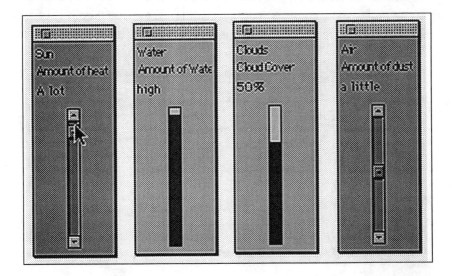

After pairs of relationships have been tested, the students increase the number of meters and may test all of them simultaneously. We begin with all of the meters set to their beginning levels. Next, the amount of heat is increased to the maximum, which causes the amount of water vapor to become high and the amount of cloud cover to increase. The students were not sophisticated enough to enter a numerical scale. The 50% is treated as a word rather than a number. Students read it as more than 50%. Many students expect the cloud cover to become 100% because the heat and water vapor are at their maximum values. Upon reconsideration and following a discussion led by a teacher, they realize that the amount of dust affects the cloud cover. In this very simple, beginning model, the amount of cloud cover does not move in a direct relationship to the amount of water vapor. The amount of dust particles also influences the amount of clouds. In the last set of meters, the dust particles have been increased and the cloud cover increased correspondingly. From this beginning level model, created by sixth-grade students, we can see that the construction and testing of models using Model-It as a tool allow students to discover the complexity of scientific systems.

Group Presentations

The culminating activity is a group presentation to the rest of the class. Often presentations are viewed only as assessments. But students are more motivated to create an accurate, complex model when they know that they will be presenting it to their peers. Each student is required to participate actively in the

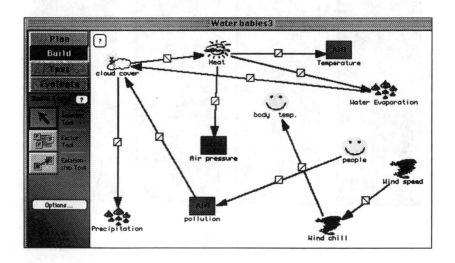

presentation by explaining at least one relationship. The screen shot is from a model presented by a group of seventh-grade students. The model and their presentation showed that they understood the water cycle, the very elusive concept of wind chill, and the difficult relationship between air pressure and temperature. Model-It gives students the freedom to express their knowledge using their own creativity and the movement of graphics. During the modeling activity, therefore, most students are actively engaged in learning science for nearly the whole class period each day. This model took several days to construct. Just imagine, seventh-grade students spending several hours willingly discussing such topics!

MORE COMPLEX MODELS

These examples of examining weather lack one essential specification, the times at which the events of the model take place. In real conditions, not all of the events of the water cycle occur at one time. Model-It can accommodate the more complex question of whether the events happen immediately, quickly or slowly.

During our examination of the quality of our river water, the students use the weighting factor for each of nine water quality tests. The amount of dissolved oxygen in a river is many times more important than the turbidity. Model-It allows you to assign values that vary with the relationship between the two. This is indicated by the thickness of the relationship arrow. In the water quality model here, the students were given the nine weighted water quality relationships. We wanted them to concentrate not on the numerical values they worked with but on the factors in the environment of our river that affect each of the parameters: for example, the events that determine the amount of dissolved oxygen rather than the mathematical importance of dissolved oxygen. They were therefore given a template upon which they built their model. One of the problems was a lack of screen space for everything they wanted to include.

MODELS DESIGNED BY TEACHERS

After using a number of model templates constructed by university personnel, I decided to construct one of my own. In the eighth grade, students review electrical circuits, learn about Ohm's Law, and study the relationships between electricity and magnetism. It was easy for me to draw a paper model that defined the objects needed for Model-It. Unable to find appropriate clip art, I had either to scan images or to take digital photos. I did the latter. A benefit to the digital pictures is that the objects are readily recognizable by the students. Getting the artwork was the most difficult part. When the images were ready, they were pasted into Model-It in the Object Editor.

Many of our models have local pictures as one of the objects and as the background for the others. This helps to make projects tangible and relevant to the students. The air quality objects here are plants, vehicles, man-made objects, weather, people, and the background is a picture of the park and the back of our school. The background picture was taken with a digital camera. For other projects, I have asked the art students to draw the clip art I need. These were scanned and saved as clip art.

CHOOSING THE QUESTION OR MODELING PURPOSE

Much of science does not fit into a complex system of cause and effect suitable for modeling. The question "What is the quality of the air in our community?" is the driving question for an authentic project-based science unit. Most of the objectives we wish to accomplish address differences between physical and chemical changes, word and formula equations, and the conservation of matter. The unit emphasizes the chemistry of air pollution. The environmental science aspects are secondary objectives and a means of making the study of chemistry applicable to the life and health of the students. However, the models address the environmental issues with an occasional very general reference to a chemical aspect. Model-It did not foster deep understanding of the chemistry underlying the environmental science. "Do I really need to wear a helmet when riding my bike?" This is the driving question for another authentic project. Acceleration, velocity, inertia, and gravity have linear relationships to injury to a biker's head, and only seldom will more than one cause point to an effect. Some of us argue that the construction of each relationship is a valuable learning experience because the students must convert their collection of knowledge into the graphic format. Others believe that there must be a network of relationships to make the testing phase meaningful and the modeling experience valuable. These arguments occur only because we are constantly mindful of the need in modeling projects to use every class minute to address efficiently the objectives of our curriculum.

Model-It is adaptable to any network of cause and effect. Social studies and history would be perfect subjects. Students could build models of the factors that brought about the Bill of Rights and the effects of the bill. Other topics that come to mind are slavery, westward expansion, and labor unions. One of our social studies teachers attended a series of Model-It training sessions, and I am looking forward to working with her to infuse Model-It into one of her units of study.

STUDENTS AND TEACHERS ENJOY MODEL-IT

Students and teachers both enjoy their experiences with Model-It. Unlike many software applications, Model-It offers creativity to both. Students do not quickly tire of a process that actively involves them in the creation of a model representing them. While there are guidelines for the general content of the model, students are in control of the objects and factors chosen to represent relationships. Because they have control, they must make decisions. The decision making process requires them to use both vocabulary and reasoning. They have both the frustration and the support of working with a partner. They begin with a blank workspace that they must fill. Contrast this to a worksheet on which they fill in some empty spaces hoping to read the mind of its designer. Think about it. Why do we enjoy teaching? We enjoy the creative efforts of designing units of instruction and creative ways of presenting materials to our students. Model-It lets students experience the creative aspect of learning. They discover the teacher as a facilitator of learning, as another partner in the creation of their model, rather than as a taskmaster constantly making demands of them. The teacher therefore benefits too, changing from provider of information to facilitator of learning.

TEACHER AS FACILITATOR

Most teachers are familiar with the emerging role of teachers as facilitators of learning. The classroom culture, however, prepares students, administrators, and parents to expect the teacher to tell the students what they need to know and to provide test results documenting their achievement. Model-It is an entirely new environment for all of the participants and an opportunity to break out of the old culture. It happens naturally. As students are constructing their models, the teacher is moving around the room solving technical problems or looking over shoulders and offering suggestions. The true facilitator does not tell the students what to do, but asks leading questions that bring the students to reconsider what they are doing. If students have a blatant error in their model, the teacher facilitator may ask them to explain the erroneous segment. As they do so, they often realize that there is an error. The teacher may ask questions to break the misconception into small segments, and further questioning will help the students assemble an appropriate component.

Once teachers have established this rapport with the class, the demand for their time becomes unmanageable. We have established the "Three Before Me!" rule. Students are to ask three people sitting near them before seeking the teacher's help. This rule helps students to appreciate one another and raises the self-esteem of classmates who give assistance. Another technique that lowers everyone's impatience for immediate help is the red and green paper cup signaling system. When the cup on the top of the computer is green, everything is going well. A red cup is a call for help. It does not take long for the students and teachers to come to appreciate this working atmosphere.

COMPUTERS AS MOTIVATORS

Because the models are built on the computers, students are much more willing to meet the challenges of model building. It is commonplace to observe unmotivated or disruptive students actively participating in the modeling process. At these times, the discipline problems become minor or disappear. The computer also helps to address the differing kinds of intelligence found in any student group. The student who cannot comprehend the science may type well or have computer expertise and become valuable in the construction of the group's model. Rejection of them for their limitations gives way to respect for their skill. In every class, there are fewer problems during the modeling phase than at any other time.

PARENTS ARE FRUSTRATED

Parents find the modeling sessions beyond comprehension. Some are irate because they cannot understand factors and relationships. They want more written material so that they can work with their student. They want to know, "What page is it on?" The students readily adapt to changes involving computing, but computers intimidate many parents. We have printed screen shots of their models for them to take home and explain to their parents. Also, teachers invite parents to attend the science class and the final presentations. The best solution to this problem would be software for home, but the memory requirements are beyond those of lower priced computers.

CONCLUSION

For more than fifteen years, I looked for software tools to enhance learning and teaching of science and mathematics. The computer is not a subject to be taught, but a tool to be used to facilitate a task. Over the years, I have had my favorite software titles, but none of them directly facilitated learning scientific principles. They were excellent tools for teaching problem solving, the organization of material in a presentation, and science facts devoid of deep understanding. Students could be successful with the computer activities and fail to learn curriculum objectives.

Model-It is the first and only piece of software I have found that supports the existing curriculum. It is a technology that can be productive throughout a curriculum. Initially I thought that the students should be experts before designing a model. But as I have indicated, the students' knowledge grows as the model is constructed. Modeling motivates the students to extend their understanding through added research, discussions, testing, and revising. An added benefit is their growth in their ability to write explanations and verbally to express their understanding of scientific concepts.

REFERENCES

Kids as global scientists. University of Michigan, Ann Arbor.

Model-It. Cogito Learning Media, Inc., San Francisco, CA.

National Research Council. 1996. *National science education standards*. Washington, DC: National Academy Press.

Metacognitive Facilitation: An Approach to Making Scientific Inquiry Accessible to All

Barbara Y. White and John R. Frederiksen

INTRODUCTION

Science can be viewed as a process of creating laws, models, and theories that enable people to predict, explain, and control the behavior of the world. Our objective in the ThinkerTools Inquiry Project has been to create an instructional approach that makes this view of understanding and doing science accessible to a wide range of students, including lower-achieving and younger students. Our hypothesis is that this objective can be achieved by facilitating the development of metacognitive knowledge and skills: students need to learn about the nature and utility of scientific models as well as the processes by which they are created, tested, and revised. More specifically, we want to help students acquire:

- self-knowledge, including awareness of what expertise they have, the forms that expertise can take, and when and why their expertise might be useful;
- self-regulatory skills, including skills for planning and monitoring such as determining goals and developing strategies for achieving those goals and then evaluating their progress to see whether their plan needs to be modified;

▶ self-improvement expertise, including expertise in reflecting on their knowledge and its use to determine how to improve both of these.

We believe that developing such metacognitive expertise is the key to acquiring inquiry skills and to "learning how to learn" in general. For further discussions regarding the nature of metacognition and the central role that it plays in learning, see Brown, 1987; Brown, Collins, and Duguid, 1989; Bruer, 1993; Collins and Ferguson, 1993; Nickerson, Perkins, and Smith, 1985; and Resnick, 1987.

METACOGNITIVE FACILITATION FOR STUDENTS

We created the ThinkerTools Inquiry Curriculum to test our hypothesis about making scientific inquiry accessible to all students by focusing on the development of metacognitive expertise. In this curriculum, students engage in constructing and revising theories of force and motion. The curricular activities and materials are aimed at developing the knowledge and skills that students need to support this inquiry process. At the beginning of the curriculum, students are introduced to a metacognitive view of research, called "The Inquiry Cycle," and a metacognitive process, called "Reflective Assessment," in which they reflect on their inquiry (see Figure 1 and Table 1). The Inquiry Cycle consists of five steps: Question, Predict, Experiment, Model, and Apply. This cycle is repeated with each module of the curriculum and provides a goal structure that students use to guide their inquiry. The curricular activities focus on enabling students to develop the expertise needed to carry out and understand the purpose of the steps in the Inquiry Cycle, as well as to monitor and reflect on their progress as they conduct their research. This is achieved via an approach in which metacognitive knowledge and skills are learned through a process of scaffolded inquiry, reflection, and generalization. We call this process "Metacognitive Facilitation."

Scaffolded Inquiry

Initially, the meaning and purpose of the steps in the Inquiry Cycle may be only partially understood by students. We therefore designed ways to provide scaffolding for their inquiry until they are able to design their own experiments and to construct their own laws to characterize their findings. We created both scaffolded activities and environments that enable students to carry out a sequence of activities corresponding to steps in the Inquiry Cycle. The scaffolded activities guide them as they do real-world experiments and help them to learn about

FIGURE 1. A METACOGNITIVE VIEW OF THE SCIENTIFIC INQUIRY PROCESS, WHICH STUDENTS USE TO GUIDE THEIR RESEARCH

THE INQUIRY CYCLE

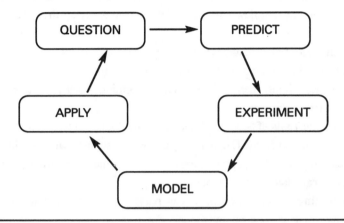

TABLE 1. THE CRITERIA FOR JUDGING RESEARCH THAT STUDENTS USE IN THE REFLECTIVE ASSESSMENT PROCESS

CRITERIA FOR JUDGING RESEARCH

Understanding

- **Understanding the Science.** Students show that they understand the science developed in the curriculum and can apply it in solving problems, in predicting and explaining real-world phenomena, and in carrying out inquiry projects.
- **Understanding the Processes of Inquiry.** Students can talk about what approach they or others have taken in exploring a research topic. For instance, they can explain what types of scientific models and inquiry processes have been used in carrying out investigations and in reaching conclusions.
- **Making Connections.** Students see the big picture and have a clear overview of their work, its purposes, and how it relates to other ideas or situations. They relate new information, ideas, and experimental results to what they already know.

Performance: Doing Science

▶ **Being Inventive.** Students are creative and examine many possibilities in their work. They show originality and inventiveness in thinking of problems to investigate, in coming up with hypotheses, in designing experiments, in creating new laws or models, and in applying their models to new situations.

▶ **Being Systematic.** Students are careful, organized, and logical in planning and carrying out their work. When problems come up, they are thoughtful in examining their progress and in deciding whether to alter their approach or strategy.

▶ **Using the Tools of Science.** Students use the tools and representations of science appropriately. The tools they choose to use (or create) may include such things as lab equipment, measuring instruments, diagrams, graphs, charts, calculators, and computers.

▶ **Reasoning Carefully.** Students can reason appropriately and carefully using scientific concepts and models. For instance, they can argue whether or not a prediction or law that they or someone else has suggested fits with a scientific model. They can also show how experimental observations support or refute a model.,

---

Social Context of Work

▶ **Writing and Communicating Well.** Students clearly express their ideas to each other or to an audience through writing, diagrams, and speaking. Their communication is clear enough to allow others to understand their work and reproduce their research.

▶ **Teamwork.** Students work together as a team to make progress. Students respect each others' contributions and support each others' learning. Students divide their work fairly and make sure that everyone has an important part.

---

the processes of experimental design and data analysis, the nature of scientific argument and proof, and the characteristics of scientific laws and models. The scaffolded environments include computer simulations, which allow students to create and interact with models of force and motion. They also provide analytic tools that help students analyze the results of their computer and real-world experiments. These scaffolded activities and environments make the inquiry process as easy and productive as possible at each stage in learning.

Reflective Assessment

In conjunction with the scaffolded inquiry, students engage in a reflective process in which they evaluate their own and each other's research. This process employs a carefully chosen set of criteria, such as "Being Systematic" and "Reasoning Carefully," that characterize expert scientific inquiry (see Table 1). Students use these criteria to evaluate their work at each step in the Inquiry Cycle, which helps them to see the intellectual purpose and properties of the inquiry steps and their sequencing. Students also employ the criteria to evaluate their own and each other's work when they finish their research projects and present their work to the class. By engaging in these evaluations in which they talk about and reflect on the characteristics of expert scientific inquiry and the functions of each inquiry step, students grow to understand the nature and purpose of inquiry as well as the habits of thought that are involved.

Generalized Inquiry and Reflection

The students employ the Inquiry Cycle and the Reflective Assessment Process repeatedly as the class addresses a series of research questions. With each repetition of the cycle, some of the scaffolding is removed so that eventually the students are conducting independent inquiry on questions of their own choosing (as in the scaffolding and fading approach of Palincsar and Brown [1984]). These repetitions of the Inquiry Cycle in conjunction with Reflective Assessment help students to refine their inquiry processes. Carrying out these activities in new research contexts also enables students to learn how to generalize the inquiry and reflection processes so that they can apply them to learning about new topics in the future.

FACILITATING INQUIRY WITHIN THE
CLASSROOM RESEARCH COMMUNITY

The project has established Classroom Research Communities in seventh, eighth, and ninth-grade science classrooms in middle schools in Berkeley and Oakland. In these classes, inquiry is the basis for developing an understanding of the physics. Physical theories are not directly taught, but are constructed by students themselves as they engage in the scaffolded inquiry and reflection. The idea is to teach students how to carry out scientific inquiry, and then have them discover the basic physical principles for themselves by doing experiments and creating theories.

The process of inquiry follows the Inquiry Cycle, shown in Figure 1, which is presented to students as a basis for organizing their explorations into the physics of force and motion. Inquiry begins with finding research questions, that is, finding situations or phenomena students do not yet understand that become new areas for investigation. Students then use their intuitions, which are often incorrect, to make conjectures about what might happen in such situations. These predictions provide them with a focus as they design experiments that allow them to observe phenomena and test their conjectures. Students then use their findings as a basis for constructing formal laws and models. By applying their models to new situations, students test the range of applicability of their models and, in so doing, identify new research questions for further inquiry.

The social organization of the research community is similar to that of an actual scientific community. Inquiry begins with a whole-class forum to develop shared research themes and areas for joint exploration. Research is then carried out in collaborative research groups. The groups thereupon reassemble to conduct a research symposium in which they present their predictions, experiments, and results, as well as the laws and causal models they propose to explain their findings. While the results and models proposed by individual groups may vary in their accuracy, in the research symposium a process of consensus building increases the reliability of the research findings. The goal is, through debate based upon evidence, to arrive at a common, agreed-upon theory of force and motion.

Organization of the Curriculum

The curriculum is organized around a series of investigations of physical phenomena that increase in complexity. On the first day, students toss a hacky sack around the room while the teacher has them observe and list all of the factors that may be involved in determining its motion (such as how it is thrown, gravity, air resistance, and so forth). As an inquiry strategy, the teacher suggests the need to simplify the situation, and this discussion leads to the idea of looking at simpler cases, such as that of one-dimensional motion where there is no friction or gravity (an example is a ball moving through outer space). The curriculum accordingly starts with this simple case (Module 1), and then adds successively more complicating factors such as introducing friction (Module 2), varying the mass of the ball (Module 3), exploring two-dimensional motion (Module 4), investigating the effects of gravity (Module 5), and analyzing trajectories (Module 6). At the end of the curriculum, students are presented with a variety of possible research topics to pursue (such as orbital motion or collisions), and they carry out research on topics of their own choosing (Module 7).

For each new topic in the curriculum, students follow the Inquiry Cycle:

1. **Question**. As described previously, the inquiry process begins with developing a research question such as, "What happens to the motion of an object that has been pushed or shoved when there is no friction or gravity acting on it?"

2. **Predict**. Next, to set the stage for their investigations, students try to generate alternative predictions and theories about what might happen in some specific situations related to the research question. In other words, they engage in "thought experiments." For example, in Module 1, they are asked to predict what would happen in the following situation:

   Imagine a ball that is stopped on a frictionless surface, one that is even smoother than ice. Suppose that you hit the ball with a mallet. Then, imagine you hit the ball again in the same direction with the same size hit. Would the second hit change the velocity of the ball? If so, describe how it would change and explain why.

   In response to this question, some students might say, "the second hit does not affect the speed of the ball because it's the same size hit as the first"; while others say, "it makes the ball go twice as fast because it gives the ball twice as much force"; and others, "it only makes the ball go a little bit faster because the ball is already moving."

3. **Experiment**. After presenting their predictions to the class, students break into research groups to design and carry out experiments to test their alternative theories. These investigations make use of both computer simulations and real-world experimental materials.

a. <u>Computer activities and experiments</u>. Computer models and experiments are made possible by use of the ThinkerTools software that we developed for the Macintosh computer. This software enables students to interact with Newtonian models of force and motion. (See Figure 2 which shows an example of a computer activity that students use in studying one-dimensional motion.) The software also lets students create their own models and experiments. Using simple drawing tools, students can construct and run computer simulations. Barriers and objects (such as the large circle shown in Figure 2) can be placed on the screen. (The objects are introduced to students as generic objects, simply called "dots," which are the pictorial equivalent of variables that students can map onto different objects such as space ships or billiard balls.) Students can define and change the properties of any object, such as its mass, elasticity (it can be bouncy or fragile), and velocity. They can then apply impulses to the object to change its velocity using the keyboard or a joystick as in a video game. (Impulses are forces that act for a specified—usually short—amount of time like a kick or a hit.) Students can thus create and experiment with a "dot-impulse model" and can discover, for example, that applying an impulse in the same direction that the dot is moving increases the dot's velocity by one unit of speed. In this way, they can use simulations to discover the laws of physics and their implications.

Such software enables students to create experimental situations that are difficult or impossible to create in the real world. For example, they can turn friction and gravity on or off and can select different friction laws (such as sliding friction or gas-fluid friction). They can also vary the amount of friction or gravity to see what happens. Such experimental manipulations in which students dramatically alter the parameters of the simulation make it possible for them to use inquiry strategies, such as "look at extreme cases," which are hard to employ in real-world inquiry. This type of inquiry enables students to see more readily the behavioral implications of the laws of physics and to discover the underlying principles.

Another advantage of having students experiment with such simulations is that the software includes measurement tools that allow students to easily make accurate observations of distances, times, and velocities. These observations would often be very

FIGURE 2. THE THINKERTOOLS SOFTWARE PROVIDES A MODELING AND INQUIRY TOOL FOR CREATING AND EXPERIMENTING WITH MODELS OF FORCE AND MOTION

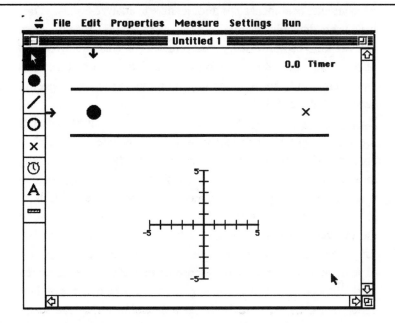

difficult to make in the corresponding real-world experiments. The software includes graphical representations of variables. As the dot moves, for example, it leaves behind "dotprints" that show how far it moved in each second and "thrustprints" that show when an impulse was applied. There is also a "datacross" that shows a dot's x and y velocity components. And, students can have the software keep a table or graph to record, say, the velocity of the dot. In addition, there are analytic tools such as "stepping through time," which allow students to pause the simulation and to proceed time step by time step so that they can better see and analyze what is happening to the motion of the dot. In this mode, the simulation runs for a small amount of time, leaves one dotprint on the screen, and then pauses again. The students have control over whether the simulation remains paused, proceeds to the next time step, or returns to continuous mode. These analytic tools and graphical representations help students determine the underlying laws of motion. They can also be incorporated within the students' conceptual model to represent and reason about what might happen in successive time steps.

Ideally, the software helps students construct conceptual models that are similar to the computer's in that they both use diagrammatic representations and they both employ causal reasoning in which the computer or student steps through time to analyze events. In this way, such dynamic interactive simulations combined with these analytic tools can provide a transition from students' intuitive ways of reasoning about the world to the more abstract formal methods that scientists use for representing and reasoning about a system's behavior (White, 1993b).

b. <u>Real-world experiments</u>. Students are also given a set of materials for conducting real-world experiments. This includes "bonkers" (a bonker is a rubber mallet mounted on a stand), balls of varying masses, and measurement tools such as meter sticks and stop watches (see Figure 3). These materials are coordinated with those in the ThinkerTools software. For instance, the bonker is similar to the joystick and is used to give a ball a standard-sized impulse. Employing such materials, students design and carry out real-world experiments that are related to those carried out in the simulated world. Students are also shown stop-motion videos of some of their experiments. Using frame-by-frame presentations, they can attach blank transparencies to the video screen and draw the position of a moving ball after fixed time intervals. These "dotprint analyses" allow them to measure the moment-by-moment changes in the ball's velocity.

4. **Model.** After the students have completed their experiments, they analyze their data to see whether there are any patterns. They then try to summarize and explain their findings by formulating a law and a causal model to characterize their conclusions. Students' models typically take the form: "If A then B because…"; for example, "if there are no forces like friction acting on an object, then it will go forever at the same speed, because there is nothing to slow it down."

The computer simulations combined with real-world experiments and the process of creating a model can help students to understand the nature of scientific models. The computer is not the real world; it can only simulate real-world behavior by stepping through time and using rules to determine how forces that are acting, like friction or gravity, will change the dot's velocity on that time step. Thus, the computer is actually using a conceptual model

FIGURE 3. AN ILLUSTRATION OF A REAL-WORLD EXPERIMENT, CALLED "THE DOUBLE-BONK EXPERIMENT," ALONG WITH THE TABLE THAT STUDENTS USE TO RECORD THEIR DATA

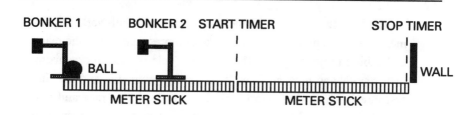

DATA TABLE

TRIAL	DISTANCE (CM)	TIME (SEC)	VELOCITY (V=D÷T)
1	100		
2	100		
3	100		
Average	100		

to predict behavior, just as the students will use the conceptual model they construct to predict behavior. In working with the computer, the students' task is to design experiments that will help them induce the laws that are used by the simulation. This is more straightforward than the corresponding real-world inquiry task. After all, objects in the real world are not driven by laws; rather, the laws simply characterize their behavior.

One example of a modeling activity, which is carried out early in the curriculum, has students explain how their computer and real-world experiments could lead to different conclusions. They might say, for instance, that "the computer simulation does not have friction, which is affecting our real-world experiments." Alternatively, they might say that "the real world does not behave perfectly and does not follow rules." Working with a computer simulation can thus potentially help students to develop metacognitive knowledge about what scientific models are, and how laws can be used to predict and control behavior. It can also enable them to appreciate the utility of creating computer simulations that embody scientific laws and idealized abstractions of real-world behavior, and then employing such simulations to do experiments in order to see the implications of a particular theory.

Based on the findings of their computer and real-world experiments, students prepare posters, make oral presentations to the class, and submit project reports. The Inquiry Cycle is used in organizing their reports and presentations. With writing, graphing, and drawing software (such as ClarisWorks) students analyze their data and prepare their reports. Then, in a whole-class research symposium, they evaluate together the findings from all the research groups, and choose the "best" laws and models to explain their data.

5. **Apply**. Once the class chooses the best laws and causal models, students try to apply them to different real-world situations. For instance, they might try to predict what happens when you hit a hockey puck on ice. As part of this process, they investigate the utility of their laws and models for predicting and explaining what would happen. They also investigate the limits of their models (such as, "What happens if the ice isn't perfectly smooth?"), which inevitably raises new research questions (such as, "What are the effects of friction?"). This brings the class back to the beginning of the Inquiry Cycle and to investigating the next research question in the curriculum.

CYCLING TOWARD INDEPENDENT INQUIRY

The Inquiry Cycle is repeated with each of the seven modules of the curriculum. As the curriculum progresses, the physics the students are dealing with increases in complexity and so does the inquiry. In the early stages of the curriculum, the inquiry process is heavily scaffolded. In Module 1, students are given experiments to do and are presented with alternative possible laws to evaluate. In this way, they see examples of experiments and laws before they have to create their own. In Module 2, students are given experiments to do but have to construct the laws for themselves. Then, in Module 3, they design their own experiments and construct their own laws to characterize their findings (see Appendix A). By the end of the curriculum, the students are carrying out independent inquiry on a topic of their own choosing.

FACILITATING REFLECTIVE ASSESSMENT WITHIN
THE CLASSROOM RESEARCH COMMUNITY

In addition to the Inquiry Cycle, which guides the students' research and helps them to understand what the research process is all about, we developed a set of criteria for characterizing good scientific research. These are presented in Table 1. They include goal-oriented criteria such as "Understanding the Science" and "Understanding the Processes of Inquiry," process-oriented criteria such as "Being Systematic" and "Reasoning Carefully," and socially-oriented criteria such as "Communicating Well" and "Teamwork." These characterizations of good work are used not only by the teachers in judging the students research projects, but also by the students themselves.

At the beginning of the curriculum, the criteria are introduced and explained to the students as the "Criteria for Judging Research" (see Table 1). Then, at the end of each step in the Inquiry Cycle, the students monitor their progress by evaluating their work on the two most relevant criteria. At the end of each module, they reflect on their work by evaluating themselves on all of the criteria. Similarly, when they present their research projects to the class, the students evaluate not only their own research projects but also each other's. They give one another feedback both verbally and in writing. These assessment criteria thus not only provide a way to introduce students to the characteristics of good research, they also help students to monitor and reflect on their inquiry processes.

In what follows, we present sample excerpts from a class's reflective assessment discussion. Students give oral presentations of their projects accompanied by a poster, and they answer questions about their research. Following each presentation, the teacher picks a few of the assessment criteria and asks students in the audience how they would rate the presentation. In these conversations, students are typically respectful of one another and generally give their peers high ratings (ratings between 3 and 5 on a five-point scale). But, within the range of high scores that they use, they do make distinctions among the criteria and offer insightful evaluations of the projects that have been presented. The following illustrates some examples of such reflective assessment conversations. (Pseudonyms are used throughout, and the transcript has been lightly edited to improve its readability.)

> **Teacher**: OK, now what we are going to do is give them some feedback. What about their "understanding the process of inquiry"? In terms of their following the steps within the Inquiry Cycle, on a scale from 1 to 5, how would you score them? Vanessa?

Vanessa: I think I would give them a 5 because they followed every-thing. First they figured out what they wanted to inquire, and then they made hypotheses, and then they figured out what kind of experiment to do, and then they tried the experiment, and then they figured out what the answer really was and that Jamal's hypothesis was correct.

Teacher: All right, in terms of their performance, "being inventive." Justin?

Justin: Being inventive. I gave them a 5 because they had complete-ly different experiments than almost everyone else's I've seen. So, being inventive, they definitely were very inventive in their experi-mentation.

Teacher: OK, good. What about "reasoning carefully"? Jamal, how would you evaluate yourself on that?

Jamal: I gave myself a 5, because I had to compute the dotprints between the experiments we did on mass. So, I had to compute every-thing. And, I double checked all of my work.

Teacher: Great. OK, in terms of the social context of work, "writing and communicating well." Carla, how did you score yourself in that area?

Carla: I gave myself a 4, because I always told Jamal what I thought was good or what I thought was bad, and if we should keep this part of our experiment or not. We would debate on it and finally come up with an answer.

Teacher: What about "teamwork"? Does anyone want to rate that? Teamwork. Nisha?

Nisha: I don't know if I can say because I didn't see them work. [laughter]

Teacher: That's fine. That's fair. You are being honest. Julia?

Julia: I gave them a 5 because they both talked in the presentation, and they worked together very well, and they looked out for each other.

There are various arguments for why incorporating such a Reflective Assessment Process into the curriculum should be effective. One is the "transparent assessment" argument put forward by Frederiksen and Collins (1989; Frederiksen, 1994), who argue that introducing students to the criteria by which their work will be evaluated enables students to better understand the characteristics of good performance. In addition, there is the argument about the importance of metacognition put forward by researchers (for example, Baird et al., 1991; Brown 1987; Brown & Campione, 1996; Collins, Brown, & Newman, 1989; Miller, 1991; Reeve & Brown, 1985; Scardamalia & Bereiter, 1991; Schoenfeld, 1987; Schön, 1987; Towler & Broadfoot, 1992) who maintain that monitoring and reflecting on the process and products of one's own learning is crucial to successful learning as well as to learning how to learn. Research comparing good with poor learners shows that many students, particularly lower-achieving students, have inadequate metacognitive processes and their learning suffers accordingly (Campione, 1987; Chi et al., 1989). Thus if you introduce and support such processes in the curriculum, the students' learning and inquiry should be enhanced. Instructional trials of the ThinkerTools Inquiry Curriculum in urban classrooms (that included many lower-achieving students) provided an ideal opportunity to test these hypotheses concerning the utility of such a metacognitive Reflective Assessment Process.

INSTRUCTIONAL TRIALS OF THE
ThinkerTools INQUIRY CURRICULUM

In 1994, we conducted instructional trials of the ThinkerTools Inquiry Curriculum. Three teachers used it in their twelve urban classes in grades seven through nine. The average amount of time they spent on the curriculum was ten and a half weeks. Two of the teachers had no prior formal physics education. They were all teaching in urban classes that averaged almost thirty students, two thirds of whom were minority students, and many were from highly disadvantaged backgrounds.

We analyzed the effects of the curriculum for students who varied in their degree of educational advantage, as measured by their standardized achievement test scores (CTBS—Comprehensive Test of Basic Skills). We compared the performance of these middle school students with that of high school physics students. We also carried out a controlled study comparing ThinkerTools classes in which students engaged in the Reflective Assessment Process with matched "control" classes in which they did not. For each of the teachers, half of the classes were reflective assessment classes and the other half were control classes. In the reflective assessment classes, the students were given the assessment framework (shown in Table 1) and they continually engaged in monitoring and evaluating their own and each other's research. In the control classes, the students were not given an explicit framework for reflecting on their research;

instead, they engaged in alternative activities in which they commented on what they did and did not like about the curriculum. In all other respects, the classes participated in the same ThinkerTools Inquiry Curriculum.

There were no significant differences in students' average CTBS scores among the classes that were randomly assigned to the different treatments (reflective assessment or control), for the classes of the three different teachers, or for the different grade levels (seventh, eighth, and ninth). (Since CTBS scores are normed for each grade level, one does not expect differences associated with grade.) Thus, the classes were all comparable with regard to achievement test scores.

AN OVERVIEW OF THE RESULTS

Our results show that the curriculum and software modeling tools make the difficult subject of physics understandable and interesting to a wide range of students. Moreover, the emphasis on creating models enables students to learn not only about physics, but also about the properties of scientific models and the inquiry processes needed to create them. Furthermore, engaging in inquiry improves students' attitudes toward learning and doing science. Below, we provide an overview of our findings with regard to the students' development of expertise in inquiry and physics. For a more in-depth presentation of all of our results, see White & Frederiksen (1998).

The Development of Inquiry Expertise

One of our assessments of students' scientific inquiry expertise was an inquiry test given both before and after the ThinkerTools Inquiry Curriculum. In this written test, the students were asked to investigate a specific research question: "What is the relationship between the weight of an object and the effect that sliding friction has on its motion?" The students were first asked to come up with alternative, competing hypotheses with regard to this question. Next, they had to design on paper an experiment that would determine what actually happens, and then they had to pretend to carry out their experiment. They had, in effect, to conduct it as a thought experiment and make up the data that they thought they would get if they actually carried out their experiment. Finally, they had to analyze their made-up data to reach a conclusion and relate this conclusion back to their original, competing hypotheses.

Scoring this test centered entirely on the students' inquiry process. Whether or not their theories embodied the correct physics was regarded as irrelevant. Figure 4 presents the gain scores on this test for both low- and high-achieving students, and for students in the reflective assessment and control classes. Notice that students in the reflective assessment classes gained more on this inquiry test, and that this was particularly true for the low-achieving students. This is the first piece of evidence that the metacognitive Reflective Assessment Process is beneficial, particularly for academically disadvantaged students.

The gain scores for each component of the inquiry test, presented in Figure 5, show that the effect of Reflective Assessment is greatest for the more difficult aspects of the test: making up results, analyzing those made-up results, and relating them back to the original hypotheses. In fact, the largest difference in the gain scores is that for a measure we call "coherence," which assesses the extent to which the experiments that the students designed address their hypotheses, their made-up results relate to their experiments, their conclusions follow from their results, and whether they compare their conclusions with their original hypotheses. This kind of overall coherence in research is, we think, a very important indication of sophistication in inquiry. It is on this coherence measure that we see the greatest difference in favor of students who engaged in the metacognitive Reflective Assessment Process.

FIGURE 4. THE MEAN GAIN SCORES ON THE INQUIRY TEST FOR STUDENTS IN THE REFLECTIVE ASSESSMENT AND CONTROL CLASSES, PLOTTED AS A FUNCTION OF THEIR ACHIEVEMENT LEVEL

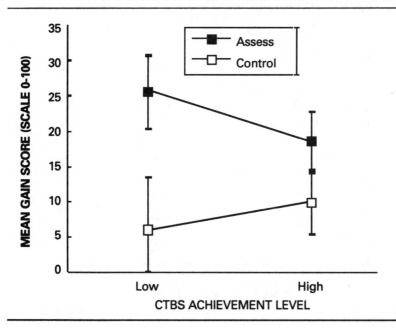

**FIGURE 5. MEAN GAINS ON THE INQUIRY TEST SUBSCORES FOR
STUDENTS IN THE REFLECTIVE ASSESSMENT AND
CONTROL CLASSES**

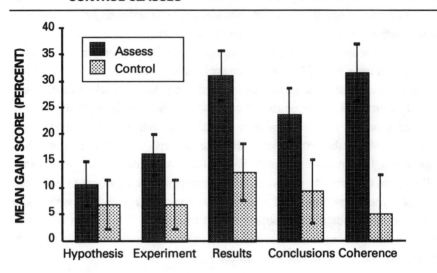

**FIGURE 6. THE MEAN SCORES ON THEIR RESEARCH PROJECTS FOR STU-
DENTS IN THE REFLECTIVE ASSESSMENT AND CONTROL CLASSES,
PLOTTED AS A FUNCTION OF THEIR ACHIEVEMENT LEVEL**

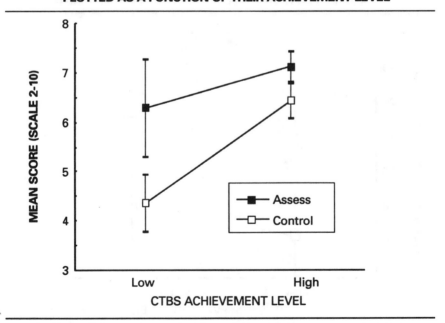

Students carried out two research projects, one about halfway through the curriculum and one at the end. For the sake of brevity, we added the scores for these two projects together as shown in Figure 6. These results indicate that students in the reflective assessment classes do significantly better on their research projects than students in the control classes. The results also show that the Reflective Assessment Process is particularly beneficial for the low-achieving students: in the reflective assessment classes, they perform almost as well as the high-achieving students. These findings were the same across all three teachers and all three grade levels.

The Development of Physics Expertise

We gave the students a General Physics Test, both before and after the ThinkerTools curriculum. This test includes items commonly used by educational researchers to assess students' understanding of Newtonian mechanics. For example, there are items such as that shown in Figure 7 in which students are asked to predict and explain how forces will affect an object's motion. On this test we found significant pretest to posttest gains. We also found that our middle school, ThinkerTools students do better on such items than do high school physics students who are taught using traditional approaches. Furthermore, on items that represent near or far transfer in relation to contexts ThinkerTools students had studied in the course, we found that there were significant learning effects on both the near and far transfer items. Together, these results show that you can teach sophisticated physics in urban, middle school classrooms when you make use of simulation tools combined with scaffolding the inquiry process. In general, this inquiry-oriented, constructivist approach appears to make physics interesting and accessible to a wider range of students than is possible with traditional approaches (White, 1993a&b; White & Frederiksen, 1998; White & Horwitz, 1988).

What is the effect of the Reflective Assessment Process on the learning of physics? The assessment criteria were chosen to address principally the process of inquiry and only indirectly the conceptual model of force and motion that students are attempting to construct in their research. Within the curriculum, moreover, students practice Reflective Assessment primarily in the context of judging their own and others' work on projects, not their progress in solving physics problems. Nonetheless, our hypothesis is that by improving the learning of inquiry skills that are instrumental in developing an understanding of physics principles, the Reflective Assessment should have an influence on students' success in developing conceptual models for the physical phenomena they have studied.

FIGURE 7. A SAMPLE PROBLEM FROM THE PHYSICS TEST*

Imagine that you kick a ball off a cliff.
Circle the path the ball would take as it falls
to the ground.

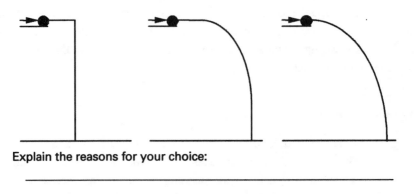

Explain the reasons for your choice:

_____

_____

*On a set of such items, the ThinkerTools students averaged 68%
correct and significantly outperformed high school physics students
who averaged 50% correct (t343 = 4.59, p = <.001).

To evaluate the effects of Reflective Assessment on students' understanding
of physics, we examined their performance on our Conceptual Model Test. In
this test, students are asked questions about the behavior of objects in the
Newtonian computer model. It assesses whether they have developed the
desired Newtonian conceptual model of force and motion. Our findings, presented in Figure 8, show that for the academically disadvantaged students, the
effects of Reflective Assessment extend to their learning the science content as
well as the processes of scientific inquiry.

The Impact of Understanding the Reflective Assessment Criteria

If we are to attribute these effects of introducing Reflective Assessment to students' developing metacognitive competence, we need to show that the students
developed an understanding of the assessment criteria and could use them to
describe multiple aspects of their work.

**FIGURE 8. THE MEAN SCORES ON THE CONCEPTUAL MODEL TEST FOR STU-
DENTS IN THE REFLECTIVE ASSESSMENT AND CONTROL CLASSES,
PLOTTED AS A FUNCTION OF THEIR ACHIEVEMENT LEVEL**

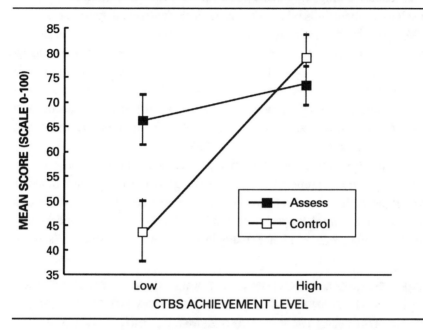

One way to evaluate their understanding of the assessment concepts is to compare their use of the criteria in rating their own work with the teachers' evaluations of their work using the same criteria. If students have learned how to employ the criteria, their self-assessment ratings should correlate with the teachers' ratings for each of the criteria. We found that students in the reflective assessment classes, who worked with the criteria throughout the curriculum, showed significant agreement with the teachers in judging their work. However, this was not the case for students in the control classes, who were given the criteria only at the end of the curriculum for use in judging their final projects. In judging Reasoning Carefully on their final projects, for instance, students in the reflective assessment classes had a correlation of .58 between their ratings and the teachers', while for students in the control classes the correlation was only .23. The average correlation for students in the reflective assessment classes over all of the criteria was .48, which is twice that for students in the control classes.

If the reflective assessment criteria are acting as metacognitive tools to help students as they ponder the functions and outcomes of their inquiry processes, then the students' performance in developing their inquiry projects should depend upon how well they have understood the assessment concepts. To evaluate their understanding, we rated whether the evidence they cited in justifying their self assessments was relevant to the particular criterion they were considering. We then looked at the quality of the students' final projects, comparing students who had developed an understanding of the set of assessment concepts by the end of the curriculum with those who had not. Our results, shown in Figure 9, indicate that students who had learned to use the interpretive concepts appropriately in judging their work produced higher quality projects than students who had not. And again, we found that the benefit of learning to use the assessment criteria was greatest for the low-achieving students.

Taken together, these research findings clearly implicate the use of the assessment criteria as a reflective tool for learning to carry out inquiry. Students in the reflective assessment classes generated higher scoring research reports than those in the control classes. And, students who showed a clear understanding of the criteria produced higher quality investigations than those who showed

FIGURE 9. THE MEAN SCORES ON THEIR FINAL PROJECTS FOR STUDENTS WHO DID AND DID NOT PROVIDE RELEVANT EVIDENCE WHEN JUSTIFYING THEIR SELF-ASSESSMENT SCORES, PLOTTED AS A FUNCTION OF THEIR ACHIEVEMENT LEVEL

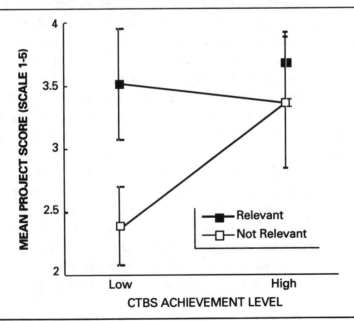

less understanding. Thus, there are strong beneficial effects of introducing metacognitive language to facilitate students' reflective explorations of their work in classroom conversations and in self assessment.

An important finding was that the beneficial effect of Reflective Assessment was particularly strong for the low-achieving students: The Reflective Assessment Process enabled them to gain more on the inquiry test (see Figure 4), and to perform close to the high-achieving students on their research projects (see Figure 6). The introduction of Reflective Assessment, while helpful to all, was thus closing the performance gap between the low- and high-achieving students. In fact, the Reflective Assessment Process enabled low-achieving students to perform equivalently to high-achieving students on their research projects when they did their research in collaboration with a high-achieving student. In the control classes, in contrast, the low-achieving students did not do as well as high-achieving students, regardless of whether they collaborated with a high-achieving student. Thus, there was evidence that social interactions in the reflective assessment classes—particularly between low- and high-achieving students—were important in facilitating learning (cf., Carter & Jones, 1994; Slavin, 1995; and Vygotsky, 1978).

THE IMPLICATIONS OF OUR FINDINGS

We think that our findings have strong implications for what curricula that emphasize inquiry and metacognition can accomplish, particularly in urban school settings in which there are many academically disadvantaged students. More specifically, we argue that three important conclusions follow from our work:

▶ To be equitable, science curricula should incorporate reflective inquiry, and assessments of students' learning should include measures of inquiry expertise.

▶ Students should learn how to transfer the inquiry and reflective assessment processes to other domains so that they learn how to learn and can utilize these valuable metacognitive skills in their learning of other school subjects.

▶ Such an inquiry-oriented approach to education, in which the development of metacognitive knowledge and skills plays a central role, should be introduced early in the school curriculum, preferably at the elementary school level.

1. **Science curricula should incorporate reflective inquiry and include assessments of students' inquiry expertise**. Our results suggest that, from an equity standpoint, curricular approaches can be created that are not merely equal in their value for, but actually enhance the learning of less-advantaged students. Moreover, adequately and fairly assessing the effectiveness of such curricula requires utilizing measures of inquiry expertise, such as our inquiry tests and research projects. If only subject-matter tests are used, the results can be biased against both low-achieving and female students. For instance, on the research projects, we found that low-achieving students who had the benefit of the Reflective Assessment Process did almost as well as the high-achieving students. And these results could not be attributed simply to ceiling effects. We also found that the male and female students did equally well on the inquiry tests and research projects. On the physics tests, however, the pattern of results was not comparable: males outperformed females (on both pretests and posttests) and the high-achieving students outperformed the low-achieving students (White & Frederiksen, 1998). Thus utilizing inquiry tests and research projects in addition to subject-matter tests not only plays a valuable role in facilitating the development of inquiry skills, it also produces a more comprehensive and equitable assessment of students' accomplishments in learning science.

2. **Students should learn to transfer what they learn about inquiry and reflection to the rest of their school curriculum**. Students' work in the ThinkerTools Inquiry Curriculum and their performance on the various inquiry assessments indicate that they acquired an understanding of the Inquiry Cycle as well as the knowledge needed to carry out each of the steps in this cycle. They also learned about the forms that scientific laws, models, and theories can take and about how the development of scientific theories is related to empirical evidence. In addition, they acquired the metacognitive skills of monitoring and reflecting on their inquiry processes. Since all of science can be viewed as a process of constructing models and theories, both the Inquiry Cycle and the Reflective Assessment Process can be applied to learning and doing all areas of science, not just physics. Understanding and engaging in the Inquiry Cycle and Reflective Assessment Process should therefore benefit students in their future science courses.

In the subsequent work of ThinkerTools students, we see evidence of these benefits and transfer to new contexts. For example, eighth-grade students who did ThinkerTools in the seventh grade were asked to do research projects that used the Inquiry Cycle. They were free to choose topics other than physics. For instance, one group of students wanted to understand how listening to music affects performance on school-work. They did an experiment in which their classmates listened to different kinds of music while taking an arithmetic test. They wrote research reports that described how they followed the Inquiry Cycle in planning and carrying out their research, and they evaluated their own and one another's research using scoring criteria shown in Table 1. Their teacher reports that their performance on these projects was equal to or better than the performance on their ThinkerTools physics projects. Moreover, at the end of the curriculum, some students were asked if the Inquiry Cycle and Reflective Assessment Process could be used to help them learn other subjects. Many of their answers involved highly creative explanations of how it could be applied to domains such as social studies, mathematics, and English, as well as to other areas of science. An example is the following observation from a student who was discussing how the Inquiry Cycle could be useful (her statement has been edited to improve its readability):

I'm sure that a lot of things will need the Inquiry Cycle, like even things like a law court. See, they have to go through a cycle. Maybe not quite the same thing, but they have a question, like why did he do it. The predictions are like possible motives. There is no real experiment, but the equipment is like the murder weapon. The analysis and models are like what did they find out from the trial.... And so almost everything has to go through sort of a cycle like this.

Furthermore, all of the teachers attest to the benefits of both the Inquiry Cycle and the Reflective Assessment Process and have chosen to incorporate nonscience subjects. In order to make the valuable skills of inquiry, modeling, and reflection apply to other experimental sciences, such as biology, as well as to the learning of nonscience subjects, various approaches could be pursued. For instance, students could be introduced to a generalized version of the Inquiry Cycle (such as: Question, Hypothesize, Investigate, Analyze, Model, and Evaluate, which represents a minor transformation of the more experimentally oriented

Inquiry Cycle that students internalize during the ThinkerTools Inquiry Curriculum). This generalization could give students a metacognitive view of learning and inquiry that can be applied to any topic in which building predictive and explanatory models is important. In addition, the students could discuss how readily the Reflective Assessment Process, which uses the criteria shown in Table 1 (such as Making Connections, Reasoning Carefully, and Communicating Well), can be generalized to learning other topics within and beyond science. Having such explicit discussions of transfer in conjunction with explicitly using versions of the Inquiry Cycle and Reflective Assessment Process in their science and other curricula should enable students and teachers to appreciate and benefit from the power of metacognition. Investigating how such generalization and transfer can be achieved is a major concern of our current research (see White, Shimoda, & Frederiksen, 1999).

3. **It is important to introduce inquiry-based learning and reflective assessment early in the school curriculum.** Another conclusion from our research is that the processes of reflective inquiry should be taught early. This would enable young students to develop metacognitive skills that are important components of expertise in learning. These skills should help the low-achieving students to overcome a major source of educational disadvantage. The findings from instructional trials of the ThinkerTools Inquiry Curriculum support the feasibility of achieving this goal. Students over a range of grades showed equal degrees of learning: We found no age differences in students' pretest or posttest scores on the inquiry test over grades ranging from seventh to ninth, nor did we find any age differences in students' gains on the physics tests. Extrapolating from these results suggests that inquiry-based science curricula could be introduced in earlier grades. Metz (1995) presents additional arguments concerning the feasibility and importance of teaching scientific inquiry to young students.

To meet this need, we are extending our work on the ThinkerTools Project to investigate how inquiry, modeling, and metacognition can be taught and assessed in earlier grades. As a first step, we are collaborating with sixth-grade teachers to develop an inquiry-oriented curriculum that utilizes the Inquiry Cycle, the Reflective Assessment Process, as well as other techniques for Metacognitive Facilitation. In this year-long curriculum, low-achieving students work in collaboration with high-achieving students to plan, carry out, and critically evaluate research on a wide variety of topics across a number of disciplines. Our hope is that such an inquiry-oriented curriculum, which focuses on the development of metacognitive skills, will enable all students to learn how to learn at a young age, regardless of their prior educational advantages and disadvantages.

METACOGNITIVE FACILITATION FOR TEACHERS

How can we enable teachers to implement such inquiry-oriented approaches to education? Our research in which we studied the dissemination of the ThinkerTools Inquiry Curriculum indicates that it is not sufficient to simply provide teachers with teacher's guides that attempt to outline goals, describe activities, and suggest, in a semi-procedural fashion, how the lessons might proceed (White & Frederiksen, 1998). We have found that teachers also need to develop a conceptual framework for characterizing good inquiry teaching and for reflecting on their teaching practices in the same way that students need to develop criteria for characterizing good scientific research and for reflecting on their inquiry processes.

To achieve this goal, we made use of a framework that we had developed for the National Board for Professional Teaching Standards (Frederiksen et al., 1998). This framework, which attempts to characterize expert teaching, includes five major criteria: worthwhile engagement, adept classroom management, effective pedagogy, good classroom climate, and explicit thinking about the subject matter, to which we added active inquiry. In this characterization of expert teaching, each of these criteria for good teaching is unpacked into a set of "aspects." For example, Figure 10 illustrates the criterion of "Classroom Climate," which is defined as "the social environment of the class empowers learning." Under this general criterion are five aspects: engagement, encouragement, rapport, respect, and sensitivity to diversity. Each is defined by specific characteristics of classroom practice, such as "humor is used effectively" or "there is a strong connection between students and teacher." Further, each of these is indexed to video clips, called "video snippets," which illustrate it. This framework characterizes good inquiry teaching and provides teachers with video exemplars of teaching practice.

Such materials can be used to enable teachers to learn about inquiry teaching and its value, as well as to reflect on their own and each other's teaching practices. For example, recently we tried the following approach with a group of ten student teachers. They learned to use the framework by scoring videotapes

FIGURE 10. AN EXAMPLE OF THE HIERARCHICAL DEFINITIONS CREATED FOR EACH CRITERION, SUCH AS CLASSROOM CLIMATE, WHICH ARE USED TO CHARACTERIZE EXPERT TEACHING

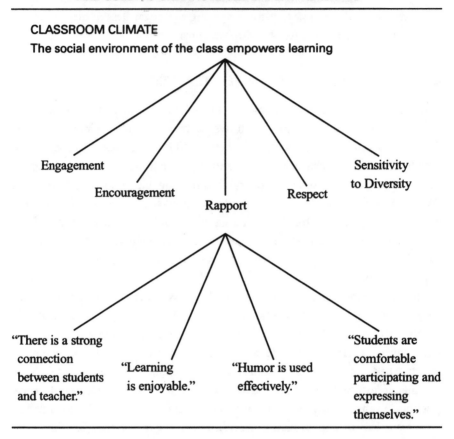

CLASSROOM CLIMATE
The social environment of the class empowers learning

Engagement

Encouragement

Rapport

Respect

Sensitivity to Diversity

"There is a strong connection between students and teacher."

"Learning is enjoyable."

"Humor is used effectively."

"Students are comfortable participating and expressing themselves."

of ThinkerTools classrooms. Then, they used the framework to facilitate discussions of videotapes of their own teaching. In this way, they participated in what we call "video clubs," which enabled them to reflect on their own teaching practices and to hopefully develop better approaches for inquiry teaching. (Video clubs incorporate social activities designed to help teachers reflectively assess and talk about their teaching practices [Frederiksen et al., 1998]). The results have been very encouraging, and our findings indicate that engaging in this reflective activity enabled the student teachers to develop a shared language for viewing and talking about teaching which, in turn, led to highly productive conversations in which they explored and reflected on their own teaching practices (Diston, 1997; Frederiksen & White, 1997; Richards & Colety, 1997).

We conclude by arguing that the same emphases on Metacognitive Facilitation that we have found is important and effective for students is beneficial for teachers as well. It can enable teachers to explore the cognitive and social goals related to inquiry teaching and to thereby improve their own teaching practices. Through this approach, both students and teachers can come to understand the goals and processes related to inquiry, and can learn how to engage in effective inquiry learning and teaching.

ACKNOWLEDGEMENT

We gratefully acknowledge the support of our sponsors: the James S. McDonnell Foundation, the National Science Foundation, and the Educational Testing Service (ETS). The ThinkerTools Inquiry Project is a collaborative endeavor between researchers at the University of California at Berkeley and ETS and middle school teachers in the public schools of Berkeley and Oakland, California. We would like to thank all members of the team for their valuable contributions to this work.

REFERENCES

Baird, J., P. Fensham, R. Gunstone, and R. White. 1991. The importance of reflection in improving science teaching and learning. *Journal of Research in Science Teaching*, 28(2): 163-182.

Brown, A. 1987. Metacognition, executive control, self-regulation, and other more mysterious mechanisms. In *Metacognition, motivation, and understanding*, edited by F. E. Weinert and R. H. Kluwe, 60-108. Hillsdale, NJ: Erlbaum.

Brown, A., and J. Campione. 1996. Psychological theory and the design of innovative learning environments: On procedures, principles, and systems. In *Innovations in learning: New environments for education*, edited by L. Schauble and R. Glaser, 289-325. Mahwah, NJ: Erlbaum.

Brown, J., A. Collins, and P. Duguid. 1989. Situated cognition and the culture of learning. *Educational Researcher* 18: 32-42.

Bruer, J. 1993. *Schools for thought: A science of learning in the class-room*. Cambridge, MA: MIT Press.

Campione, J. 1987. Metacognitive components of instructional research with problem learners. In *Metacognition, motivation, and under-standing*, edited by F. E. Weinert and R. H. Kluwe, 117-140. Hillsdale, NJ: Erlbaum.

Carter, G., and M. Jones. 1994. Relationship between ability-paired interactions and the development of fifth graders' concepts of bal-ance. *Journal of Research in Science Teaching* 31, 8, 847-856.

Chi, M., M. Bassock, M. Lewis, P. Reimann, and R. Glaser. 1989. Self-explanations: How students study and use examples in learning to solve problems. *Cognitive Science* 13: 145-182.

Collins, A., J. Brown, and S. Newman. 1989. Cognitive apprenticeship: Teaching the craft of reading, writing, and mathematics. In *Knowing, learning, and instruction: Essays in honor of Robert Glaser*, edited by L. Resnick, 453-494. Hillsdale, NJ: Erlbaum.

Collins, A., and W. Ferguson. 1993. Epistemic forms and epistemic games: Structures and strategies to guide inquiry. *Educational Psychologist* 28: 25-42.

Diston, J. 1997. *Seeing teaching in video: Using an interpretative video framework to broaden pre-service teacher development*. Unpublished master's project, Graduate School of Education, University of California, Berkeley, CA.

Frederiksen, J. (1994). Assessment as an agent of educational reform. *The Educator* 8(2): 2-7.

Frederiksen, J., and A. Collins. 1989. A systems approach to education-al testing. *Educational Researcher* 18(9): 27-32.

Frederiksen, J., M. Sipusic, M. Sherin, and E. Wolfe. 1998. Video port-folio assessment: Creating a framework for viewing the functions of teaching. *Educational Assessment* 5(4): 225-297.

Frederiksen, J.R., and B.Y. White. 1997. Cognitive facilitation: A method for promoting reflective collaboration. In *Proceedings of the second international conference on computer support for collaborative learning*. Toronto, Canada: University of Toronto.

Metz, K. 1995. Reassessment of developmental constraints on children's science instruction. *Educational Researcher* 65(2): 93-127.

Miller, M. 1991. Self-assessment as a specific strategy for teaching the gifted learning disabled. *Journal for the Education of the Gifted* 14(2):178-188.

Nickerson, R., D. Perkins, and E. Smith. 1985. *The teaching of thinking*. Hillsdale, NJ: Erlbaum.

Palincsar, A., and A. Brown, A. 1984. Reciprocal teaching of comprehension fostering and monitoring activities. *Cognition and Instruction* 1(2):117-175.

Reeve, R.A., and A.L. Brown 1985. Metacognition reconsidered: implications for intervention research. *Journal of Abnormal Child Psychology* 13(3): 343-356.

Resnick, L. 1987. *Education and learning to think*. Washington, DC: National Academy Press.

Richards, S., and B. Colety. 1997. *Conversational analysis of the MACSME video analysis class: Impact on and recommendations for the MACSME program*. Unpublished master's project, Graduate School of Education, University of California, Berkeley, CA.

Scardamalia, M., and C. Bereiter. 1991. Higher levels of agency for children in knowledge building: A challenge for the design of new knowledge media. *The Journal of the Learning Sciences* 1(1):37-68.

Schoenfeld, A. H. 1987. What's all the fuss about metacognition? In *Cognitive science and mathematics education*, edited by A. H. Schoenfeld, 189-215. Hillsdale, NJ: Erlbaum.

Schön, D. 1987. *Educating the reflective practitioner*. San Francisco, CA: Josey-Bass Publishers.

Slavin, R. 1995. *Cooperative learning: theory, research, and practice (2nd edition)*. Needham Heights, MA: Allyn and Bacon.

Towler, L., and P. Broadfoot. 1992. Self-assessment in primary school. *Educational Review* 44(2):137-151.

Vygotsky, L. 1978. *Mind in society: The development of higher psychological processes*. (M. Cole, V. John-Steiner, S. Scribner, & E. Souberman, Eds. and Trans.). Cambridge, England: Cambridge University Press.

White, B. 1993a. ThinkerTools: Causal models, conceptual change, and science education. *Cognition and Instruction* 10(1): 1-100.

White, B. 1993b. Intermediate causal models: A missing link for successful science education? In *Advances in instructional psychology. Vol. 4.*, edited by R. Glaser, 177-252. Hillsdale, NJ: Erlbaum.

White, B., and J. Frederiksen. 1998. Inquiry, modeling, and metacognition: Making science accessible to all students. *Cognition and Instruction* 16(1): 3-118.

White, B., and P. Horwitz. 1988. Computer microworlds and conceptual change: A new approach to science education. In *Improving learning: New perspectives,* edited by P. Ramsden, pgs. London: Kogan Page.

White, B., T. Shimoda, and J. Frederiksen. 1999. Enabling students to construct theories of collaborative inquiry and reflective learning: Computer support for metacognitive development. *International Journal of Artificial Intelligence in Education* 10(2):121-152.

Appendix A

This appendix contains the outline for research reports that is given to students and an example of a student's research report including data and her self-assessment.

AN OUTLINE AND CHECKLIST FOR YOUR RESEARCH REPORTS

❏ QUESTION:
  ▶ Clearly state the research question.

❏ PREDICT:
  ▶ What hypotheses did you have about possible answers to the question?
    • Explain the reasoning behind each of your hypotheses.

❏ EXPERIMENT:
  ▶ Describe your computer experiment(s).
    • Draw a sketch of your computer model.
    • Describe how you used it to carry out your experiment(s).
  ▶ Show your data in tables, graphs, or some other representation.
  ▶ Describe your real-world experiment(s).
    • Draw a sketch of how you set up the lab equipment.
    • Describe how you used the equipment to carry out your experiment(s).
  ▶ Show your data in tables, graphs, or some other representation.

❏ MODEL:
  ▶ Describe how you analyzed your data and show your work.
  ▶ Summarize your conclusions.
    • Which of your hypotheses does your data support?
    • State any laws that you discovered.
  ▶ What is your theory about why this happens?

❏ APPLY:
  ▶ Show how what you learned could be useful.
    • Give some examples.
  ▶ What are the limitations of your investigation?
    • What remains to be learned about the relationship between the mass of an object and how forces affect its motion?
    • What further investigations would you do if you had more time?

AN EXAMPLE RESEARCH REPORT ABOUT MASS AND MOTION WRITTEN BY A SEVENTH GRADE STUDENT (AGE 12)

During the past few weeks, my partner and I have been creating and doing experiments and making observations about mass and motion. We had a specific question that we wanted to answer — how does the mass of a ball affect its speed?

I made some predictions about what would happen in our experiments. I thought that if we had two balls of different masses, the ball with the larger mass would travel faster, because it has more weight to roll forward with, which would help push it.

We did two types of experiments to help us answer our research question – computer and real world. For the computer experiment, we had a ball with a mass of 4 and a ball with a mass of 1. In the real world they are pretty much equal to a billiard ball and a racquetball. We gave each of the balls 5 impulses, and let them go. Each of the balls left dotprints, that showed how far they went for each time step. The ball with the mass of 4 went at a rate of 1.25 cm per time step. The ball with the mass of 1 went at a rate of 5 cm per time step, which was much faster.

For the real-world experiment, we took a billiard ball (with a mass of 166 gms) and a racquetball (with a mass of 40 gms). We bonked them once with a rubber mallet on a linoleum floor, and timed how long it took them to go 100 cm. We repeated each experiment 3 times and then averaged out the results, so our data could be more accurate. The results of the two balls were similar. The racquetball's average velocity was 200 cm per second, and the billiard ball's was 185.1 cm per second. That is not a very significant difference, because the billiard ball is about 4.25 times more massive than the racquetball.

We analyzed our data carefully. We compared the velocities, etc. of the lighter and heavier balls. For the computer experiment, we saw that the distance per time step increased by 4 (from 1.25 cm to 5 cm) when the mass of the ball decreased by 4 (from 4 to 1). This shows a direct relationship between mass and speed. It was very hard to analyze the data from our real-world experiment. One reason is that it varies a lot for each trial that we did, so it is hard to know if the conclusions we make will be accurate. We did discover that the racquetball, which was lighter, traveled faster than the billiard ball, which was heavier.

Our data doesn't support my hypothesis about mass and speed. I thought that the heavier ball would travel faster, but the lighter one always did. I did make some conclusions. From the real-world experiment I concluded that the surface of a ball plays a role in how fast it travels. This is one of the reasons that the two balls had similar velocities in our real-world experiment. (The other reason was being inaccurate.) The racquetball's surface is rubbery and made to respond to a bonk and the billiard ball's surface is slippery and often makes it roll to one side. This made the balls travel under different circumstances, which had an effect on our results.

From the computer experiment I concluded that a ball with a smaller mass goes as many times faster than a ball with a larger mass as it is lighter than it. This happens because there is a direct relationship between mass and speed. For example, if you increase the mass of a ball then the speed it travels at will decrease.

I concluded in general, of course, that if you have two balls with different masses that the lighter one will go faster when bonked, pushed, etc. This is because the ball doesn't have as much mass holding it down.

The conclusions from our experiments could be useful in real-world experiences. If you were playing baseball and you got to choose what ball to use, you would probably choose one with a rubbery surface that can be gripped, over a slippery, plastic ball. You know that the type of surface that a ball has affects how it responds to a hit. If you were trying to play catch with someone you would want to use a tennis ball rather than a billiard ball, because you know that balls with smaller masses travel faster and farther.

The investigations that we did do have limitations. In the real-world experiments the bonks that we gave the balls could have been different sizes, depending on who bonked the ball. This would affect our results and our conclusions. The experiment didn't show us how fast balls of different masses and similar surfaces travel in the real world. That is something we still can learn about. If there was more time, I would take two balls of different masses with the same kind of surface and figure out their velocities after going 100 cm.

Overall, our experiments were worthwhile. They proved an important point about how mass affects the velocity of a ball. I liked being able to come up with my own experiments and carrying them out.

COMPUTER EXPERIMENTS

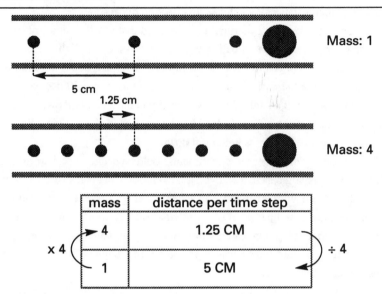

mass	distance per time step
4	1.25 CM
1	5 CM

× 4   ÷ 4

RACQUETBALL			
mass: 40 g.	**distance**	**time**	**velocity** [cm per sec]
trial 1	100 cm.	.56 sec	178.5
trial 2	100 cm.	.43 sec	232.5
trial 3	100 cm.	.53 sec	189.0
average	100 cm.	.51 sec	200.0

BILLIARD BALL			
mass: 166 g.	**distance**	**time**	**velocity** [cm per sec]
trial 1	100 cm.	.60 sec	166.6
trial 2	100 cm.	.50 sec	200.0
trial 3	100 cm.	.53 sec	188.7
average	100 cm.	.54 sec	185.1

VELOCITIES OF TWO BALLS

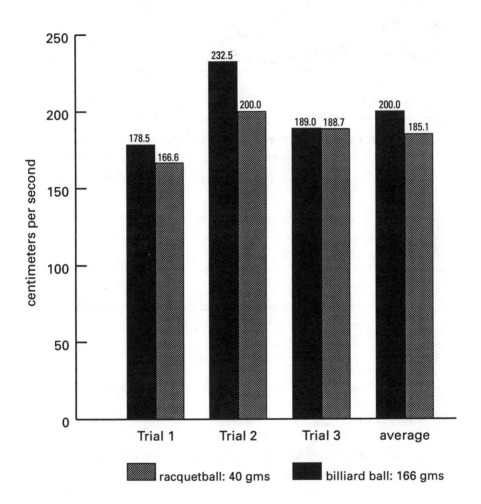

AN EXAMPLE OF A SELF ASSESSMENT WRITTEN BY THE STUDENT WHO WROTE THE PRECEDING RESEARCH REPORT

UNDERSTANDING

Understanding the Science

NA	1	2	3	④	5
	NOT ADEQUATE		ADEQUATE		EXCEPTIONAL

Justify your score based on your work. I have a basically clear understanding of how mass affects the motion of a ball in general, but I don't have a completely clear sense of what would happen if friction, etc. was taken into account.

Understanding the Processes of Inquiry

NA	1	2	3	④	5
	NOT ADEQUATE		ADEQUATE		EXCEPTIONAL

Justify your score based on your work. I used the inquiry cycle a lot in my write up, but not as much while I was carrying out my experiments.

Making Connections

NA	1	2	3	④	5
	NOT ADEQUATE		ADEQUATE		EXCEPTIONAL

Justify your score based on your work. I made some references to the real world, but I haven't fully made the connection to everyday life.

PERFORMANCE: DOING SCIENCE

Being Inventive

NA	1	2	3	④	5
	NOT ADEQUATE		ADEQUATE		EXCEPTIONAL

Justify your score based on your work. What I did was original, but many other people were original and did the same (or similar) experiment as us.

Being Systematic

NA	1	2	3	④	5
	NOT ADEQUATE		ADEQUATE		EXCEPTIONAL

Justify your score based on your work. On the whole I was organized, but if I had been more precise my results would have been a little more accurate.

Using the Tools of Science

NA	1	2	3	4	⑤
	NOT ADEQUATE		ADEQUATE		EXCEPTIONAL

Justify your score based on your work. I used many of the tools I had to choose from. I used them in the correct way to get results.

Reasoning Carefully

NA	1	2	3	④	5
	NOT ADEQUATE		ADEQUATE		EXCEPTIONAL

Justify your score based on your work. I took into account the surfaces of the balls in my results, but I didn't always reason carefully. I had to ask for help, but I did compute out our results mathematically.

SOCIAL CONTEXT OF WORK

Writing and Communicating Well

NA	1	2	3	④	5
	NOT ADEQUATE		ADEQUATE		EXCEPTIONAL

Justify your score based on your work. I understand the science, but in my writing and comments I might have been unclear to others.

Teamwork

NA	1	2	3	④	5
	NOT ADEQUATE		ADEQUATE		EXCEPTIONAL

Justify your score based on your work. We got along fairly well and had a good project as a result. However, we had a few arguments.

REFLECTION

Self-assessment

NA	1	2	3	④	5
	NOT ADEQUATE		ADEQUATE		EXCEPTIONAL

How well do you think you evaluated your work using this scorecard?
I think I judged myself fairly — not too high or too low. I didn't always refer back to specific parts of my work to justify my score.

Young Children's Inquiry in Biology: Building the Knowledge Bases to Empower Independent Inquiry[1]

Kathleen E. Metz

INTRODUCTION

Elementary science educators have long been concerned with achieving a "developmentally appropriate" curriculum and have for many years sought to use cognitive developmental stage theory to derive constraints on curricula at different age levels. While interpretations of Piaget's concrete operational thought have strongly influenced science educators' conceptualization of the inquiry processes that elementary school children *can* do, interpretations of formal operational thought have strongly influenced conceptualizations of what they can't do.

Consider, for example, recommendations from *Science for All Children: A Guide for Improving Elementary Science Education in Your School District*, a recent publication of the National Science Resources Center:

> [Piaget's] theories still provide basic guidelines for educators about the kind of information children can understand as they move through the elementary school.... Through the primary grades, children typically group objects on the basis of one attribute, such as

color.... The significance of this information for educators is that young children are best at learning singular and linear ideas and cannot be expected to deal with more than one variable of a scientific investigation at a time.... Toward the end of elementary school, students start to make inferences.... At this stage of development, students are ready to design controlled experiments and to discover relationships among variables. (1997, pp. 28-29)

The book emphasizes the dangers of failing to teach in accordance with Piagetian stages:

If these developmental steps are not reflected in science instructional materials, there will be a mismatch between what children are capable of doing and what they are being asked to do.... When this kind of mismatch happens over and over again, children do not learn as much as they could learn about science. Equally important, they do not enjoy science. (p. 29)

The *National Science Education Standards* reflects a similar perspective:

Students should do science in ways that are within their developmental capacities.... [C]hildren in K-4 have difficulty with experimentation as a process of testing ideas and the logic of using evidence to formulate explanations.... Describing, grouping, and sorting solid objects and materials is possible early in this grade range. By grade 4, distinctions between the properties of objects and materials can be understood in specific contexts, such as a set of rocks or living materials. (National Research Council, 1996, pp. 121-123)

More generally, in line with widespread interpretations of concrete operational thought, most elementary school science curricula have emphasized the "science process skills" of observation, measurement, and organization of the concrete and, conversely, have avoided abstractions and any thinking demanding hypothetical-deductive thought—including experimental design and analysis of data sets (Metz, 1995). This approach stems from science educators' assumption that cognitive developmental stages constitute largely inflexible, hard-wired constraints on children's reasoning, within which the teacher must teach.

Nevertheless, we have negligible grounds for regarding the stages identified by cognitive developmentists as hard-wired. The reasoning that children exhibit in the developmentalist's laboratory reflects their thinking *without* the benefit of instruction. Furthermore, we have robust evidence that the adequacy of individuals' reasoning is strongly impacted by the adequacy of their knowledge of the domain within which the reasoning is tested. Thus, inside the research laboratory and beyond, cognitive performance is always a complex interaction of scientific reasoning capacities and domain-specific knowledge.

Reader, think of a field in which you have negligible training—perhaps theoretical physics or evolutionary biology. If a researcher tested your reasoning capacities within this field, you could not demonstrate your best thinking and your capacities would be underestimated. In this same vein, cognitive developmental researchers focusing on preschoolers have tended to pay much more attention to using domains with which their subjects have familiarity than have the cognitive developmental researchers focusing on elementary school children. Thus, ironically, if one compares the research from these two literatures, the preschoolers frequently look *more* competent than their older peers!

As another example of the impact of knowledge on reasoning, consider the assumption—derived from interpretations of concrete operational thought—that elementary school children's thinking is bound to the concrete. Indeed, as children tend to know less than adults about most domains, their thinking is most frequently restricted to the surface, concrete features. But when by reason of particular interest or instruction children do have deeper knowledge, their thinking can be abstract. Thus, with instruction, second graders studying ecology in Brown and Campione's (in press) scientific literacy project developed a robust understanding of the abstract idea of interdependence. Conversely, adults with little knowledge of a domain can be restricted to concrete, surface features. For instance, Chi, Glaser, and Rees (1982) found that while physicists categorized physics problems on the basis of abstract physics principles, adults with little knowledge of physics categorized them on the basis of concrete features such as the kind of objects involved.

Carey has pointed out this interaction—and experimental confound—in Piaget and Inhelder's (1958) classic book, *The Development of Logical Thinking From Childhood to Adolescence*, the series of experiments upon which the elementary school science community has largely based their ideas of what elementary school children cannot do in science:

> These experiments confound knowledge of particular scientific concepts with scientific reasoning more generally. It is well-documented that before the ages of 10 or 11 or so the child has not fully differentiate weight, size, and density and does not have a clear con-

ceptualization of what individuates different kinds of metals (densi-ty being an important factor). If these concepts are not completely clear in the child's mind, due to incomplete scientific knowledge, then the child will of course be unable to separate them from each other in hypothesis testing and evaluation. (Carey, 1985, p. 498)

Carey argues that in many of the domains in which Piaget and Inhelder examined the elementary school children's reasoning, the children could not make the differentiations and coordinations Piaget and Inhelder were looking for, due to their weak conceptual knowledge.

A vicious cycle has emerged here. Children's performance in the laboratory is frequently handicapped by weak knowledge of the domain within which they are tested. This weak knowledge has resulted in poor reasoning and thus an underestimation of their reasoning capacities. This underestimation of their reasoning capacities, interpreted as a ceiling on age-appropriate curricula, has resulted in unnecessarily watered-down curricula. The watered-down curriculum has led to less opportunity to learn and thus weaker domain-specific knowledge, again undermining children's scientific reasoning. In short, given the impact of adequacy of domain-specific knowledge on adequacy of scientific reasoning, this approach to defining appropriate curriculum simply maintains children's scientific reasoning more or less at the level with which they enter the classroom.

The American Association for the Advancement of Science's (AAAS) *Benchmarks for Science Literacy* (1993), while trying to delineate a frame for age-appropriate science, also acknowledges the complexity of the relation between developmentalists' descriptions of the stages of cognitive developmental and appropriate science instruction; in particular, the danger of interpreting the stages reported in the developmental literature as hard and fast limits that cannot be modified by effective instruction. *Benchmarks* states:

> Research studies suggest that there are some limits on what to expect at this level of student intellectual development [grades 3rd -5th]. One limit is that the design of carefully controlled experiments is still beyond most students in the middle grades. Others are that such students confuse theory (explanation) with evidence for it and that they have difficulty making logical inferences. However, the studies say more about what students at this level do not learn in today's schools than about what they might possibly learn if instruction were more effective. (pp.10–11)

In summary, to regard the scientific reasoning that children manifest in the developmentalist's research laboratory as indicative of the level of scientific reasoning children can reflect in the classroom assumes the scientific reasoning manifested in that laboratory is largely hard-wired, an assumption that is invalidated by the profound interaction of adequacy of reasoning with adequacy of domain-specific knowledge. The interpretation of cognitive deficiencies manifested in the developmentalist's laboratory as reflections of an interaction of the adequacy of the children's knowledge with stage-based constraints demands a fundamentally different approach to curriculum design. The rest of this chapter identifies the author's conceptualization of knowledge bases needed to empower children's inquiry in biology, and then describes a curriculum aimed at scaffolding these knowledge bases and the learning the curriculum supported.

KNOWLEDGE BASES TO EMPOWER CHILDREN'S INQUIRY

We are working toward the elaboration of this teaching approach and examination of the scientific inquiry it supports at the elementary school level, through classroom-based teaching experiments and parallel laboratory studies. Following the lead of Ann Brown, we think of this tactic as an "educational design experiment." Brown (1992) explained, "As a design scientist in my field, I attempt to engineer innovative educational environments and simultaneously conduct experimental studies of those innovations." While the classroom context enables us to refine the learning environment and study the thinking it supports, the laboratory studies allow us to research key cognitive and instructional issues under more controlled conditions.

While there is a broad spread belief that elementary school children are incapable of independently formulating researchable questions or designing and implementing empirical studies, even at the high school level most science curriculum is not designed to foster independent inquiry (Tobin, Tippins, & Gallard, 1994). As educational psychologist Lauren Resnick (1983) noted, "we do not recognize higher-order thinking in an individual when someone else 'calls the plays' at every step" and yet this level of external control characterizes most science teaching and learning at both the elementary and high school level. In the words of *Benchmarks*:

> The usual high-school science "experiment" is unlike the real thing: The question to be investigated is decided by the teacher, not the investigators; what apparatus to use, what data to collect, and how to

> organize the data are also decided by the teacher (or the lab manual).
> (AAAS, 1993, p. 9)

Nevertheless, the gap between school science and children's science will remain immense unless we scaffold the knowledge that supports their control over the inquiry process. Narrowing the gap constitutes the top-level goal of the project.

We began with the value of maximizing the children's control over the inquiry process, stemming from our view that having significant control and responsibility over the line of inquiry constitutes a *sine qua non* of the delight and essence of doing research. At this point, given children's knowledge-base handicap, the scope of the inquiry within elementary school children's reach is unclear. We are investigating inquiry within their reach by seeking to provide instruction in the knowledge bases most important to effective data-based inquiry in biology.

The research project described in this chapter aims to develop the knowledge bases and associated metacognitive knowledge needed for the children, working in teams of two or three, to undertake largely independent scientific inquiry; conceptualizing, implementing, and revising studies in various domains of biology. We view a strong *meta* focus as crucial for this goal of educating children to take control of their own studies. Our aim is to permeate the scaffolding of the knowledge bases with a concern for metacognition (Brown, 1987), both in the sense of self-regulation and the sense of reflecting on the state of their knowledge—what they know, what they don't know.

Scaffold Domain-Specific Knowledge

Analysis of a range of cognitive literatures leads us to conclude that if we are to support relatively powerful empirical inquiry, we need to concentrate children's science study in a relatively small number of domains. Although the idea that the depth of one's domain-specific knowledge would strongly influence the adequacy of one's reasoning seems so commonsensical that it would have failed Robert Siegler's test for a worthwhile research question (Wouldn't your grandmother have assumed that thinking is best when one has knowledge of the domain?), the idea has frequently not been reflected in children's science curricula. As Brown, Campione, Metz, and Ash (1997) stated, in the majority of children's science texts, "There is a striking lack of cumulative reference (volcanoes following magnets, following a unit on whales, etc.). This lack of coherent themes or underlying principles all but precludes systematic knowledge building on example, analogy, principle, or theme or theory; it does not encourage sustained effort after meaning." (pp. 20-21)

Given the deep connection between adequacy of domain-specific knowledge and adequacy of scientific reasoning, a curriculum structure which emphasizes coverage over depth handicaps inquiry. Speaking of the importance of specialization in the biologist's thinking, evolutionary biologist, and historian of science Stephen Jay Gould writes:

> No scientist can develop an adequate "feel" for nature (that undefinable prerequisite of true understanding) without probing deeply into minute empirical details of some well-chosen group of organisms. (1985, p.168).

Similarly, how can children transcend the most superficial scientific inquiry—or even be in a position to formulate questions worthy of investigation—in a domain in which they have little knowledge? If we are to support children's thinking in the cognitively demanding task of scientific inquiry, we need to avoid the survey curriculum structure. By concentrating children's science in a relatively small number of domains, we can support the depth of knowledge that begins to create the conditions for knowledge-building and effective scientific reasoning.

Scaffold Knowledge of the Enterprise of Empirical Inquiry

In most of the research literature examining the impact of knowledge on scientific reasoning, "knowledge" connotes level of expertise in the field as a whole (e. g., of physics or biology) or, more specifically, one's knowledge of that aspect of the field involved in the task at hand (e. g., Newtonian mechanics or natural selection). However, as Brewer and Samarapungavan (1991) have asserted, in their shared knowledge of the culture of science scientists draw upon knowledge that extends far beyond the bounds of their particular specialty. "Clearly," Brewer and Samarapungavan argue, "the individual child does not yet share the enormous body of knowledge that is part of the institution of science" (p. 220). These authors identify experimental methodology and associated institutional norms as one of the fundamental knowledge bases that differentiate the scientific reasoning of adult scientists from that of children. Our project aims to scaffold the knowledge fundamental to empirical inquiry, avoiding a simplistic portrayal of some singular scientific method.

The distinction between observation and inference, for instance, undergirds the research enterprise. Thus we begin the curriculum with a focus on this distinction and at a subsequent point in the children's investigations introduce the parallel distinction between theory and evidence (and counter-evidence). Similarly, carefully designed experiments with appropriate controls are fundamental to experimental research. The curriculum scaffolds the idea of a "fair" experiment, extraneous variables, and different means of controlling them. More subtle, we are seeking to help children grasp distinctions between the norms of scientific argumentation and everyday classroom argumentation, such as the importance of recognizing and codifying uncertainty and the acceptability of a failure to reach consensus.

Scaffold Knowledge of Domain-Specific Methodologies

A classic distinction within the cognitive science research tradition is the difference between weak and strong methods. "Weak methods" denote domain-general problem-solving strategies (such as generate and test and means/ends analysis) that one without any expertise in the domain at hand can draw upon. "Strong methods" denote the problem-solving strategies specific to the domain, and powerfully suited to its particular characteristics, purposes, and challenges. Domain experts are defined and identified in part by their knowledge of these strong methods. We conceptualize the intersection of the two knowledge bases delineated above—domain-specific knowledge and empirical inquiry—as the specialized methodological techniques of the domain under study.

To empower children's inquiry, we need to identify a repertoire of methodologies that will enable them to conduct effective investigations in the field under study. For example, while in a preliminary statistics module we focus the children's investigations on ideas of sampling, later in the study of animal behavior we teach children the domain-specific sampling techniques of time-sampling of behavior and time-sampling of location. In the study of botany, they learn other domain-specific sampling methods; including relevé sampling (continuing to double the area sampled until they study an area in which they find no additional new species) and random quadrat methods (involving random selection of which squares in the grid are to be sampled).

Scaffold Data Representation, Data Analysis, and Fundamental Constructs of Statistics and Probability

Knowledge of how to represent and analyze data, in conjunction with fundamental constructs of statistics and probability, is fundamental for data-based scientific inquiry. We seek to develop knowledge of a repertoire of ways of representing data; including the meta-knowledge of when to use each one, what kinds of patterns each makes salient, and what each obscures. Similarly, we aim to develop elementary heuristics of data analysis, to help the children identify patterns and variability within them. Here, in particular, the heuristics vary by grade level. For example, at the second-grade level the curriculum includes analysis of the mode, median, and range, and identification of the biggest clump in the distribution of their data, whereas at the fifth-grade level children can also consider such aggregate measures as means and quartiles, and the symmetries and asymmetries in the distributions.

Even when science is taught from the perspective of active investigations, the challenges of data analysis and the uncertainty in data interpretation are rarely confronted. However, real data sets frequently are messy and doing science involves consideration of uncertainty and chance, from the point of research design to data analysis. Some fundamental knowledge of statistics and probability, at the level of ideas of chance variation, randomness, the Law of Large Numbers, and sampling, are needed for the practice of data-based research. We aim to foster these ideas through investigations focused on statistical ideas, as well as investigations focused on science domains in which children use the statistics.

Scaffolding Knowledge of Tools

Tools such as external memory aids, calculators, data representation software, and flexible knowledge bases are considered fundamental to the scientists' work. Such tools are assumed to profoundly affect the reach of their inquiry. From the perspective of information-processing theory, these kinds of tools can fundamentally expand the processing capacities available to the human problem-solver.

Indeed, historians of science frequently identify a new tool as crucial in the development of new constructs. The computer constitutes a preeminent example of a tool fundamentally changing the bounds of human cognition along multiple parameters. In his seminal article, "The Science of Patterns," mathematician Lynn Steen examines the impact of the advent of computers on the field of mathematics:

> [T]he computer is now the most powerful force changing the nature of mathematics. Even mathematicians who never use computers may frequently devote their entire research careers to problems generated by the presence of computers. Across all parts of mathematics, computers have posed new problems for research, provided new tools to solve old problems, and introduced new research strategies. (1988, p. 612)

In his analysis of the mathematics that made the computer possible and, in turn, the mathematics made possible by the computer, Steen documents a complex interaction of technological and conceptual advancements. Wiser and Carey's (1983) historical case study of the differentiation of heat and temperature constitutes another example of the power of tools and, as in the case of the computer, the interaction of conceptual and technological advancements. Wiser and Carey analyze changes in the understanding of heat and temperature in conjunction with an analysis of the refinement of the thermometer.

Whereas the invention of either computers or thermometers demanded significant breakthroughs, the naïve—including young children—can successfully use either tool. Thus while the thermometer supports the novice's measurement of temperature as well as the differentiation of heat and temperature (still a nontrivial distinction), the computer can support a broad range of cognitive functions. We view the scaffolding of children's knowledge of relevant tools as key to the empowering of their inquiry. For example, in our project the children learn to use visual tools such as binoculars and microscopes, in accordance with the needs of their sphere of study. They learn to use the computer for a variety of functions; including knowledge-base reference sources, simulations, vehicles to contribute to national databases, compiling their own data sets, data representation, and data analysis.

A CURRICULUM MODULE IN ANIMAL BEHAVIOR

The author developed curriculum modules in ornithology, animal behavior, and ecology. This has been supplemented by curricula on the mathematical and statistical aspects of inquiry, including lessons drawn from the TERC curriculum *Investigations in Number, Data and Space* (e.g., Russell, Corwin, & Economopoulus, 1997; Tierney, Nemirovsky, & Weinberg; 1995) and other units developed by the author. In the first year, the fourth/fifth-grade class focused their project-related science in ecology, while the second graders focused on animal behavior.[2] In the second year, the fourth/fifth-grade class studied animal behavior through a modified version of the second-grade curriculum. In the third year, second graders, third graders, and fourth/fifth graders studied ornithology in the fall and animal behavior in the spring. In the upcoming year, the foci will be botany and ecology in three classrooms, including grades one, three, and five. The three project teachers and the author are currently engaged in developing the botany curriculum.

This chapter describes how we implemented the design principles in the animal behavior module. There are a number of reasons why animal behavior constitutes a strategic domain to concentrate young children's scientific investigations. This domain of inquiry is well suited to children's analysis of patterns and variability in data, along with their reflection on ideas of uncertainty, chance, and causality. The domain is amenable to both observational and experimental research. Finally, this sphere of investigation holds high interest for children.

Children began their study of animal behavior by observing rodents in a large enclosed area; focusing on issues of behavior from a scientific perspective, scientific note taking, observation versus inference, and stimulus and response. (See Table 1.) In accordance with our *meta* focus, in those instances where there was not consensus about attribution of observation or inference, the children reported the basis of their differing attributions. This supported rich discussions of the continuum from low to high inference and the elements of interpretation that can enter in, even at this relatively basic level of the research process. Despite claims to the contrary, children even at the second-grade level, after one or two sessions, were *all* successful at the task of distinguishing inference from observation and providing appropriate justifications for their attributions. These initial observations were later elaborated by the exploration of such ideas as stimuli and response (in which the children proposed objects that might prove of interest to the rat), the challenge of trying to affect minimally what is being observed, social behavior, and the potential of variability in recorded observations between different observers.

TABLE 1. SUMMARY OF ANIMAL BEHAVIOR CURRICULUM MODULE

COGNITIVE OBJECTIVES	INSTRUCTIONAL ACTIVITIES	2nd	4/5th
Scientists' view of "behavior" Distinction between observation and inference (high and low).	Observe an animal, taking notes that differentiate and link observations and inferences. Collaboratively examine and critique attributions of observations and inferences.	x	x
Knowledge about the anatomy, life cycle, needs, and behavior of insects in general and crickets in particular. How is knowledge of the animal behavior used to protect and preserve species?	Directed drawings. Analysis of videos, books, knowledge-bases on computer software, and thought experiments based on scientists' accounts of resolving conceptual and/or methodological dilemmas.	x	x
Concepts of stimuli and response. Social behavior Goal of observation with minimal disturbance of the observed.	Observe and analyze behavior in response to different stimuli. Observe two animals together, focusing on social behaviors. Attention to the way in which they observe (cf. noisily) can unintentionally constitute an additional stimulus.	x	x
Conceptualization of interpretative and subjective element in doing science, even at the level of observations.	Each "research team" (mostly dyads) observes a cricket; thereafter identifying a behavior they both noticed but described differently; then exploring possible explanations for difference.		x
Purpose and process involved in constructing coding categories from notes regarding behavior.	Teacher helps children construct categories of behavior for time-sampling from their compiled lists of observations. Children may develop their own categories.	x	x

COGNITIVE OBJECTIVES	INSTRUCTIONAL ACTIVITIES	2nd	4/5th
Time sampling of behavior, multiple representations of these data, with idea that different representations highlight and obscure different patterns and relationships.	Children collect time sampling of behavior data, represent in multiple ways, analyze, and formulate questions. Children enter categorical data into computer; analyze data using different representations.	x	x
Conceptualization and analysis of social behavior.	Reiterate with multiple animals to examine social behaviors.	x	x
Time sampling of location as data collection technique and analysis thereof .	Children develop maps of their crickets' terrarium. Collect and analyze time sampling of location data under solitary and social conditions; formulate questions.	x	x
Research design. Experimental controls.	Research teams select a variable for experimental manipulation and identify other variables they will need to hold constant. Collect time sampling of behavior data.	x	x
Development of experimental procedure. Representation and interpretation of numerical data.	Class collaboratively develops procedure for studying how far a cricket hops. Children enter their numerical data into computer; analyze data using different representations.	x	x
Conceptualization of a menu of research methodologies and their utility.	Class collaboratively develops a table of questions they have researched, methodologies employed, and steps involved (including methods used in prior domains).	x	x
Conceptualization of a menu of data representations and their utility.	Class collaboratively develops a table of data representations and what they are good for.		x

COGNITIVE OBJECTIVES	INSTRUCTIONAL ACTIVITIES	2nd	4/5th
Understanding and application of the distinction between researchable and unresearchable questions.	Presented with heuristic of "Can you [begin to] answer this question through collecting data?", children brainstorm questions and discuss whether or not they are researchable.	x	x
Question formulation. Heuristics for question generation.	Throughout module children record their questions. Class explores heuristics for question generation; e.g., comparisons of same crickets under different conditions or different classes of crickets under the same conditions.	x	x
Conceptualization and implementation of a research study. Possibility of competing theories to account for same data sets.	Research teams develop question, select methodology, formulate experimental procedure, construct experimental apparatus, collect, represent and analyze data, prepare and present research poster.	x	x
Cyclical nature of science: revision of research designs and ideas. Distinction between science of discovery & science of verification.	Research teams think through additional studies needed, in order to be more confident in their findings. Research teams identify uncertainties and weaknesses in their studies and plans to address.	x	x

We made the transition from observational research to experimental research by presenting the challenge of thinking through the categories that could capture most of the behaviors on the class's compiled list of observations or, alternatively, to generate a classification schema that focuses in on the sphere(s) of behaviors on which they would personally like to concentrate their observations. Given our goal of independent inquiry, the teacher did not identify the categories, but rather presented the task to the children. As philosopher of science Karl Popper (1972) has written, the observation and classification each presume "interests, points of view, problems"; without focus on purpose, the task becomes "absurd" (p. 46). We aim to have children's categorization-construction be problem-driven. More generally, our goal here and throughout the curriculum is for children to learn how to assume responsibility and control over the decision-making process, as opposed to simply carrying out the activity as specified by the teacher.

To build toward more independence, instead of having each child in the class contribute to a compiled data set, at this point in the curriculum children were assigned to "research teams" of two children at the second-grade level and two or three at the fourth/fifth-grade level. Assignments to teams were formed largely on friendship lines, in order to facilitate smooth functioning of the collaboration. At this point where the children begin to function as research teams, we substituted crickets for the rodents, in order to provide sufficient organisms for all.

In parallel with these and subsequent empirical instructional activities, we sought to develop the children's knowledge of the anatomy, needs, life cycle, and behavior of insects in general and crickets in particular. We employed directed drawings, books the children read independently, books the teacher read to the children, videos of animal behaviorists at work, computer software, teacher presentations, and thought experiments. In our version of thought experiments, the teachers would present a question that a real animal behaviorist had faced, have the children to try to formulate strategies to address the question, and then examine the scientist's thinking. For instance, we framed one thought experiment in terms of Stephen Emlen's (1975) curiosity, theory-building, and experimentation concerning how birds found their way in the course of migration.

At first all the teams conducted the same studies, designed to promote exploration and reflection on different methodologies and different data representations. As the class moved from one research approach to the next, the teacher led the children's collaborative analysis on what the different methodological, analytical, and representational techniques elucidated and obscured.

After completion of these studies, the teacher helped the class construct two tables, one of data representations, the other of methodologies, that explicitly defined the range of techniques they had employed (including ones from prior modules), the purpose(s) for which they had used the technique, and more generally what it was good for. We viewed these menus as external memory aids that we hoped would eventually become internalized. Through the process of constructing these "menus" of possibilities and the end-product of the two tables, we aimed to support the students' independent decision-making in the context of their own studies.

The next phase involved considerably more independence, as the teams formulated their own research project. This phase of the design and implementation of the teams' research studies was the most challenging and exciting to children and teacher alike, which we attribute to the fact that the question they investigate is their own and the pathway they take to investigate it of their own design. This project drew intensively on the knowledge—conceptual, procedural, epistemological, and metacognitive—that the children had developed earlier in the unit. A recurrent theme in the transfer literature is the difficulty of knowing when knowledge learned in one situation is relevant in another (Pea, 1987). Transfer becomes a particularly challenging issue in the context of experimental research, as children need to be able to decide not only what technique may be well suited to a particular question, but also how to implement or adapt the technique for the particular characteristics of the task at hand.

The first challenge involved formulation and refinement of the question for investigation. We had tried to engineer the classroom learning environments as cultures that fostered generation of questions above and beyond requests for clarification. Throughout the module the children had been recording, in their science notebooks, questions they had about cricket behavior. At this point in their study, we also explored with them different heuristics for generating researchable questions. One crucial consideration is whether a question is amenable to empirical investigation. The heuristic we used here is "Can you [begin] to answer this question through collecting data?" In this context, we also scaffolded the children's exploration of limits in the power of a single study to resolve a question. Additional heuristics used for brainstorming questions included identifying particularly interesting behaviors or spheres thereof (for example, chirping, hopping, or aggressive behaviors); conceptualizing conditions which might affect cricket behavior; listing classes of crickets for which they had reason to anticipate variability in behavior; and, at a more aggregate level of question formation, comparing different crickets under the same condition or of the same crickets under different conditions.

Following formulation of their research question, the research teams started the process of designing their study by identifying the most appropriate methodology. While most children selected a method from the menu of possibilities, some invented approaches outside of it (for instance, planning to study the competence of crickets in working their way through a maze to get food, by building a maze and charting the crickets' progress through the maze over time). After checking back with the teacher and obtaining her agreement that their question was researchable and the methodology they had selected was well suited to the question, the children planned their study in more detail. At the second-grade level, this included a reiteration, frequently in more refined form, of the question to be investigated, a listing of materials they would need, and a step-by-step description of their experimental procedure. At the fourth/fifth-grade level, the teams' elaboration of their study included: (a) specification of the question under investigation; (b) if they conceptualized their study, not as an exploratory study, but as one testing an hypothesis, then their hypothesis and their reasoning underlying it; (c) step-by-step delineation of their experimental procedure, such that another investigator could replicate their study; (d) delineation of the extraneous variables that they thought might affect their results; (e) their plan for coping with these extraneous variables (with the caveat that it might not be feasible to control for all of them); (f) a sketch of the data sheet on which they planned to record their data; and (g) how their research plan would help them reduce the possibility that their results were simply due to chance.

The children's subsequent work—building their experimental apparatus, collecting their data, entering their data on the computer, selecting two data representations that suited their purposes—was implemented largely independently, with the teacher helping children think through occasional stumbling blocks. We structured this process through specification of the end-state, in the form of a spatial representation of a research poster and its constituent parts. Components of the poster for second graders were:

(a) Question: Write your question here. If you had an idea about what would happen, explain what you thought would happen and why. You could say why you thought this question was interesting or how you came up with the question.

(b) Organisms: Tell what crickets you used (what kinds, how many of each).

(c) Materials: Make a list of materials you used.

(d) Experimental Procedure: Describe step-by-step your procedure. Include what you kept the same and what you varied. Say enough about how you did it, so another scientist could repeat your study.

(e) Experimental Set-Up: Draw a picture of how you set up your materials to collect data.

(f) Data representations: Include two representations of your data.

(g) Data Analysis: What do your data tell you about your question? What is your evidence AND counter-evidence for your interpretation?

(h) Conclusions: What did your study tell you about your question? What are you sure of? What are you not sure of? If you think there is more than one way of explaining your results, include both. What new questions does your study raise?"

The fourth/fifth-grade template consisted of a somewhat more elaborated version of the second-grade version. The teachers found this template very useful in scaffolding independence, as any team of children could go back to the template to see what needed to be done next.

In the last phase of the project, each team critiqued their study and designed a new one in a group interview with the author (a component we are now incorporating into the curriculum itself). The children were asked to think through how sure they were of their results and how they could be more sure. Subsequently, they are challenged to formulate changes that would improve their study and to explain why these changes would strengthen it. They were also posed the thought experiment: given the chance to do another study of cricket behavior, what question would they research and what method would they use? This activity helped the children to think further about issues connected with the science of discovery ("What would I like to find out? What am I curious about?"), as well as the science of verification ("Have I been fooled?"). These questions also encouraged the children to think more deeply about issues of research design.

EXAMPLES OF CHILDREN'S RESEARCH PROJECTS
AND THEIR REVISIONS THEREOF

Across the age span from second through fifth grade, all the teams—including those comprised of two academically weak children—formulated a researchable question and corresponding research plan, and subsequently implemented their study. Second and fifth graders alike were concerned about "fair tests"; they identified variables that could affect their results that they needed to hold constant or, where they couldn't or didn't keep them the same, needed to keep in mind as they sought to interpret their findings. All children formulated ideas of how to improve their research design that would indeed have strengthened their study. Our laboratory studies of the children's understanding of scientific inquiry indicate a much higher level of sophistication than that generally attributed to the elementary school child.

At the second-grade level, studies that pairs of children have formulated include: (a) "Do crickets act differently inside and outside?" relying first on careful observation to construct their coding categories and then time-sampling of behavior; (b) "Do crickets stay away from other bugs?" using time-sampling of location; (c) "Do crickets act differently in different colors?" using time-sampling of behavior; (d) "Do crickets do more in the night or in the day?" using time-sampling of behavior; and (e) "What size do female crickets get their ovipositor (the tube through which the female lays her eggs and deposits them in the ground)?" by measuring under a microscope (after slowing them down by putting them in the refrigerator). At the fourth/fifth-grade level, studies that pairs or triads have conducted include: (a) "What are the crickets' defenses in defending their territory and their mate?" using first careful observations to identify the defenses and subsequently time-sampling of behavior to research their relative frequency; (b) "How does a hissing roach and cricket nymph (at different times) [sic] affect a cricket's behavior?" using time-sampling of behavior; (c) "Does the gender affect how much the crickets eat?" using measurement; (d) "Where would crickets go on the playground?" using time-sampling of location; and (e) "Would two females fight over a male?" using time-sampling of behavior.[3] To provide a more detailed picture of the children's work, a case from the second grade and the fourth/fifth combined class is included here.

Second graders Ashley and Maria studied "What food do crickets like the most, dog food or apple?" On their poster, Ashley and Maria state that on the first day they put the same size of apple and dog food in their terrarium, and on the next two days checked to see how much food was missing from each. In their final interview with the author, they contended that the crickets like apples more than dog food, basing their argument on their data:

> **Ashley:** We found out which one they liked because they ate the apple before they ate the dog food.
>
> **Maria:** We put the dog food and the apple in [the terrarium] and it was the same size. And then, the second day the apple was half gone. The dog food was still there.
>
> **Ashley:** The dog food had just like two little bites.
>
> **Maria:** Two little bites. And then the next day the apple was all gone and the dog food had three bites.

When the students are subsequently asked "Can you think of a way you could change how you did your study to make it even a better study?", the girls have several ideas. Maria suggests manipulating the amounts of food, providing only a small amount of the seemingly less desirable food, in combination with moving to a "no choice" situation of sequential food presentation.

> **Maria:** First we put the apple in, the most; and then the dog food a little bit and then we'll see if he eats it all. And if he eats it [the apple] all and the dog food only a little bit, then we'll be really really sure, because the apple of course was bigger.
>
> **Author:** I want to understand your ideas and how your ideas right now will make it stronger than it was before.
>
> **Maria:** If we put both [kinds of food] then maybe the cricket will choose. But if we put the dog food first, maybe it, it <u>can't</u> choose. So maybe it won't eat and just a few bites if it gets hungry. And then maybe when we put the apple in, it'll eat all of it. It won't get no choice. No choice.

Thus, within Maria's logic, if the cricket chooses hunger or minimal bites when presented the dog food and then consumes all the apple, they will have stronger evidence for their initial conclusions.

Ashley suggests another strategy for improving their research, based on the experimental confound she identifies between eating (and associated behavior of approaching the food) and social aversion.

> **Ashley:** I've got another idea! I think we should put the dog food on one side [of the terrarium] and the apple on the other.

Author: How did you do it?

Ashley: We had them like this [indicating, with the distancing of her hands, a space of about two inches].

Author: Why do you think that would make it better?

Ashley: Because maybe they didn't want to be that close and they didn't eat that much. I think we should have put them [two kinds of food] on sides like that [far ends of the terrarium]. Because I think that the female or the male would be scared of the female to get close to them and it may scare them away from each other and they won't want to eat.

In the first implementation of the animal behavior curriculum at the second-grade level, we had not introduced the idea of experimental confound since we assumed that it would be too sophisticated an idea for these children. To our surprise, several children independently introduced it. In one context, the teacher had modeled how to think through the development of an experimental apparatus and experimental procedure, using the question of whether temperature affected the crickets' behavior. As part of this discussion, the class formulated a list of how to make the terrarium hot: heating pad, hot rock, or heat lamp. The next day when one team began to construct an experimental apparatus using a heat lamp to create hotter conditions, the author heard (and recorded) a second grader from another team advising, "Better not use the lamp. Then you'll have three things: Cold, hot, and light." Although the idea and concern of experimental confound was not shared with the class at large, it still was independently invented by two other research teams, including Ashley and Maria's. We now incorporate in the second-grade version of the curriculum the issue of extraneous variables and the importance of trying to hold constant all the variables outside of the one under investigation. The children have interpreted the idea as a way to make a "fair test" of the variable under investigation.

Kimberly and Matthew are second graders in the second year in which we implemented the animal behavior curriculum. They had the benefit of our curriculum revision, which included greater emphasis on research design, adequacy of sample size, and the use of computer tools—software encyclopedias and *Data Explorer*, a computer-based tool for recording of data and data representation. (We did not use the program's data analysis function at the second-grade level.) Kimberly and Matthew were curious about when the ovipositor first appears on the female cricket. They checked their reference sources, using the computer-based encyclopedia and classroom cricket books. When they found no source that provided the answer, they decided on this question for their investigation.

Their research poster description of their experimental procedure includes: "4) Put nymphs under microscope. 5) Observe carefully measure size of nymph and ovipositor if it has one [sic]. 6) Keep accurate records." To maximize the number measured, Kimberly and Matthew searched all three project classrooms for nymphs, finally measuring sixty-four individuals (ranging in length from five to eighteen millimeters). They used *Data Explorer* to enter their data. Their data representations were the data sheet itself and a scatterplot. As they collected their data, they developed an intuitive feel for the emergent relationship between nymph length and ovipositor length, as revealed in Kimberly's comment, "This one has got to be a male. If it was a female, its ovipositor would be about ten millimeters."

In the context of the interview, Kimberly and Matthew reflect on the issue of when the ovipositor first emerges in the female, including the uncertainty of their results, the source of the uncertainty, and the strategy for addressing it:

> **Author:** What did you find out?
>
> **Kimberly:** That usually it has to be at least eight millimeters to get an ovipositor, because the littlest one that had an ovipositor was eight millimeters.
>
> **Author:** How sure are you that a cricket has to be at least eight millimeters before it starts to grow?
>
> **Kimberly:** Not very sure, because just on one study, that can't tell you anything about it. Because we only had one that was eight millimeters. And we measured it. And it was eight millimeters. But that was only one.
>
> **Author:** Matthew, how sure are you that crickets have to be at least eight millimeters long before they get their ovipositor.
>
> **Matthew:** I'm not so sure. Maybe there might be one that, probably, has seven millimeters that has one.

When the two are asked "Is a way to improve your study to be more sure that crickets have to be eight millimeters long before they get their ovipositor?" Kimberly immediately addresses the need for including more crickets in their study:

> **Kimberly:** Study more nymphs. You have to look at um more different nymphs. And lots more.

Matthew: [Nodding agreement]

The children subsequently address the question of changes that they could make in their study to make it even better seemingly in biological engineering terms, trying to conceptualize changes in the environment which might result in earlier appearing ovipositors:

> **Kimberly:** Maybe grass in the habitats. It might change it a little.
>
> **Matthew:** Probably if we [pause]. Maybe if we add more food, they'll get theirs quicker. If you put them in hot, maybe they get an ovipositor quicker.

In the first year of the animal behavior curriculum at the fourth/ fifth-grade level, two fourth graders, Shayla and Leah, and a fifth grader, Jena, designed a study to get at the question "Do crickets react differently when they are in different colors of light?" using time-sampling of location. The team collected time-sampling data for six female crickets and six males; under conditions of red, clear, green, and black light (as produced by light bulbs of these colors in a dark closet). They found considerable inter-variability of behavior in the various light conditions (for example, "half of the crickets stayed still in the red light. Four of the six crickets stayed still in the green light").

When asked if there were ways they could be more sure of their findings, like most children at all grade levels they think extending the number of individuals tested and retesting them would help:

> **Author:** Is there a way you could be more sure of your results?
>
> **Shayla:** Yeah.
>
> **Leah:** If we tested more crickets and tested the same crickets over again.
>
> **Author:** And why would that help?
>
> **Leah:** 'Cause they might do different things.

The team subsequently addresses the issue of a possible order effect; that is, that their results may have been due in part to the order in which the different colored lights were presented. To address this problem, they propose varying the order.

> **Shayla:** If maybe you'd change the routine of the light colors. 'Cause in this one the whole time we did red, clear, green and black. Maybe red, green, black, clear.

Author: And why would you want to do that?

Shayla: 'Cause maybe it like maybe it would be on time. Like maybe they're like, maybe they could be hungry and you wouldn't know that and maybe they didn't feed them or something.

Jena: If it went like clear, red, black, green. Say if the green was last and they might be tired and they might stay still, so….

Leah is concerned that they have tested an insufficient range of colors and suggests adding more different light conditions:

Leah: If we use more lights…. Because it might make different if there's different colors of light. Like they might just not be different in those colors of light and the other colors of light they might react the exact same as the clear.

The girls raise the possibility that their results may be affected by where the experiment has been conducted:

Jena: It also might affect them, like what their surroundings are. It's like if we did it in here, they'd be surrounded by….

Leah: Books.

Jena: Books, yeah, and if we did it outside like in their normal habitat, then they'd probably act different.

Author: So what, so what's your suggestion there for what you might do differently to be sure, more sure?

Jena: Like we could do it in two different places. We could do it like in here and in a like a grass field or something.

Asked how they might improve their study, the girls have several ideas: increasing the number of organisms tested, controlling the experience of those tested, extending the range of light color conditions, constructing purer light conditions (where all stimuli in the testing area reflect the targeted light color), adding additional species of crickets, counterbalancing the order to address potential order effect, keeping constant a newly identified extraneous variable or, alternatively, redesigning the experimental procedure to study the crickets' behavior under conditions of experimental manipulation of the extraneous variable. For example:

Author: When scientists finish their study, they frequently think, "Hmm, how could I change my study to make it even better." It's a good study already, but how could you make your study an even better study?

Jena: Test more crickets maybe.

Author: Why would you want to do that, Jena?

Jena: To be more accurate, 'cause we like tested only six and tested each cricket four times, well like five times on each light, but there's four [lights]. You might want to test, like instead of European House Crickets, you could test like, yeah, like a tree cricket or something.

Leah: And we had to um make sure that they weren't nymphs, 'cause we started out with grown-ups and then we get new crickets and they were all nymphs. So we had to borrow some and those ones were used to other cages. So if we got them all the same time and they were all grown-ups and we didn't have to get somebody else's. Maybe, since they have been through the same thing,…maybe they'd act the same.

Jena: 'Cause if we had just like get them out of the container thing, then maybe, and put them in the cage and tested them like a minute or a couple of minutes later, they might be exploring more….

Shayla: They had never been in our cage before, ever…. Maybe, so they were kind of scared.

Conversation continues.

Leah: We could do an experiment on that. Are they more active in this [cage] or this [other] cage?

Shayla: In a different arrangement.

Jena: It would give us more of a variety to see the crickets. 'Cause the crickets [that the team borrowed] had never been in there and we changed the cage [introduced the novel cage] and the crickets. I think we should've maybe like tested, we should have left the cage that way and took the crickets from the first cage we had and then we should have tested them in this and then take the other four crickets [their own crickets] and test them in this cage. We should have tested all the crickets in here, that cage, and all the crickets in that cage.

Shayla: Because—

Leah: Maybe they would act different. We would have two tests, two tests on each. So if like somebody didn't know anything about our experiment, they could look at it and say, oh, they did it in two different kinds of cages to see if they reacted differently in different colors of light and different cages and they could look at each one.

Also in the first year of implementation of the animal behavior curriculum, two fourth graders, Bryant and Douglas, researched "Will smaller or larger cages affect the crickets defensive fighting methods?" using time-sampling of behavior. The data representation on their research poster represented changes in level of intensity of aggression over time. The y axis indicated intensity of aggression as derived from an ordering of their coding categories; the x axis indicated time in one-minute increments (as coded in the time-sampling), across the four experimental conditions. The four experimental conditions consisted of: (a) large container, no female; (b) small container, no female; (c) small container with female; and (d) large container with female. A change over time graph is included for both Trial One and Trial Two. The conclusions on their research poster reflect a value on aggressive behavior and consequently disappointment with its absence:

> By looking at our data, you can see that small with female got the most action. If you look at trial two, you can see we had a big disappointment. The reason we think that the fighting level went down is because they already had their territory marked. The question our experiment raises is why did the crickets fight less in trial two.

In their interview with the author, Bryant and Douglas again raise the issue of how they can account for why the intensity of the fighting decreased over time and, more generally, why crickets fight:

> **Bryant:** Well, our first trial they fought a lot more. The second trial, they didn't fight as much. And we think that because they had already marked their territory.
> **Douglas:** And claimed it.
> **Bryant:** And claimed it so the other crickets knew to stay away. 'Cause look at our results on this other one [Trial One]. Mostly all like um walking over, no movement, climbing, like drinking....
> **Douglas:** It [the graph] shows what was made up of how intense the fighting was.
> **Bryant:** We put headbutting at the top because we thought that one was the most intense.... Then put kicking, wrestling.

Note that in their research poster and interview, the boys clearly differentiate between their theory and evidence.

The author then brings Bryant and Douglas back to the question they initially posed and asks them how sure they are about their results.

> **Author:** So will smaller or larger cage affect the crickets' defensive fighting method?
> **Bryant:** We really never answered that.

> **Author:** Do you think you can answer your question?
> **Bryant:** But if we— Yeah, it does affect it....
> **Author:** How sure are you?
> **Bryant:** We're in the middle of sure, sure, and sure.

> **Douglas:** Well, I'm pretty certain, but I won't say that because we were using the same crickets on both trials.

Douglas then shifts back to the question of what the crickets were fighting over: was it simply territory or something else?

> **Douglas:** So I'm thinking that if we. What I'd like to do is switch around the whole cage and make something new in it.
> **Bryant:** Yeah!
> **Douglas:** And see if they do the same thing. But if they don't, that um it really shows that they might um, it might not affect. Still know where it is, but it might not affect them. They might have just fought for territory.

> **Author:** What do you mean, change it all around?
> **Douglas:** Like move everything.
> **Bryant:** Like put our big water dish in the small one and put more sticks in the little one and put different kinds of sand or something.

When probed for other ways in which they could be more sure of their conclusions, the boys get into ideas of replication, increasing the number of crickets tested, and the need to restrict their study to adults. Their rationale for using only adults reflects the key issue (and methodological correlative) of different needs at different stages of the life cycle:

> **Author:** Is there another way you could be more sure of your conclusion?
> **Douglas:** And conducting other experiments the same way.

Bryant: With different crickets. The same way but with different crickets. We couldn't find enough adult crickets. 'Cause this is mostly with adults....

Douglas: Nymphs won't fight because they're immature and they didn't have a need for territory. But the adults needed for their mates and stuff.

Subsequently, when asked how they might improve their study, Douglas proposes further experimental manipulations to test whether the crickets' fighting is indeed over territory.

Author: When scientists finish a study, they frequently think about "How could I have done the study so it would be even a better study?" How might you have done it differently to make it even better.

Douglas: Maybe no food, because they might be fighting over the food.

This question concerning the impetus for crickets fighting stays with the boys throughout the year. Eight months later, upon hearing that as fifth graders they will be designing new animal behavior studies, Douglas came up to the author to request that he and Bryant again work together, on the basis that they hadn't finished their research project—questions still were left unanswered. Douglas, Bryant, and Sarah (with the class now divided into triads) designed and conducted a study to research the primary stimulant of the aggression: female, food, or territory.

In parallel with our first implementation of the curriculum at the second-grade level and at the fourth/fifth-grade level, we observed child-initiated scientific reasoning above our expectations and thus in subsequent years have scaffolded, for all the fourth and fifth graders, more sophisticated research design. At the fourth/fifth grade, we now introduce more strategies for coping with extraneous variables; both holding constant variables or the experimental manipulation of variables through which one can begin to explore interactions. More generally, as an iterative process, the educational design experiment enables us to continue to refine our model of the scientific reasoning within the children's reach and, correlatively, to refine a curriculum that can more effectively empower their thinking.

THE TEACHERS' PERSPECTIVE: VALUE AND CHALLENGES
OF THIS APPROACH

Clearly, effective implementation of this way of teaching places new demands on the elementary school teacher. I interviewed the three project teachers to elicit their perspective on the value and challenges of this instructional approach. Their comments reflected the dramatic, indeed systemic, change this approach had involved for them.

One central issue they themselves had addressed in the context of project participation and also anticipated as an issue for other teachers concerned the discrepancy between the model of science reflected in this curriculum and the model of science reflected in their college science courses and prior school curricula. Conceptualizing science as a way of knowing was fundamentally new for the teachers. In the teachers' words:

> We grew up thinking of science as factoids. It's a whole retraining, because people weren't taught science this way. We had to learn a new way of thinking. (Fourth/fifth-grade teacher)

> There is a real simple word built into this that keeps on coming to mind: That's to do science. You're not doing much science if all you ever do is just go through the steps and procedures that someone else set up. So in the college science courses I took we were doing someone else's science. And science was things to learn about. You could learn about something. Like the wind or the weather. (Third/fourth-grade teacher; her emphases)

In the same vein, all three teachers remarked on the scaffolding of scientific inquiry, in the form of the students' design and implementation of their own studies, as a core aspect of the curriculum's power:

> They're trying to get more of "discovery" science in schools. This curriculum goes beyond discovery, in that the children ask the questions to discover instead of you. It forces them to think. (Third/fourth-grade teacher)

> Where they got to work on their own science projects, I was just the facilitator. It was really neat to see how they would go back to the chart and see which methodology they would use. And they knew exactly what they would use on a science question. It's a lot easier to get

involvement. They <u>are</u> involved! When they get to the point of doing their own research studies, when they're in charge of their own research, they can do it the whole day! (Second-grade teacher; her emphasis)

It's kid-driven rather than teacher-driven. In most other science curricula, the teacher controls everything; "Here's the graph we're going to do. Here's the materials we need." Our goal is always that they'll be able to come up with their own investigation, come up with the methodology, the materials—everything. The <u>learning responsibility</u>, that was a really key thing. I think the specific skills we teach them, like how to think like a scientist thinks, that's the part that's missing from canned programs. (Fourth/fifth-grade teacher; her emphasis)

The relatively large degree of the research teams' independence also raised issues of classroom organization. All of the teachers spoke of the need to foster the children's responsibility from the beginning of the year—responsibility for use of time, materials, following through on a complex task, and more generally for their own learning—to the point at which you could eventually "turn your back on groups." Related to the classroom management issue, the teachers also addressed the importance of positive social interactions. "The social interactions," the fourth/fifth-grade teacher contended, "can really destroy everything. If they don't have the background of how to work together and value each other, what do you do when three kids want to do something different in a study?"

The teachers reported continuing to struggle over how best to engineer the formation of research teams. While pairs seemed to support the most concentrated involvement and responsibility of all members, this arrangement combined with absenteeism sometimes resulted in a child working alone—which for many children proved difficult. Similarly, whereas assigning a relatively weak student to work with two stronger students tended to result in a low level of contribution on the part of the low student (despite preambles intended to encourage honoring everyone's perspective), assigning weak students to work together typically resulted in their needing much more support than other groups. Friends frequently formed productive teams, but the teachers also noted that they needed to be on the lookout for how the friends interacted over academic issues, since some friends reflected patterns of asymmetry and dominance.

Finally, the teachers all spoke of the importance of resources in making the curriculum work. For example:

It helps to have books, <u>lots</u> of books that give kids information. (Second-grade teacher; her emphasis)

A really big part for me is having the resources we need to teach. The knowledge, the stuff, you. Most teachers are alone with the textbook. For people who haven't majored in science, the support is really important. (Fourth/fifth-grade teacher)

The curriculum materials were developed for the project teachers, for the purpose of exploring the possibilities of children's relatively independent data-based inquiry in biology. We are beginning to think through how we might change the curriculum materials to make them friendly to other teachers with less direct support. These issues identified by the teachers—including shifts in vision of what science is and what it means to do science, shifts in student and teacher roles, background knowledge, the nuts and bolts of running a classroom with pairs of students assuming responsibility for their own projects—are each a critical part of the systemic change involved in teaching science this way, any of which left unattended can undermine its power.

CONCLUSIONS

We began this project skeptical about some fairly broad spread assumptions about children's scientific reasoning abilities, together with a concern with narrowing the gap between scientists' inquiry and the inquiry of children in science lessons. We viewed maximizing the students' control over the inquiry process as crucial. Stemming from the deep connection between adequacy of knowledge and adequacy of scientific reasoning, our tactic has been to try to empower young children's relatively independent scientific inquiry by scaffolding those spheres of knowledge most fundamental to inquiry: (a) domain-specific knowledge; (b) knowledge of the enterprise of empirical inquiry; (c) domain-specific methodologies; (d) data representation, data analysis, and fundamental constructs of statistics and probability; and (e) relevant tools. The curriculum aimed to permeate the teaching of each of these knowledge components with a metacognitive perspective, involving keen attention to reflections on the adequacy of their knowledge (What do I know? What do I not know?), as well the meta knowledge needed for independent inquiry (e. g., When would I want to use this? What is this good for? How can I adapt it for different situations?).

Given the advantage of a number of relevant knowledge bases, we have found the children's inquiry to extend beyond that reflected in the developmental literature and most elementary science classrooms. In short, the project children's thinking is neither restricted to the singular, linear ideas, nor tied to the concrete. They successfully engage in the complex task of designing controlled, albeit imperfect, experiments. Their thinking about the meaning of their data and how to improve their studies reflect inferences and hypothetical-deductive thought.

We caution science educators not to rely on the cognitive developmental literature to derive schemas of age-appropriate science curriculum. Knowledge, of various forms, can dramatically extend children's scientific inquiry. Our challenge is to further explore the science within children's reach by means of a variety of teaching experiments emphasizing different aspects at the core of thinking scientifically.

ENDNOTES

1. The preparation of this chapter was supported by the National Science Foundation (NSF) under Grant No. REC-9618871. The ideas expressed herein are the author's and do not necessarily reflect those of the NSF.
2. Each year teachers use science kits mandated by the district for subjects outside of the life sciences.
3. We were intrigued that while most teams at the upper elementary level choose to research some aspect of social behavior, we have had only one second-grade team formulate such a project; a tendency we tentatively attribute to the increasing concerns with issues of social interactions across this age span. Curriculum elaboration of the unit now underway includes strengthening this social behavior component to empower more adequately the children's research projects in this sphere.

REFERENCES

American Association for the Advancement of Science. 1993. *Benchmarks for science literacy*. New York: Oxford University Press.

Brown, A.L. 1990. Domain-specific principles affect learning and transfer in children. *Cognitive Science* 14: 107-133.

Brown, A.L. 1992. Design experiments: Theoretical and methodological challenges in evaluating complex interventions in classroom settings. *The Journal of the Learning Sciences* 2(2): 141-178.

Brown, A.L. 1997. Metacognition, executive control, self-regulation, and other mysterious mechanisms. In *Metacognition, motivation, and understanding*, edited by R.H. Kluwe, 65-116. Hillsdale, NJ: Lawrence Erlbaum.

Brown, A., D. Ash, M. Rutherford, K. Nakagawa, A. Gordon, and J.C. Campione. 1996. Distributed expertise in the classroom. In *Distributed cognitions*, edited by G. Solomon. New York: Cambridge University Press.

Brown, A. L., J. Campione, K.E. Metz, and D. Ash. 1997. The development of science learning abilities in children. In *Growing up with science: Developing early understanding of science*, edited by A. Burgen and K. Härnquist, 7-40. Göteborg, Sweden: Academia Europaea.

Brewer, W., and A. Samarapungavan. 1991. Children's theories versus scientific theories: Differences in reasoning or differences in knowledge? In *Cognition and the symbolix processes: Applied and ecological perspectives*, edited by R.R. Hoffman and D.S. Palermo. Hillsdale, NJ: Lawrence Erlbaum.

Carey, S. 1985. Are children fundamentally different kinds of thinkers than adults? In *Thinking and learning skills*, edited by S. Chipman, J. Segal, and R. Glaser, 2. Hillsdale, NJ: Lawrence Erlbaum.

Chi, M., R. Glaser, and E. Rees. 1982. Expertise in problem solving. In *Advances in the psychology of human intelligence*, Vol. 1, edited by R. Sternberg, 7-75. Hillsdale, NJ: Lawrence Erlbaum.

Emlen, S. T. 1975. The stellar-orientation system of a migratory bird. *Scientific American* 233(2): 102-111.

Gould, S. J. 1985. *The flamingo's smile: Reflections on natural history.* New York: W.W. Norton & Company.

Inhelder, B., and J. Piaget. 1958. *The growth of logical thinking from childhood to adolescence*, translated by A. Parsons and S. Milgram. New York: Basic Books. (Original work published in 1955.)

Metz, K.E. 1995. Re-assessment of developmental assumptions in children's science instruction. *Review of Educational Research* 65(2): 93-127.

National Research Council. 1996. *National science education standards*. Washington, DC: National Academy Press.

National Science Resources Center. 1997. *Science for all children: A guide to improving elementary science education in your school district*. Washington, DC: National Academy Press.

Pea, R. 1987. Putting knowledge to use. In *Education and technology in 2020*, edited by R. Nickerson and P. Zodhiates. Hillsdale, NJ: Lawrence Erlbaum.

Popper, K. 1972. *Conjecture and refutations: The growth of scientific knowledge*. 4th ed. London: Routledge, Kegan & Paul, Ltd.

Resnick, L. 1983. *Education and learning to think*. Washington, DC: National Academy Press.

Russell, S. J., R. Corwin, and K. Economopoulus. 1997. *Sorting and classifying data: Does it walk, crawl or swim?* Palo Alto: Dale Seymour Publications.

Rutherford, F.J., and A. Ahlgren. 1990. *Science for all americans*. New York: Oxford University Press.

Steen, L.A. 1988. The science of patterns. *Science* 240: 611-614.

Tierney, C., R. Nemirovsky, and A.S. Weinberg. 1995. *Changes over time*. Palo Alto: Dale Seymour Publications.

Tobin, K., D.J. Tippins, and A.J. Gallard. 1994. Research on instructional strategies for teaching science, In *Handbook of research on science teaching and learning*, edited by D. L. Gabel, 43-93. New York: MacMillan Publishing Company.

Wiser, M., and S. Carey. 1983. When heat and temperature were one. In *Mental models*, edited by D. Gentner and A. L. Stephens. Hillsdale, NJ: Lawrence Erlbaum.

Inquiry Learning as Higher Order Thinking: Overcoming Cognitive Obstacles

Anat Zohar

Inquiry learning is a complex activity that requires much higher order thinking and a variety of cognitive performances. But school children typically have difficulties with thinking strategies that are necessary for the practice of sound inquiry (for details, see "Literature Review" below). Ignoring such difficulties may hinder successful inquiry learning.

Science teachers are trained to invest much thought in preparing detailed lessons plans when they launch on the goal of teaching a complex science topic. Teaching experience and research findings repeatedly show that after they have gone through a unit of instruction, some students still hold on to their preconceptions. Often this happens even after much time and thought have been devoted to designing a learning sequence. Teaching scientific reasoning strategies is at least as difficult as teaching scientific concepts. But few lesson plans in science are designed specifically for the purpose of inducing a change in students' scientific reasoning strategies. Most science teachers do not devote their pedagogical skills to structure learning activities that are specifically designed to foster particular thinking skills. In fact, science teachers often do not consciously think of thinking skills as explicit educational goals (Zohar, 1999). Accordingly, in order to foster thinking we first need to consider it a distinct, explicit educational goal.

The aim of this chapter is to describe common cognitive difficulties that children encounter when they engage in inquiry learning and to suggest instructional means for coping with them. We shall start with several examples to illustrate the problem, proceed to a brief review of the relevant literature, and conclude by suggesting some practical instructional means.

ILLUSTRATIONS OF THE PROBLEM

One illustration of the problem may be based on a "confession" recently made by a colleague, now a university professor. Over twenty years ago when he took accelerated biology in high school, his teacher introduced the class to hypothesis testing. The "if A then B..." algorithm was written neatly in his notebook. He remembers his efforts to study it by heart, failing to find any meaning in the neatly written words. Whenever he had to write a lab report, he leafed through his notebook, searching for that mysterious looking algorithm in order to be able to make a match between the particular experiment conducted in class and the correct way of writing a hypothesis. Although he wrote numerous lab reports in this way (and usually got good marks), it wasn't until college that the mystery was solved, and he finally understood the logic of hypothesis testing.

Another illustration is taken from my current work with science teachers. Towards the end of a recent school year, a seventh-grade biology class conducted an experiment to investigate whether various parts of living organisms contain water. Flowers, stems, leaves, and a piece of meat were put into four glass containers sealed with glass covers. After heating the containers, little drops of water accumulated on all four covers. When asked about the conclusions from the experiment, several students responded by describing the experimental results: little drops of water accumulated on the glass covers. The teacher, Jane, then asked a number of questions.

1. **Teacher:** The little drops on the glass cover—what is it, a result or a conclusion?
2. **Student:** (Several students answer at the same time.) A result.

3. **Teacher:** What is the difference between a result and a conclusion?
4. **Student:** A result is what's out there, what we could see. A conclusion is what we can learn from the result.
5. **Student:** Conclusions are like a summary of all the results.
6. **Student:** The conclusion is what you can conclude from the results.

7. **Teacher:** We can't explain a word by using the very same word. A conclusion is what you can conclude. What does it mean?

8. **Student:** A conclusion is just like a result. They are the same.

9. **Teacher:** If it is the same, why do we need two separate words?

10. **Student:** Results are like, facts.... We saw the drops of water. Conclusions are our ideas, what we think.

None of the students drew the correct conclusions from this experiment—that all the parts of living organisms that were examined contained water. Some children, however, revealed an ability to explain the difference between results and conclusions (see lines 4 and 10), while others missed the difference between the two concepts at both the operational and the procedural level (see lines 5, 6, and 8).

After the lesson, I sat down with Jane to watch parts of the videotaped lesson. She said that at the beginning of the year she had taught about scientific inquiry, explaining the meaning of several concepts including "conclusions." Conclusions were discussed in general terms and defined as "what one learns from an experiment." Yet later on during the year, when students had to describe results and conclusions in the context of their experimentation, many were unable to complete the task successfully.

"We discussed it theoretically, in general terms," Jane said, "but then in each experiment, they kept telling me, 'But Jane, it's the same thing' [results and conclusions]. Each time anybody said so, we immediately discussed whether results and conclusions are indeed the same thing or not.... And it happened several times during the year. That's why I was a little surprised [to see that they still found it so hard by the end of the year]."

These two illustrations have a common feature. In both cases, students rote-learned some definition that pertains to a thinking skill required for inquiry learning. The professor learned a definition by heart and could use it correctly under certain circumstances, without understanding either its meaning or how it relates to the experiments conducted in class. In Jane's class, even students who could cite the definition of conclusions and experimental results were unable to distinguish between them in specific instances. This disability is especially striking because the teacher was sensitive to her students' difficulty and had devoted repeated attention to that issue.

These two cases are drawn from inquiry learning but not from lessons that consisted of <u>open</u> inquiry. Because thinking is dependent on context and content, it may be supposed that students' reasoning difficulties are an artifact generated by inauthentic learning environments, and therefore do not represent genuine difficulties. Students engaging in open inquiry, however, encounter the same kind of reasoning problems. Table 1 describes a sample of problems detected in a ninth-grade classroom in an urban, middle-class school, where students were engaged in open inquiry:

TABLE 1. EXAMPLES OF REASONING PROBLEMS DETECTED IN AN OPEN INQUIRY CLASSROOM

1. Inadequate hypotheses

▶ A group of students that conducted a survey to investigate whether health food improves students' school achievements, formulated several hypotheses that were phrased as research questions: "Do sweet foods improve one's energy and concentration before classes?"; "Is a student who wakes up early and eats a healthy breakfast, more alert during school than a student who had no time for breakfast?"; "Does food improve students' general feeling?"

▶ A second group of students who investigated the difference between the level of vitamin C in fresh and in conserved orange juice, formulated their hypothesis as "Is the amount of vitamin C in fresh orange juice higher than in conserved juice?"

▶ Another frequent problem regarding hypotheses is the formulation of hypotheses that are irrelevant to the research question. The hypothesis formulated by a group of students who wanted to find out whether various periods of storage affect the level of vitamin C in orange juice was: "The level of vitamin C in the juice turns to a different state" (i.e., from liquid to solid).

2. No control of variables.

▶ A group of students formulated their research question as: "Does the level of vitamin C decrease when orange juice is heated to several temperatures, for various periods?" The students' experimental design included heating the orange juice up to several temperatures for different periods of time.

▶ Another group of students defined their research question as "Does adding sugar affect the level of vitamin C in lemon juice?" Their experimental design included manipulation of several variables at once: the amount of sugar added, the temperature and state of the juice, and the type of juice.

▶ A third group of students investigated the effect of several temperatures on the level of vitamin C. They made a sound experimental design, but when they carried out their plan they did not pay attention to the fact that the test tubes were heated for different periods.

3. Mismatch between research problem and experimental design

▶ Students defined the following research problem: "Does adding sugar to lemon juice affect the level of vitamin C?" But their experimental design was based on a comparison between home-made and frozen lemonade.

4. Problems in the processing of experimental results
(Students had trouble in translating their experimental results into graphs and charts.)

▶ One group who conducted a survey to investigate the frequency of vegetarians among different age groups had tried to draw a pie chart of the results. Participants did not understand that the pie represented one hundred percent, and did not know what corresponded to the whole population in their own survey.

5. Confusion between dependent and independent variables

▶ In the course of conducting a survey about diets, students tried to represent their results in a graph. They were observed while discussing which of their variables is a dependent and which an independent variable. Finally, they gave up and decided to guess (they had studied about variables before but apparently were unable to transfer their knowledge to their own research).

(These examples are taken from classrooms that used the Thinking in Science Classrooms project learning materials.)

Several studies that took place in classrooms (e. g., German et al., 1996) show us that the type of problems demonstrated in Table 1 are indeed prevalent in inquiry learning. A look into some of the theoretical studies that investigated this issue may teach us about the possible cognitive source of such problems.

LITERATURE REVIEW

A vast literature describes studies about children's scientific thinking. For the purpose here, a distinction can be made between studies that describe deficiencies in science process skills and studies that describe students' thinking from the perspective of the relationships between experimental evidence and scientific theories and hypotheses.

Science process skills are derived from a list of activities that were used traditionally, according to a positivist paradigm of science, to describe the work of scientists. Numerous inventories include somewhat different lists of science process skills (e. g., Tamir & Lunetta, 1978; Tamir, Nussinovitz, & Friedler, 1982; Lawson, 1995). Typical items on such lists include: defining a research problem, formulating hypotheses, testing hypotheses, designing experiments (including the design of adequate controls), performing experiments, collecting data, analyzing experimental data, and drawing conclusions.

The classic work of Inhelder and Piaget (1958), as well as many studies that stemmed from their work, documented children's deficiencies in abstract thinking. These deficiencies were explained by arguing that before the ages of thirteen or fifteen, children's thinking is concrete and they are incapable of performing formal logical operations. Even after fifteen, many children are still not able to carry out such abstract thinking. These findings point to a likely source of the problems many children experience in inquiry learning. Apparently, they are incapable of employing the logical thinking necessary for successful scientific inquiry. After all, several components of what Piaget termed "logical operations" are the foundations of thinking patterns applied during inquiry. The logic of hypotheses testing is necessary for sound formulation and testing of hypotheses. The ability to manipulate variables is necessary for the design of sound experiments, including the identification of variables, the differentiation between independent and dependent variables, and the control of variables. Analysis of experimental data may also require complex procedures such as the understanding of probability and correlation or the ability to make multiple representations of data, as in coding data in tables or graphs.

Clearly, even one deficiency in any of these abilities may prevent children from drawing valid conclusions. For several decades, therefore, the predominant view among science educators was that elementary school science curricula must be constrained to activities that children of the age can handle. The activities include observing, classifying, comparing, categorizing, measuring, and the drawing of inferences on the basis of these limited activities. This view, however, has been seriously criticized. The research literature following Piaget fails to support the assumption that elementary school children cannot grasp abstract ideas. Although abstract ideas tend to be more elaborated and more prevalent in subsequent ages, some abstract ideas do emerge in the elementary school years and even earlier. Young children are capable of abstract thinking especially when they engage in inquiry that is conducted in authentic contexts (see Metz, 1995 and entries in this volume by Metz and Lehrer et al.). Moreover, the literature that has been used to derive constraints on instruction has typically been based on research that describes competencies based on alienating testing conditions, apart from instruction. More and more studies indicate that suitable instruction may bring children's thinking abilities to higher ceilings than have previously been assumed possible.

Among studies that describe students' thinking from the perspective of the relationships between experimental evidence and scientific theories is that of Kuhn, who views the coordination of theories and evidence as the heart of scientific thinking (Kuhn, Amsel, & O'Loughlin, 1988; Kuhn, 1989). Kuhn's findings suggest that children (and many lay adults) do not differentiate between theory and evidence. Instead, they meld the two into a single representation of the "way things are." When the two are discrepant, children exhibit strategies for maintaining their alignments—either adjusting the theory or adjusting the evidence by ignoring it altogether, or by attending to it in a selective, distorting manner.

Carey and her colleagues (1989; 1993) investigated the understanding of seventh-grade students regarding the nature of scientific knowledge and inquiry in contrast to proper scientific inquiry. Although there are multiple views among scientists about the nature of scientific work, they tend to agree on the basic distinction between hypotheses and the experimental evidence supporting or refining it. Scientific ideas are not simply copies of the facts, but rather distinct, constructed, and manipulable entities. The seventh-grade students, it was discovered, do not appreciate this, nor do they understand that scientists' ideas affect their experimental work and vice versa. Instead, ideas are confused with experiments, and there is no acknowledgment of the theoretical motivations behind scientific experiments. Students do not know what a hypothesis is, explaining it as an idea or a guess. They also do not know where scientists get their hypotheses from, what experiments are, why scientists perform them, and how they choose which experiment to perform. Students also do not understand when and why scientists change their ideas and what they do when they get unexpected results. Most of the seventh-grade students in this study simply saw the goal of science to be the gathering of specific facts about the world.

This brief review of the literature about children's scientific reasoning difficulties is much too short to be comprehensive. But it shows that many cases seen in classrooms are embedded in thinking difficulties that are widely documented and analyzed in theoretical cognitive studies. Several of the classroom examples described in the previous section may be caused by children's difficulty in logical thinking: the mystery of the algorithm for hypothesis testing, the lack of variable control, and the confusion between dependent and independent variables. Difficulties portrayed in other classroom examples may be in accord with the findings in the studies into children's limitations in understanding the nature of scientific experimentation. Children's prevalent inability to differentiate between experimental results and conclusions may be an instance of the difficulties Kuhn defined in distinguishing between evidence and theories, since experimental results correspond with evidence while conclusions relate to scientific theories. Students' difficulties to formulate adequate hypotheses and the mismatch found between their research questions and their experimental design, may also be related to the findings of Carey et al., who speak of the inability among young students to understand the nature of scientific knowledge and inquiry (see also Klahr & Dunbar, 1988; Sodian, Zaitchik, & Carey, 1991; and Klahr & Fay, 1993).

Although there is no consensus among researchers about the sources and developmental phases of children's difficulties in scientific reasoning, it is obvious that these difficulties will have a considerable effect on how children construct their knowledge while learning by inquiry. It is therefore essential to address them during instruction.

INSTRUCTIONAL MEANS

Several Approaches to Instruction

Can higher order thinking skills be taught? An accumulation of evidence indicates an affirmative answer. Many recent studies show that higher order thinking skills in general and scientific inquiry skills in particular can indeed be taught, leading to considerable gains in students' reasoning abilities.

How should higher order thinking be taught? The general literature about instruction of higher order thinking contains methods that may be adaptable to instruction of scientific inquiry skills. A primary distinction in that literature concerns the difference between the "general" and the "infusion" approaches to teaching thinking (Ennis, 1989). The general approach attempts to teach thinking abilities and dispositions generally, separately from any specific curricular content. Thinking skills are typically taught through some content that is conscripted for the purpose of teaching reasoning—local or national political issues, problems in the school cafeteria, or some school subject—but instruction aims at general thinking skills and not deep conceptual understanding of contents. The infusion approach involves the integration of thinking into the regular school curriculum. According to this approach students learn subject matter in a deep and thought provoking way, from which general principles of thinking are made explicit.

Educators working toward instruction of inquiry skills have been embracing both methods. Friedler and Tamir (1986) taught basic concepts of scientific research to high school students by designing a module that included invitations to inquiry and a variety of exercises that lead the students gradually from simple to more complex and some highly sophisticated experiences in solving problems. An evaluation of this unit showed that students who have used this module demonstrated substantial gains in applications of inquiry skills that were measured in inquiry-oriented practical laboratory tests taken from the matriculation examinations.

Adey and colleagues Shayer and Yates (1989; 1993) applied a different approach in CASE (Cognitive Acceleration through Science Education). The project aims at improving the ability to use thinking skills across multiple subjects and topics. CASE addresses ten of Piaget's formal operations by designing a set of special lessons, replacing regular science lessons every two weeks for two years. Some of the operations are directly related to inquiry: for instance, the notion of variables, dependent and independent variables, control of variables, and probability. The CASE program is firmly grounded in the cognitive literature about children's learning. Each unit utilizes several means that are proposed for bringing about long-term effects in the general ability of learners:

- ▶ **concrete preparation**—concrete activities are used to introduce the terminology and the context in which a problem is presented;
- ▶ presentation of problems in ways that will induce a **cognitive conflict**;
- ▶ special activities to foster **metacognition**;
- ▶ explicit activities that induce the **transfer of thinking strategies** to novel situations.

CASE lessons are taught as special lessons, involving topics that are not part of the regular science curriculum. To this extent they exemplify the strategy of teaching skills in the general approach. But CASE teachers are encouraged to practice transfer activities, including specific ideas for applying the reasoning strategies taught in CASE lessons to other parts of the curriculum. An extensive evaluation program of CASE indicated that the intervention led both to gains in Piagetian measures of cognitive development and to gains in subject-matter knowledge (in science, mathematics, and English).

A third approach to teaching scientific thinking skills is applied in the Thinking in Science Classrooms (TSC) project that supplements the regular science curriculum with learning activities designed to foster scientific reasoning skills, scientific argumentation, and knowledge of scientific concepts (Zohar, Weinberger, & Tamir, 1994; Zohar, 1996). Although fostering scientific thinking skills is among the project's explicit goals, skills are not taught as context-free entities. The TSC approach is based on the assumption that scientific reasoning cannot be taught by developing discrete, decontextualized skills but must always be deeply embedded in specific contents. The contents of the learning activities always match topics from the regular science syllabus. Therefore, teachers may incorporate these activities in the course of instruction whenever they get to a topic that is covered by one of the activities. The project produces a set of opportunities, calling for "thinking events" to take place in multiple scientific topics. Thus, the project is designed according to the infusion approach to teaching higher order thinking (Ennis, 1989).

Instruction always begins with concrete problems regarding specific scientific phenomena that students are asked to solve. During the learning process, students are active. Much of their work takes place in small groups with rich scientific argumentation. After students have used the same reasoning skill in various concrete contexts, they are encouraged (usually through class discussion) to engage in metacognitive activities that include generalization, identification of skills, and formulation of rules regarding those skills. Learning of reasoning skills is therefore achieved through an inductive process in which generalizations are made by the learners themselves. In order to avoid fixed patterns of learning activities which might eventually train students to deal with them in a merely algorithmic way, varied types of learning activities were designed:

- inquiry and critical thinking skills learning activities;
- investigation of microworlds;
- learning activities promoting argumentation skills about bioethical dilemmas in genetics;
- open-ended inquiry learning activities.

Evaluation studies investigating the effect of the first, second, and third types of learning activities indicate gains in both reasoning skills and scientific knowledge (Zohar, Weinberger, & Tamir, 1994; Zohar, 1996; Zohar & Nemet, submitted). An evaluation study investigating the effect of the fourth type of learning activities is currently under progress.

Teaching Inquiry Skills: Detailed Description of One Unit

To illustrate how inquiry skills are taught as a distinct educational objective, let us turn in some detail to one of the TSC learning activities, Investigation of Microworlds.

Description of Task and Typical Students' Performance. The learning activities for investigation of microworlds consist of a computerized simulation, a set of worksheets that students employ individually or in small groups, and an instructional sequence that is taught to a whole class (Zohar, 1994, 1996). Several similar microworlds were developed in various biological topics such as photosynthesis, plant germination, ecology, and nutrition, so that teachers can choose to use a microworld that matches a particular topic they teach. The idea for these learning activities originated in tasks used in a set of theoretical studies designed to investigate the development of scientific reasoning skills (Schauble, 1990; Kuhn, Schauble, & Garcia-Mila, 1992). Rapid and universal progress in thinking skills were observed in these studies, suggesting that the tasks they used might be applied to practical educational purposes (Zohar, 1994). In order to adapt tasks from research purposes to classroom use, several major changes were introduced. The topics of the tasks were changed to match topics that are part of the science curricula and means for class management—among them worksheets and computerized databases—were added.

In a learning activity related to photosynthesis, for example, students are asked to investigate five variables—light intensity, temperature, species of plant, natural growth area, and carbon dioxide—to determine which of them affect the rate of photosynthesis as measured by the amount of oxygen released in a fixed period. Students' investigation consists of defining the variables they wish to investigate, planning a combination of variables they want to examine, conducting the simulated experiment on the computer, making inferences and justifying them.

When students begin their inquiry, their investigation is often unsystematic, characterized by the use of invalid reasoning strategies such as: ignoring evidence, conducting experiments without controlling variables, or constantly changing the focus of their inquiry. An unsystematic investigation is illustrated by Alice, an eighth-grade student, in her first few experiments with the photosynthesis problem.

Attempting to investigate the effect of light intensity, Alice conducted these two experiments:

VARIABLE	EXPERIMENT #1	EXPERIMENT #2
light intensity	1 light bulb	2 light bulbs
temperature	17 C	25 C
carbon dioxide	added	added
species of plant	R	R
natural growth area	Y	Y
Results	1/4 test tube of oxygen	3/4 test tube of oxygen

After completing the second experiment, the following exchange took place between Alice and the experimenter :

Experimenter: So what did you find out?
Alice: That adding more light is good. It makes more photosynthesis.
Experimenter: How do you know that?
Alice: Because of what I did. When I added another light bulb, there was more oxygen.
Experimenter: Did you find out anything else?
Alice: Yeah. Adding more temperature is also good.
Experimenter: And how do you know that?
Alice: Because I saw what happened when the temperature was 25. There was more oxygen.

Alice's inferences are clearly invalid because she did not control variables. In the second experiment she changed both the light intensity and the temperature. Alice's confusions are by no means exceptional. An evaluation study indicated that approximately 90% of the inferences made by junior high school students when they start investigating the microworlds are invalid. A special learning sequence was designed to advance their thinking.

Description and Rationale of Learning Sequence. The sequence consists of three stages. At the first, students conduct independent investigations of the problem. The second includes a whole class discussion structured around the issue of variable control. At the third stage, students once again resume their independent investigations of the microworld.

Stage One. The main goal of Stage One is to expose the thinking strategies that students employ before instruction. Since the microworld represents a simplification of an actual experiment, the results of students' unsystematic experimentation often lead to a cognitive disequilibrium, undermining the confidence in their initial unsound strategies for solving the problem (Zohar, 1996). A conversation that took place in the classroom between George, a ninth-grade student and an experimenter, who acted as a passive observer, illustrates the point.

During a lesson, George approached the experimenter with a puzzled look on his face.

> **George:** I must ask you a question. I have a problem. I don't know whether what I'm proving is correct.
> **Experimenter:** ??????
>
> **George:** Look at what I just did.
> (George shows his work sheet, pointing to two experiments in which he failed to control variables. He changed the levels of both light and temperature: in the first experiment he used two light bulbs and a temperature of seventeen degrees Celsius while in the second experiment he used one light bulb and a temperature of twenty-five degrees. The resulting amount of oxygen was the same in both experiments—half a tube of oxygen, because the two variables were compensating for each other.)
> **George:** These results show me that both the temperature and the light make no difference. But I'm not really sure.... Because it may be both. It may be both the temperature and the light. Because both of them are being changed. So I don't know what to do.

An examination of this student's work sheet showed that during his initial experiments he had concluded that increasing the level of either light or temperature, increases the rate of photosynthesis. This student's initial experiments and inferences, then, were similar to Alice's. Although these conclusions are correct, the student drew them through the use of invalid reasoning strategies—he did not control variables. But when variables are not controlled, students often draw contradictory inferences. Indeed, this is what George did. Although in his initial experiments he had concluded that both the light and the temperature affect the rate of photosynthesis, he now concluded the opposite. The conflict with his initial findings induced a state of cognitive dissonance in his mind. As the final two lines in the excerpt show, he is now unsure about his findings and puzzled about how to carry on his investigation.

When George resumed his independent investigation, he once again changed his mind. This time he concluded that the light does make a difference but the temperature does not. Nevertheless, he was still not satisfied and rightly so, because his conclusion regarding the temperature is incorrect. He therefore concluded by saying that he would have to continue his experimentation. According to conceptual change learning theory, an undermining of one's initial thinking is an important stage in learning. It creates dissatisfaction with existing reasoning strategies and induces the motivation for adopting alternative strategies.

Stage Two. The next stage of learning presents two main alternatives. Students may be allowed to construct new thinking strategies on their own, or they may be guided in this process. Previous studies have shown that even without any guidance, the percentage of valid inferences students make increases with time, indicating the acquisition of new thinking strategies. But these studies were conducted under research conditions and not in real classrooms where it seems unlikely that students will engage in one task for a period sufficiently long to induce change in reasoning strategies. Another study that compared independent with guided discovery learning in the context of our microworlds indicated that guidance contributes to improved performance as measured by the frequency of valid inferences. It was therefore decided to design guidance that will assist students in constructing the control of variables thinking strategy.

To begin with, it cannot be assumed that all students went through the same process as George, undermining their initial, invalid thinking strategies. So the teacher first directs all students to a set of experiences that are similar to the ones George had generated on his own. The teacher thus generates sets of uncontrolled experiments that naturally lead to contradicting conclusions. The contradictions produce hot debates among students about the correct conclusions and about the correct means for drawing these conclusions. Such a debate among peers serves to expose students' initial thinking strategies, to bring those strategies up in the public domain of the class, and to lay open conflicting views that may lead students towards cognitive dissonance with respect to suitable strategies for solving the problem.

With those conflicts in the background, the teacher then turns to help students construct new reasoning strategies. It is assumed that the same students who did not control variables in the context of the relatively complex photosynthesis problem, do have an intuitive understanding of variable control in everyday type of problems that consist of only two variables. The teacher then tells the following story:

John and Susan enter a room and Susan turns on the light. The light doesn't go on. John says: "Oh, it's the plug. It's not plugged in properly."

Susan says: "No, it's the bulb. The bulb is burnt." Then Susan changes the bulb and tightens the plug and the light goes on. So John says: "See—I was right. It was the plug." Susan says: "No, you were wrong. It was the bulb."

The teacher asks who is right, John or Susan? This question is followed by another class discussion. Then, depending on students' responses, the teacher might continue to ask one or more of these questions: "Can you tell for sure which of them is right? Why not? What would John and Susan have to do in order to know for sure who is right?"

Most junior high school students can explain it in this way: "We can't tell for sure who is right, because John and Susan changed two things at once." Our goal is to use this intuitive understanding as the basis for construction of a more general understanding of the rule of variable control. This is done by asking students whether they see any similarity between their thinking in the case of the photosynthesis problem and in the case of the story about the light bulbs. Although some students always answer that the similarity is that there are light bulbs in both stories, many detect the common thinking strategy and generate a response reflecting the need in both situations for changing only one factor at a time. Otherwise, students say, we can never know for sure what is the right answer. Thus, in their own words, students formulate their own rule of variable control. The teacher summarizes the discussion by introducing the formal term <u>control of variables</u>.

The aim of the subsequent, final stages of our learning sequence is to stabilize the use of the control of variables thinking strategy. Students are asked to bring examples of other incidents in which the same strategy should be used. Then several examples taken from their previous experiences with other scientific investigations as well as from other school topics and from everyday experiences are discussed. The goal of this stage is to prevent the welding of the strategy to the specific circumstances in which it was acquired and to enhance transfer.

Stage Three. Soon afterwards, students once again resume their independent investigations of the microworld. Consequently, they have multiple opportunities to practice the new rule and to stabilize its use.

During the second and third stages, special attention is given to metacognition. In the course of Stage Two, students are encouraged to reflect on their reasoning strategies. The product of this process is the generalization or rule regarding the necessity of changing only one feature at a time. Whenever students fail to control variables during their independent investigation in the third stage, teachers once again direct them to reflect on the thinking strategies they have used, to compare them with the rule for the control of variables and thus to evaluate them.

While this learning sequence may be useful for most junior high school students, it is superfluous for some of them. Our assessment showed that approximately one tenth of the junior high school students tested had already mastered the thinking strategies for controlling variables before instruction took place. Those students certainly do not need three lessons on this subject, and are given instead an alternative assignment that consists of investigating interactions among variables in the photosynthesis microworld. The details of this assignment are beyond the scope of this chapter. It is, however, important to note that variability among students in initial reasoning abilities should be acknowledged and heterogeneous instruction may be designed to satisfy that variability. Another variation of the task designed for younger students consists of a similar but simplified microworld with only three variables. This version was used successfully in fourth, fifth and sixth grade.

Assessment. Students' progress was assessed through the comparison of a set of individual interviews conducted before learning took place, with a set conducted at the end of learning. An interviewer followed students' independent investigation of a problem, asking them to explain their inferences. Interviews were audiotaped and then transcribed and analyzed. Inferences were coded as either valid or invalid, according to the key of inference forms described by Kuhn, Schauble & Garcia-Mila. (1992). Altogether, sixteen eighth-grade students and seventeen ninth-grade students were interviewed.

In order to investigate transfer and retention, some students were interviewed twice more. For transfer, students were asked to investigate a new, logically equivalent problem in a different topic. In order to test retention of the acquired thinking skills across time, the eighth-grade students were interviewed again in the following school year, when they were in ninth grade, approximately five months after instruction had taken place.

The findings from this study showed that the rate of valid inferences increased from eleven percent in the early interviews to seventy-seven percent in the late interviews. Students were able to transfer their newly acquired reasoning strategies to a new problem taken from a new biological topic. They were also able to retain their newly acquired strategies across time, and to transfer them to yet another biological topic five months after instruction took place (Zohar, 1996).

SUMMARY AND CONCLUSIONS

This chapter focuses on the observation that inquiry learning involves reasoning patterns that many students have not yet mastered. Consequently, when they engage in inquiry learning they often fail to understand some of the processes included in sound investigation. Rather than give up inquiry altogether or present algorithms that children must follow in a meaningless way, it is suggested to address reasoning difficulties as an explicit instructional goal and to teach accordingly.

Conversations with many science teachers reveal that they are well aware of their students' reasoning incapabilities and occasionally address them in class. But they rarely think of them as distinct, explicit educational goals that must be addressed repeatedly and systematically. They usually plan lessons according to content objectives, and almost never plan lessons to address specific reasoning goals (Zohar, 1999). It may be naïve to expect that a single series of three lessons, such as the ones described here can indeed induce a significant change in the reasoning of students. It should be remembered that this series of lessons is part of a larger set of learning activities that science teachers who participate in the TSC project can incorporate constantly into the course of their instruction. Not all learning activities require three whole lessons. Some may consist of less time consuming means, such as a few questions, or a guided reflection upon reasoning processes that had taken place during inquiry. Addressing students' reasoning by pre-planned means, however, should become a routine part of instruction that is integrated constantly into science learning.

Conceptual change theories of learning, are based on a diagnosis of students' alternative concepts, followed by well-planned, structured, remedial instruction. Accordingly, sensitivity to students' reasoning difficulties should lead to a diagnosis of particular thinking problems that may then lead to instruction aimed to help students overcome those problems. Integrating such instruction into inquiry learning will help students to improve their understanding of scientific inquiry processes and thus to find them more meaningful.

REFERENCES

Adey, P. S., M. Shayer, and C. Yates 1989. *Thinking science: Student and teachers' materials for the CASE intervention*. London: Macmillan.

Adey, P., and M. Shayer. 1993. An exploration of long-term far-transfer effects following an extended intervention program in the high school science curriculum. *Cognition and Instruction* 11: 1-29.

Carey, S., R. Evans, M. Honda, E. Jay, and C. Unger. 1989. An experiment is when you try it and see if it works: A study of grade 7 students' understanding of the construction of scientific knowledge. *International Journal of Science Education*. Special Issue 11: 514-529.

Carey, S., and C. Smith. 1993. On understanding the nature of scientific knowledge. *Educational Psychologist* 28: 235-251.

Ennis, R.H. 1989. Critical thinking and subject specificity: Clarification and needed research. *Educational Researcher* 18: 4-10.

Friedler, Y., and P. Tamir. 1986. Teaching basic concepts of scientific research to high school students. *Journal of Biological Education* 20: 263-270.

German, P. J., R. Aram, and G. Burke. 1996. Identifying patterns and relationships among the responses of seventh-grade students to the science process skill of designing experiments. *Journal of Research in Science Teaching* 33: 79-99.

Inhelder, B., and P. Piaget. 1958. *The growth of logical thinking from childhood to adolescence*. New York: Basic Books.

Klahr, D., and K. Dunbar. 1988. Dual space search during scientific reasoning. *Cognitive Science* 12: 1-48.

Klahr, D., and A.L. Fay. 1993. Heuristics for scientific experimentation: A developmental study. *Cognitive Psychology* 25: 111-146.

Kuhn, D., E. Amsel, and M. O'Loughlin. 1988. *The development of scientific thinking skills*. Orlando, FL: Academic Press.

Kuhn, D. 1989. Children and adults as intuitive scientists. *Psychological Review* 96: 674-689.

Kuhn, D., L. Schauble, and M. Garcia-Mila. 1992. Cross-domain development of scientific reasoning. *Cognition and Instruction* 9: 285-327.

Lawson, A. E. 1995. *Science teaching and the development of thinking*. Belmont, CA: Wadsworth Publishing Co.

Metz, K. E. 1995. Reassessment of developmental constraints on children's science instruction. *Review of Educational Research* 65: 93-127.

Schuable, L. 1990. Belief revision in children: The role of prior knowledge and strategies for generating evidence. *Journal of Experimental Child Psychology* 49:31-57.

Sodian, B., D. Zaitchik, and S. Carey. 1991. Young children's differentiation of hypothetical beliefs from evidence. *Child Development* 62: 753-766.

Tamir, P., and V. N. Lunetta. 1978. An analysis of laboratory inquiries in the BSCS yellow version. *The American Biology Teacher* 40: 353-357.

Tamir, P., R. Nussinovitz, and Y. Friedler. 1982. The design and use of practical tests assessment inventory. *Journal of Biological Education* 16: 45-50.

Zohar, A., Y. Weinberger, and P. Tamir. 1994. The effect of the biology critical thinking project on the development of critical thinking. *Journal of Research in Science Teaching* 31:183-196.

Zohar, A. 1994. Teaching a thinking strategy: Transfer across domains and self learning versus class-like setting. *Applied Cognitive Psychology* 8: 549-563.

Zohar, A. 1996. Transfer and retention of reasoning skills taught in biological contexts. *Research in Science and Technological Education* 14: 205-219.

Zohar, A. 1999. Teachers' metacognitive knowledge and the instruction of higher order thinking. *Teaching and Teacher Education* 15: 413-429.

Zohar, A., and F. Nemet. Submitted. Fostering students' argumentation skills through bioethical dilemmas in genetics.

Teaching Science as Inquiry for Students with Disabilities

J. Randy McGinnis

INTRODUCTION

Currently, 5.4 million students in American public schools are identified as having disabilities: impairments such as speech, hearing, motor or orthopedic, and visual difficulties, and conditions eligible for special education services such as learning disabled (LD), developmental delay or mental retardation (MR), autism, traumatic brain injury, and seriously emotionally disabled (EH). Data reported by the Department of Education in 1991 indicate that over half of students with documented disabilities receive instruction in regular education classes. This figure will increase as more school districts come into compliance with Public Law 101-476, the Individuals with Disabilities Education Act (IDEA) of 1997, by placing students with disabilities in the mandated "least restrictive environment."

Teachers have customarily identified science classes as especially suited for inclusion of students with disabilities (Atwood & Oldham, 1985). They note the relevance of the content, the possibility of practical experience, and the opportunity for group learning with typical peers (Mastropieri et al., 1998). This perspective does not mean that most science teachers are sanguine about including students with disabilities in their classrooms. Instead, as reported by Katherine Norman, Dana Caseau, and Greg Stefanich in1998, both elementary and secondary school teachers identify as one of their primary concerns teaching students with special needs.

Recent standards for instruction as F. James Rutherford and Andrew Ahlgren (1989), the American Association for the Advancement of Science (1993), and the National Research Council (1996) present them put inquiry at the center. What could inquiry-based science instruction be for students with disabilities? For a beginning to an answer, I want to review four relevant fields of the literature: present-day portrayals of inquiry learning by writers addressing the teaching of science to students with disabilities; reasons for science instruction of that sort for such students; evidence that the approach is appropriate for them; and implications for teachers of science to students with disabilities.

PORTRAYALS OF INQUIRY LEARNING BY PROFESSIONALS CONCERNED WITH TEACHING STUDENTS WHO HAVE DISABILITIES

Among writers considering science instruction based in inquiry for students with disabilities are Thomas Scruggs et al., who in 1993 linked inquiry-based instruction with specific science curriculum projects such as the *Full Option Science System* (Lawrence Hall of Science, 1992) that emphasize student activities in small groups. Scruggs and Margo Mastropieri in 1994 identified such instruction with strategies endorsed by F. James Rutherford and Andrew Ahlgren in 1989: emphasis on concrete meaningful hands-on experiences in place of vocabulary acquisition and textbook learning. In 1995 Joyce Sasaki and Loretta Serna described such instruction in science as an inductive process that proceeds in a sequence referred to as "knowing and doing" (p. 14): establish what students already know, ask questions about what is to be observed, investigate, and obtain new knowledge. The appropriate role for the science teacher is that of facilitator.

In one of the most current studies reported, Mastropieri, Scruggs, and Butcher in a study published in 1997 identify instruction founded in inquiry as a type of inductive thinking that requires drawing general rules after observing a number of specific observations. They associate it with "constructivism" (p. 199), in which every learner constructs meaning individually in classrooms where the teacher engages in coaching and in promoting thinking activities. Bridget Dalton et al., (1997) describe this variety of science instruction as informed by the constructivist principle of social interaction. Support by the teacher they view as critical to the process. This is so important a tenet that they name "supported inquiry science" their version of instruction by inquiry. Paramount is their idea of learning as conceptual change. A central assumption is that hands-on activities are not sufficient to change a learner's thinking. The students should be asking questions, designing experiments, observing, predicting, manipulating materials, keeping records, and learning from mistakes (p. 670).

REASONS FOR GIVING STUDENTS WITH DISABILITIES SCIENCE INSTRUCTION BASED IN INQUIRY

In the *National Science Education Standards*, issued in 1996 by the National Research Council (NRC), references to teaching students with disabilities unequivocally support including them in science classrooms that teach by inquiry and having them participate. A central principle guiding the development of the *Standards* is "Science for all students" (p. 19), defined as a principle of "equity and excellence" or fairness (p. 20). All students are also assumed to be included in "challenging science learning opportunities" (p. 20).

This equity principle is reflected in Teaching Standard B of the *Standards*. Teachers should recognize and respond to student diversity and encourage all students to participate fully in science learning (p. 32). Students with physical disabilities might require modified equipment; students with learning disabilities might need time to complete science activities (p. 37). The equity principle is also contained in Program Standard E: All students in the K-12 science program must have equitable access to opportunities to achieve the *Standards* (p. 221). Actions to promote this include bringing in students who have not customarily been encouraged to do science, among them students with disabilities, and making adaptations responsive to their needs (p. 221). This equity principle is further reflected in Assessment Standard D: Assessment practices must be fair (p. 85). Assessment tasks must be appropriately modified to accommodate the needs of students with physical disabilities (and) learning disabilities (p. 85). In particular, fairness requires students with disabilities to "demonstrate the full extent of their science knowledge and skills" (p. 86).

The ethical position taken in the *Standards* is in compliance with the constitutional reasons expressed in Public Law 101-336, the Americans with Disabilities Act of 1990, and in IDEA. IDEA is the most encompassing legislative victory by advocates for students with disabilities who have long fought for appropriate educational opportunities for all students. It is a telling repudiation of a mindset that got expression in 1903, when the Committee on Colonies for

Segregation of Defectives influenced the National Conference on Charities and corrections to campaign for the exclusion of students with disabilities in American schools (Gilhool, 1998). Of particular relevance in IDEA is the mandate to base on the content of the regular science curriculum all instruction for students with disabilities. This means that the Individualized Education Plans (IEPs) for students with disabilities must now describe curricular adaptations and accommodations based on the regular curriculum. So as the recommendation by the NRC to base science instruction on inquiry permeates the nation's science curricula, federal law increasingly supports inquiry for all students.

EVIDENCE THAT INQUIRY IS APPROPRIATE FOR STUDENTS WITH DISABILITIES

The passage in 1975 of Public Law 94-142, the Education for the Handicapped Act intensified research on science instruction for students with disabilities (Bay et al., 1992). The provision mandated by the law to place handicapped students in the least restrictive environment meant suddenly that significantly more students with disabilities were in the general education science classroom. The chief concern was whether students with disabilities would benefit from their "selective placement...in one or more regular education classes" (Rogers, 1993, p. 1), and if so, which type of instruction is the most effective. Findings from these studies typically relate exclusively to students described with mild disabilities and concentrate on the impact of "mnemonic instruction, free study, direct questioning, and direct instruction" (Bay et al., 1992, p. 556).

In a comprehensive review published in 1992 of the literature from the 1950s until the early 1990s on science education for students with disabilities, Margo Mastropieri and Thomas Scruggs observe that from the early 1950s to the passage of P.L. 94-142 in 1975, research centered on "the effectiveness of developmentally oriented, hands-on curriculum to improve the content knowledge and cognitive functioning of students with disabilities" (Scruggs et al., 1993, p. 2). But, concurrently with the passage of P.L. 94-142 and the emergence of the back-to-basics movement, and extending until the 1980s, published studies on the developmental science activities for students with disabilities gradually stopped. Instead, increasing numbers of studies focused exclusively on basic skill acquisition by students classified as having learning disabilities.

Only in the 1990s, when the "commitment to educate each child, to the maximum extent appropriate, in the school and classroom he or she would otherwise attend" (Rogers, 1993, p. 1) became the norm, did research into the impact of instruction by inquiry for students with disabilities get published again. But, as Thomas Scruggs and Margo Mastropieri reported in 1994, the research in the inclusion in the science classroom of students with disabilities had little to say of students with developmental, emotional, or behavioral as opposed to physical disabilities (p. 805). A review of recent studies looking into the effects of science instruction by inquiry on selective students with certain disabilities indicates that when the lessons are appropriately structured, students with learning disabilities benefit by the acquisition of knowledge; they also express greater satisfaction with hands-on science activities rather than with textbook activities (Scruggs et al., 1993). Students with visual, physical, auditory, or learning disabilities were evaluated by their teachers as successfully participating in all aspects of their elementary school science classes, including science activities, classroom discussion, and completion of adapted assignments (Scruggs & Mastropieri, 1994a). Students with various disabilities benefit from the use of technology to solve problems and to acquire and analyze data (Alcantra, 1996, as reported by Woodward & Rieth, 1997). For mastering concepts, students with and without learning disabilities profit from inquiry (Dalton et al., 1997). In courses centered in inquiry, students with learning disabilities, mental retardation, or emotional problems make academic gains comparable to those of their classroom peers and superior to the gains of most peers without disabilities who take courses based in textbooks (Mastropieri et al., 1998). In construction of scientific knowledge and in learning, remembering, and comprehending, inquiry aids more than direct provision of the same information (Scruggs & Mastropieri, 1994b). Students with learning disabilities may be able to participate in inquiry based on constructivist principles and benefit from it, but it is suggested that mentally retarded students may not benefit to a similar degree (Mastropieri, Scruggs, & Butcher, 1997).

Still needed to inform this research are many thoughtful case studies that examine specific disabilities in science learning based on inquiry. In their absence, there is no preponderance of evidence to indicate whether or not instruction by inquiry as guided by the *Standards* remains for students with disabilities a promising pedagogical initiative.

IMPLICATIONS FOR INSTRUCTORS WHO TEACH SCIENCE TO STUDENTS WITH DISABILITIES

Students with disabilities are currently taught science in either self-contained or inclusion classrooms. Research based on survey methodology indicates that in both of these contexts teachers believe they are ill-prepared for the task (Holahan, MacFarland, & Piccollo, 1994; Norman, Caseu, & Stefanich, 1998). In both contexts, the *Standards* recommend instruction based in inquiry. The primary implication for teachers who teach science to students with disabilities is therefore to develop a vision for instruction by inquiry for all students. The emerging literature on science instruction by means of inquiry for students with disabilities provides an additional essential source of information from which to forge personal visions of inquiry. Here I present two examples.

In 1995 Sasaki and Serna reported on an inductive science program for middle school students with mild disabilities that they evaluated as effective. The program is titled Foundational Approach to Science Teaching I (FAST I). FAST I is described as a "hands-on, practical, inquiry approach" (p. 14) that teaches physical science, ecology, and relational topics. It features work by students in groups as well as experiments and public data charts. Students solve anomalies and are expected to make interpretations of the data, which includes making extrapolations and interpolations and drawing conclusions. A nonnegotiable portion of the instruction is clear direction on appropriate student participation. Recommendations are made for warnings and time-out procedures. Points are awarded for appropriate participation. The two teachers also recommend from experience requiring a notebook with a format; generation by students of presentations or hypothesis on what will be learned from the experiment; participation in experimental activities; oral presentations by the learning groups on the data they collected; elicitation of summary statements and group construction of conclusions. The authors assert that this approach develops critical thinking, enhances self-esteem, and furthers academic success.

For science classrooms employing inquiry and including students with disabilities and others without, a collaborative relationship with special educators is recommended (McGinnis & Nolet, 1995; Stefanich, 1994). A collaborative pairing makes for exchanging pedagogical and scientific knowledge. A model for bridging the space between science instruction and special education proposed by Victor Nolet and Gerald Tivnan as described by J. Randy McGinnis and Nolet in 1995 is directed toward achieving a specifically defined content. This model also includes problem-solving formats that emphasize concepts, not rote memorization of facts. Expected of students are the intellectual operations of description and problem solving and the acquisition of an identified assemblage of facts, concepts, and principles.

For a lesson on fossil fuels, for example, a middle school science teacher, for example, might identify "acid preparation" as an essential concept. Inquiry would be guided by recommendations in the *Standards:* making observations; posing questions; examining books and other sources of information to see what is already known; planning investigations; reviewing what is already known in light of experimental evidence; using tools to gather, analyze, and interpret data; proposing answers, explanations, and predictions; and communicating the results. The teacher will have established expectations for student assessment differing according to the IEP of the student. A student whose IEP targeted the intellectual descriptive operation would be expected minimally to summarize the defining attributes and provide illustrative examples of "acid rain." A typical student in the same lesson might be expected to demonstrate intellectual evaluation in a task that requires taking a positive stance with a well-developed rationale toward the continued use of fossil fuels.

CONCLUSION

This is a time of national reform in science education. The *Standards* published by the NRC urge that for all students instruction in science be by inquiry. While much is not settled on realizing such instruction for students with disabilities, intellectual ferment and practitioner initiative are in ample supply. The answer to what science instruction by inquiry could be for students with disabilities are multiple resplendent visions with more still to emerge during this time of remarkable opportunity.

REFERENCES

American Association for the Advancement of Science. 1993.
   Benchmarks for science literacy. New York: Oxford University Press.

Atwood, R.K., and B. R. Oldham. 1985. Teachers' perceptions of
   mainstreaming in an inquiry oriented elementary science program. *Science Education* 69:619-624).

Bay, M., J. R. Staver, T. Bryan, and J. B. Hale. 1992. Science instruction for the mildly handicapped: Direct instruction versus discovery teaching. *Journal of Research in Science Teaching* 29 (6), 555-570.

Dalton, B., C. C. Morocco, T. Tivnan, and P.L.R. Mead. 1997. Supported inquiry science: Teaching for conceptual change in urban and suburban science classrooms. *Journal of Learning Disabilities* (November/December) 30(6):670-684.

Gilhool, T.K. 1998. Advocating for our children: Will it every stop? *Down Syndrome News* (November) 22 (9):115-120.

Holahan, C., J. McFarland, and B.A. Piccollo. 1994. Elementary school science for students with disabilities. *Remedial and Special education.* 15(2): 86-93.

Lawrence Hall of Science. 1992. *Full option science system.* Chicago: Encyclopedia Britannica Educational Corp.

Mastropieri, M.A., and T.E. Scruggs. 1992. Science for students with disabilities. *Review of Educational Research* 62(4): 377-412.

Mastropieri, M.A., T.E. Scruggs, and K. Butcher. 1997. How effective is inquiry learning for students with mild disabilities? *The Journal of Special Education* 31 (2):199-211.

Mastropieri, M.A., T.E. Scruggs, P. Mantziopoulos, A. Sturgeon, L. Goodwin, and S. Chung. 1998. A place where living things affect and depend on each other: Qualitative and quantitative outcomes associated with inclusive science teaching. *Science Education* 82:163-179.

McGinnis, J.R., and V.W. Nolet. 1995. Diversity, the science classroom and inclusion: A collaborative model between the science teacher and the special educator. *Journal of Science for Persons with Disabilities* 3:31-35.

National Research Council. 1996. *National science education standards.* Washington, DC: National Academy Press.

Norman, K., D. Caseau, and G. Stefanich. 1998. Teaching students with disabilities in inclusive science classrooms: Survey results. *Science Education* 82:127-146.

Rogers, J. 1993. The inclusion revolution. *Phi Delta Kappan* 11:1-5.

Rutherford, F.J., and A. Ahlgren. 1989. *Science for all Americans*. New York: Oxford University Press.

Sasaki, J., and L. Serna. 1995. FAST science: Teaching science to adolescents with mild disabilities. *Teaching Exceptional Children* 27 (4):14-16.

Scruggs, T.E., and M.A. Mastropieri. 1994a. Successful mainstreaming in elementary science classes: A qualitative study of three reputational cases. *American Educational Research Journal* 31(4): 785-811.

Scruggs, T.E., and M. A. Mastropieri. 1994b. The construction of scientific knowledge by students with mild disabilities. *The Journal of Special Education* 28: 307-321.

Scruggs, T.E., M.A. Mastropieri, J. P. Bakken, and D. F. Brigham. 1993. Reading versus doing: The relative effects of textbook-based and inquiry-oriented approaches to science learning in special education classrooms. *The Journal of Special Education* 27(1):1-15.

Stefanich, G. 1994. Science educators as active collaborators in meeting the educational needs of students with disabilities. *Journal of Science Teacher Education* 5(2): 56-65.

U.S. Department of Education. 1991. *Thirteenth annual report to Congress on the implementation of the education of the handicapped act*. Washington, DC: US Government Printing Office.

Woodward, J., and H. Rieth. 1997. A historical review of technology research in special education. *Review of Educational Research* 67(4):503-536.

Appropriate Practical Work for School Science— Making It Practical and Making It Science1

Brian E. Woolnough

This chapter is partly a cautionary tale and in part exhortation. As a cautionary story, it looks back on the long tradition in the United Kingdom (UK) of practical work in science lessons and that too much of such practice trivialises science. But it is also an exhortation to continue personally motivating and fulfilling labor that introduces students to genuine, authentic science activity.

In this chapter I will be talking about the types of practical work that engages students, either individually or in small groups, with scientific or quasi-scientific apparatus. There is no single name for such activities. What is called an inquiry-based classroom in one country is called a practical laboratory in another; what some call hands-on experimentation others call inquiry learning and others still practical work, explorations or investigations. I hope that in the discussion below I will make clear what type of practical work or inquiry learning I am talking about, what its purposes are and what is effective.

I should say by way of introduction that I believe passionately in the importance of practical work in science teaching. In the UK much practical work is done in teaching science, many science teachers feel guilty if they are not teaching science through practice. Yet I think that much of the practical work that is done there in science lessons, and I would suggest in the science lessons of most other countries, is ineffective and detrimental to an appreciation and enjoyment of science.

Do we have problems with the practical work currently being done? If so, and I believe that we do, what is the cause? What type of practical work is appropriate and is it possible to do a type of science in schools that is authentic, introducing students to the way that many scientists actually work?

PROBLEMS WITH THE CURRENT SITUATION

A recent leader in the 1995 British Council Newsletter said that much practical work is ineffective, unscientific, and a positive deterrent for many students to continue with their science. It is ineffective in helping students understand the concepts and theories of their science. It is unscientific in that it is quite unlike real scientific activity. And it is boring and time wasting for many students who find it unnecessary and unstimulating.

I agree with that: I wrote it! But I am not alone in my worries about the effectiveness of practical work in school science. Practical work as many schools employ it, observed Derek Hodson in 1992, "is ill-conceived, confused and unproductive. It provides little of real educational value. For many children, what goes on in the laboratory contributes little to their learning of science or to their learning about science. Nor does it engage them in doing science in any meaningful sense."

An article of 1996 by Hodson describes the development of practical work since the 1960s as " three decades of confusion and distortion." In the words of OFSTED (the UK government's Office for Standards in Education) (1994), quoting Paul Black and Rosalind Driver and their group of leading science educators, "...there is a large body of evidence, both in this country and many others, that the understanding of the ideas of science which pupils develop within and at the end of their courses is alarmingly poor. . .."

THE CAUSES OF THE PROBLEMS

Among the reasons why, I believe, much practical work is ineffective, is that many teachers and students appear to be uncertain about the aims and objectives for doing practical work in science lessons, and therefore turn inquiry into a mere exercise in busyness. We need to be clear about what the reason is for doing any particular practical task and alert the students to it. If we hope that a single practical work will fulfill a whole range of vaguely specified objectives at the same time we will probably be disappointed.

Another caveat is that two particular aims often get confused in a way that does serious damage to both. Is the aim of practical work to give an increased understanding of some theory or is it to develop some practical skills? Is the experiment being done for the sake of theory, to discover, to verify or to clarify it, or is it being done for the sake of developing ways of working like a scientist? If we try to do an experiment which fulfills both of those aims at the same time we will not succeed in either. To get the experiment to clarify the right theory we will have so to direct and constrain it that we kill any freedom that the students may have for developing their practical skills. The old Nuffield theory of guided discovery just did not work (Stevens, 1978; Hodson, 1996).

There has been confusion here for a long time. A recent research investigation that I carried out with teachers and students in Oxford asked, among other things, for teachers to rank a series of aims for practical work in order of importance and also to rate how frequently they did different types. The most important objectives in their judgement related to developing practical skills; using practice to discover or clarify theory they rated very low. And yet the type of practice which teachers said that they used the most were "structured practical linked to theory." This finding confirms earlier research (Kerr, 1964; Thompson, 1975; Beatty et al., 1982). Another recent survey (Watson, 1997) asked not only the teachers but the students what they thought the purpose of doing practical work in science was. The teachers split their aims almost evenly between the development of theoretical understanding and the crafting of practical skills. But all the students believed that the only reason for doing practical work was to increase their theoretical knowledge and understanding! We must help our students, as well as our teachers, to distinguish between learning to understand scientific knowledge and learning to do scientific activity.

A further cause of confusion in the use, or misuse, of practical work is the information overload which bombards students, of which we are often unaware. As teachers, we know why we are doing a practical inquiry and see clearly the theory underlying the experiment. But students will have to sort out the apparatus, remember how it works, select the relevant data from the experiment, ignore the irrelevant, take measurements to the appropriate degree of accuracy, handle those inevitably imprecise and messy data, analyse into a meaningful conclusion the information gathered and then, after so much information has come in, try to remember what the experiment was all about in the first place—what question were they trying to answer. Often the underlying principle—say, Newton's laws of motion—is elegantly simple but by the time we have wrapped it up with carts, springs, and dotted ticker tape we have produced a highly complicated and distracting mass of clutter which hides it.

Another reason why much practical work is ineffective is that the preconceptions the students bring with them to the experiment determine what they will see. Just as we, the teachers, see the correct theory shouting at us through the experiment, so the students will see what they think is correct. In an electric circuit with two light bulbs in series, we would anticipate that the bulbs are lit equally brightly, and if one is slightly brighter than the other we conclude that the bulbs are not identical. But if we, as students, expected the current to be used up going round a circuit, we would expect one bulb to be brighter than the other and would convince ourselves that it was so. The POE (Predict, Observe, Explain) strategy (Gunstone, 1991) is useful here, forcing students first to predict what will happen, then observe and then explain the observation. This compels the students to engage with the experiment. As Joan Solomon said in 1980 when analysing why some of her experiments were successful for her science students and others were not, "Imaginative insight was not a sequel to successful experiments. On the contrary, it was an essential prerequisite."

A final and perhaps the most important reason why so many students gain so little from the practical work is that they come to the experiments without any intellectual curiosity, purpose or motivation. They come casually to the lesson, do what the teacher tells them to do and go away. I believe that almost any experiment can be effective if the student is genuinely motivated to find out what is going on and is determined to succeed. This entails ensuring that the students have a sense of ownership, that it is their experiment and not the teacher's.

A RATIONALE FOR PRACTICAL WORK IN SCIENCE

So far I have been negative about much practical work. Let me now be positive and suggest what I believe is a rationale for practical work which is thoroughly constructive and also involves the students in doing real science. Fifteen years ago, when sorting out my own thinking on practical work in science, I established three clear principles. I wanted greater clarity for the reasons for doing practical work, and a matching of project to aim. I wanted to separate theory from practice, employing practice not to discover or elucidate theory, but for its own sake. And I wanted a threefold rationale for practical tasks: experiences to develop a feel for the phenomena being studied; exercises to develop practical skills; and investigations to develop experience and expertise at working like a problem-solving scientist, which involves planning, performing, interpreting, and communicating. I still think that was a useful starting place but would now want to develop the framework a little more.

My rationale for practical work in school science would now be based on this framework:

- exercises to help develop practical skills;
- experiences to give students a feel for the phenomena;
- scientific investigations that include problem solving to gain experience of being a problem-solving scientist and often using technological, open-ended investigations and hypothesis testing to develop the way of working as a hypothesis tester and often using pure science and closed investigations;
- demonstrations to develop a theoretical argument and to arouse interest and to make an impact;
- recipe experiments to keep pupils occupied and to personalize some theory with POE.

There are certain scientific skills, sometimes called the processes of science, which need to be learnt, such as taking measurements, using scientific equipment, analysing data, tabulating and interpreting graphs, and these need to be developed with appropriate **exercises**. Sometimes these exercises will form an integral part of a larger scientific experiment, and be developed as occasion arises, but it should be made clear to the student what the aim actually is. Some educators talk about process skills, such as planning, hypothesising, observing, analysing, and interpreting, and suggest that these can be taught and assessed separately, out of context, and then put together to form the whole process through which a mature scientist works. But as Millar and Driver so clearly argued in 1987 and Millar in 1991, such processes are dependent on context. Being good at observing scientifically is not the same as being observant. The broad process skills of science can be properly developed only in the context of an authentic scientific activity

Experiences are invaluable for gaining a feel for the phenomena being studied, as opposed to developing the underlying theory. Such experiences will often be very simple, short and straightforward; stretching an elastic band to get a sense of elasticity; burning magnesium ribbon to get a sense of combustion; dissecting a plant to get a sense of the size, the shape, and the texture of the component parts; moving your finger around and above the lighted flame of a candle to get a sense of convection. These experiences can then be the basis for thinking about the underlying theory—but that will have to be done theoretically, by question and answers, discussion and exposition.

Investigations come in different forms. They may be closed or open-ended, involving finding what factors affect the strength of an electromagnet, which material makes the best sole for a shoe or how to make a parachute which lowers an egg from the top of a science block roof onto the ground in the quickest time without breaking the egg. They may take a short time or be spread over a week or so. They may fit into the normal science curriculum or be extracurricular. They may be based in a lab or the environment. But all should start with a question or a problem and get the students to make their own plans for tackling it. This very act of having responsibility for the planning of the investigation gives the students ownership of it and a determination to make it work. They will then perform the investigation, analysing and displaying their results as appropriate, and then evaluate and possibly modify their experiment. There is no single way of doing a scientific investigation. It involves the affective as well as the cognitive; it demands creativity, guesses, hunches, experience, a whole range of tacit knowledge and perseverance—and all these can be developed through investigations. Creativity cannot be taught, but it can be encouraged. Scientific investigations encourage creativity in a way little else in science teaching does. I have written more about scientific investigations elsewhere (Woolnough, 1994) and the ways that these can be incorporated in school science, both within the normal science syllabus and as extracurricular science activities. Suffice to say here that I believe that the opportunities for students to do their own investigations as a science project do more to transform a school's science course, and the students' enjoyment of it, than any other single thing.

A particular form of investigation which we, through the English National Curriculum for Science, have had experience of is a form of hypothesis testing. The original version in 1991 of the practical Attainment Target, Sc1, was called Scientific Investigation and took a topic from the physics, chemistry or biology programme of study and set up a rather closed investigation in which the student had first to set up an hypothesis as to what was likely to happen, to do the investigation and then to evaluate how far the hypothesis was proved correct. This led to some closed verification experiments—what factors, for example, affect the rate of reaction of metals and acids, what factors affect the strength of an electromagnet—which tightly prescribed what would happen and prevented genuine scientific activity. I have little time for such restrictive investigations, which seem to me to represent a very poor model of what scientists really do. The latest version of our national curriculum, produced in 1995, allows a rather wider interpretation of scientific activity which includes both investigations and experimentation based on the processes of planning, obtaining evidence, analysing evidence and drawing conclusions and evaluating evidence. This encourages a far wider range of practical activity, developing the skills and processes, and the conceptual understanding, in the most appropriate way.

Demonstrations, which I believe are more popular in the science classes of other countries than in England, are an excellent way of developing a theoretical argument in which the teacher can establish the structure of the relationships in a way which would not be possible for students left to sort out the ideas for themselves. Class demonstrations, often with student involvement, are an ideal way for the teacher to link together the theoretical and the practical,

Another class of experiment which I would refer to as **recipe experiments** has students merely follow instructions. I have little enthusiasm for these, which so often put students to just doing what they are told without understanding, insight or ownership. If such experiments are to be useful then they need to be focused. Somehow the students need to be motivated and appreciate what the experiment is about. Utilising the POE technique to force the students to think and become committed to their results, will perhaps give them a clearer impression of the significance of the experiment.

AUTHENTIC SCIENCE, PERSONAL KNOWLEDGE, AND SCHOOL SCIENCE

Much of scientific knowledge comes in the form of public knowledge. Knowledge that is public is written down in books and journals, is described in syllabuses, and provides a basis for public examinations. Public knowledge is explicit and makes possible a community of effort.

But there is another type of knowledge which is personal, gained through private experience and practice, and less easy to define. It relates both to understanding aspects of the physical world and also to the procedures required to tackle problems in it. It is tacit, and involves our senses as well as our intellect, our emotional commitment along with our intelligence. In Polanyi's words in his seminal book on personal knowledge published in 1958, "We know more than we can tell." We assert that a piece of music is by Bach even though we have never heard it before, we know how to ride a bicycle even though we cannot explain its stability and why we don't fall off, we recognise a suitable substance and a sensible course of action in tackling a scientific problem even though we cannot always explain why. Guy Claxton argued in 1997 that "we should trust our unconscious to do the thinking for us." Scientists in their professional life rely on their personal knowledge, their intuition, and their creativity as much as their public knowledge, if not more. Indeed it is only as the public knowledge has become personalised that it can be used. But personal knowledge is more than just personalised public knowledge: it is deeper, more sensual, inarticulate, and yet most useful.

School science in most countries is increasingly dominated by accountability and assessment and thus the form of science which is taught is predominantly public knowledge. Personal knowledge, because the individual cannot make it reliably explicit, gains little credit or status. But I would argue that unless school science teaching allows some place for the development and expression of personal knowledge in science, we will be misleading students about the authentic nature of scientific activity. I believe that we need also to stress the difference between the two types of knowledge and stress that both have value and importance. In 1971 J. Ravetz spoke of science as "a craft activity, depending on a personal knowledge of particular things and a subtle judgement of their properties." Learning in science W. M. Roth described in 1995 as "an enculturation into scientific practices." Millar asserted in 1996 that "learning science is a process of coming to see phenomena and events through a particular set of spectacles, and the intention of science education is to bring the learners inside the particular, and peculiar, view of things which scientists, by and large, share."

Including authentic science activity of a practical problem-solving kind enables students to develop and use their personal knowledge through experience. It allows them to experience doing authentic science and thus partake in one of our principal cultural activities. It provides students with skills and attitudes, the personal knowledge, that are useful to employers. And motivates them towards science and hence increases their propensity to learn public knowledge of science too.

Employers, as well as society at large, want school leavers to have personal life skills and self-confidence enabling them to work as autonomous, self-motivated citizens. A survey by M.C. Coles in 1997 asking employers in industry, most of it based in science, what they wanted in their recruits, found a very large consensus. He found that they wanted:

- Commitment and interest (the most important criteria),
- Core skill capabilities such as communication, numeracy, and IT (Information Technology) capability,
- Personal effectiveness, relationships and team work,
- Self reliance, resourcefulness,
- Initiative, creativity,

▶ The skill of analysis,
▶ Good general knowledge including understanding of the world of work,
▶ Professional integrity.

Employers were not just interested in their applicants' having a large amount of scientific knowledge or understanding, or specific scientific skills, though a basic amount of both was required. They asked for broad personal core skills and attitudes. These are exactly the type of skills and attitudes that are developed when students do personal investigations in their science practical work. A recent evaluation of the engineering education schemes being done in schools in Scotland demonstrated that the student research projects within them greatly enhanced the skills of communication, interpersonal relations, and problem solving of the students involved.

CREST,[2] AN EXAMPLE OF REAL PRACTICAL WORK

One of the most impressive and successful initiatives in the United Kingdom encouraging students to become involved in genuine scientific and technological activity has been the national CREST award scheme (CREST = CREativity in Science and Technology). CREST's aim is to stimulate, encourage, and excite young people about science, technology, and engineering through project work centered in tasks. It stimulates and supports projects connected to industry or community that draw on students' creativity, perseverance, and application of scientific and technological challenges. The projects cover a wide range of topics. Typical efforts might study the pollution of a local river, design and make an auto pilot for a sailing boat, or investigate the efficiency of alternative energy supplies. CREST gives bronze, silver, and gold awards. I have recently had the task of evaluating CREST after ten years of its existence and it quickly became apparent, through the response of both the teachers who organised it and the students who took part in it, that it was immensely impressive and highly successful in fulfilling its aims. Though CREST is an optional activity, involving only about 25,000 students—it is done by a minority of students in a minority of schools—those schools and students who are involved in it gain an enormous amount both in improving students' attitudes towards science and in developing their communication, interpersonal, and problem-solving skills.

My earlier researches in FASSIPES (a research project on the Factors Affecting Schools' Success In Producing Engineers and Scientists)(Woolnough, 1994b) showed that one of the factors which switched many students onto science and technology was involvement with student research projects, and such extracurricular activities in science. One way of doing this was through CREST. It and similar ways of doing projects in and out of the curriculum are described in the book *Effective Science Teaching* (Woolnough, 1994a). Others have written of the effectiveness of such investigational projects in different countries. R. Tytler's account in 1992 of the Australian Science Talent Searches is especially convincing. In the States there is a similar Science Talent Search for many years sponsored by Westinghouse and now by Intel; South Africa has Young Scientists; Scotland has its Young Engineers competitions. Many countries have science clubs, science workshops, and "great egg races." I would rate this type of practical work as being the best way of developing all the skills, educational and vocational, that could be beneficial to responsible citizens, whether or not they were to enter scientific and technological industry—and doubly so if they were!

KNOWLEDGE, SKILLS OR ATTITUDES?

Much of what I have said about effective practical work has centred on the affective domain, on the principle of motivation. Motivation is important purely in personal terms, in developing and expressing a sense of self-worth. But motivation is also vital for gaining any understanding and appreciation of science, both as a body of knowledge and as a way of working. It matters not what students know, understand, and can do; what is important is what they *want* to do. Motivation, challenge, ownership, success, and self-confidence: These are much more vital outcomes than the acquisition of specific knowledge or skills. Many are skeptical of transferable skills which may be useful in broader aspects of life. Derek Hodson shared that scepticism but said in 1988

> What may be transferable are certain attitudes and feelings of self worth...successful experience in one experiment may make children more determined and more interested in performing another experiment. The confidence arising from successfully designing an experiment might be a factor in helping children to stay on task long enough to design a new experiment successfully. (p. 260)

The Nuffield A level Physics curriculum of 1985 has a practical investigation as part of its course, and this is often the highlight of students' academic work. Students chose their own investigation, which typically lasts twenty hours. It could involve an investigation of water flows from pipes, an investigation of the acceleration of athletes, or the effect of stirring on the viscosity of non-drip paint, or the factors affecting the lift on an airfoil. Such practical investigations are not always easy, but almost without exception they enable students to produce work of a higher quality than in any other aspect of their school work. For when motivated, students are without limits in what they can achieve in their scientific and technological projects (Woolnough, 1997).

ENDNOTES

1. In the United Kingdom (UK) we refer as practical work to the type of activity done in laboratories based in inquiry. As much of this chapter will be a reflection of the UK experience (with the implications to be drawn for other countries left to the reader!), the term "practical work" will be used.
2. CREST stands for CREativity in Science and Technology. Details about the program are available from the CREST National Centre, 1 Giltspur Street, London EC1A 9DD.

REFERENCES

Beatty, J.W., and B.E. Woolnough. 1982. Practical work in 11-13 science: The context, type and aims of current practice. *British Educational Research Journal* 8(1): 23-30

Claxton, G. 1997. *Hare brain, tortoise mind—why intelligence increases when you think less.* London: Fourth Estate.

Coles, M.C. 1997. Science education—vocational and general approaches. *School Science Review* 79 (286): 27-32.

Gunstone, R.F. 1991. Reconstructing theory from practical experience. In *Practical science*, edited by B. Woolnough, 67-77. Buckingham: The Open University Press.

Hodson, D. 1988. Experiments in science and science teaching. *Educational Philosophy and Theory* 20: 253-66.

Hodson, D. 1992. Redefining and reorienting practical work in school science. *School Science Review* 73(264): 65-78.

Hodson, D. 1996. Laboratory work as scientific method: Three decades of confusion and distortion. *Journal of Curriculum Studies* 28(2): 115-135.

Kerr, J.F. 1964. *Practical work in school science.* Leicester: Leicester University Press.

Millar, R., and R. Driver. 1987 Beyond processes. *Studies in Science Education* 14: 33-62.

Millar, R. 1991. A means to an end: the role of processes in science education. In *Practical Science* edited by B. Woolnough. Buckingham: The Open University Press.

Millar, R. 1996. In pursuit of authenticity. *Studies in Science Education* 27: 149-165.

Nuffield. 1985. *Advanced level physics. Examinations and investigations.* Rev. London: Longman.

Office for Standards in Education. 1994. *Science and mathematics in schools: A review.* London: Her Majesty's Stationery Office.

Polanyi, M. 1958. *Personal knowledge.* London: Routledge and Kegan Paul.

Ravetz, J. 1971. *Scientific knowledge and its social problems.* New York: Oxford University Press

Roth, W.M. 1995. *Authentic school science: Knowing and learning in open-inquiry science laboratories.* The Netherlands: Kluwer Academic Publishers.

Solomon, J. 1980. *Teaching children in the laboratory.* London: Croom Helm.

Stevens, P. 1978 On the Nuffield philosophy of science. *Journal of Philosophy of Education* 12: 99-111.

Thompson, J. (Ed.) 1975. *Practical work in sixth form science.* Oxford: Oxford University Department of Educational Studies.

Tytler, R. 1992. Independent research projects in school science: Case studies of autonomous behaviour. *International Journal of Science Education* 14(4): 393-412.

Watson, J. 1997. Unpublished internal paper. London: Kings College.

Woolnough, B.E. 1994a. *Effective science teaching.* Milton Keynes: Open University Press

Woolnough, B.E. 1994b. Factors affecting students' choice of science and engineering. *International Journal of Science Education* 16(6): 659-676.

Woolnough, B.E. 1994c. Why students choose physics, or reject it. *Physics Education* 29:368-374.

Woolnough, B.E. 1995. Switching students onto science. *British Council Science Education Newsletter.* London: British Council.

Woolnough, B.E. 1997. Motivating students or teaching pure science? *School Science Review* 78(285): 67-72.

Assessing Inquiry

Audrey B. Champagne, Vicky L. Kouba, and Marlene Hurley

A cross the nation, science teachers are being called upon to use inquiry as a method of teaching and to enhance their students' ability to inquire. The American Association for the Advancement of Science (AAAS) and the National Research Council (NRC) have made inquiry an essential goal of the contemporary standards-based reform movement in science education. The AAAS *Benchmarks for Science Literacy*, published in 1993, and the NRC *National Science Education Standards*, which appeared in 1996, advocate the engagement of students in science inquiry so that they can develop both the ability to inquire and understanding of scientific principles and theories. The *Benchmarks* and *Standards* also contain descriptions of knowledge about inquiry and abilities to conduct inquiries that all students should gain as a result of their school science experiences. While the *Benchmarks* and the *Standards* differ somewhat in the emphasis placed on inquiry as a goal of science education,[1] each makes inquiry central to students' opportunity to learn[2] and the primary goal of contemporary science education reform. The *Benchmarks* and *Standards* serve as important guides to teachers who have the responsibility to meet national, state, or local science education standards.

This chapter describes assessment practices appropriate for science inquiry. It also describes how teachers can use assessment to increase their students'

opportunity to learn to inquire. These practices have their foundation in the principles of sound assessment practices contained in the assessment standards in
the *National Science Education Standards*.

INTRODUCTION

"Assessment," "testing," "evaluation," and "grading" are often used synony-
mously. The meaning of assessment as we use it in this chapter is much broader.
For our purposes, assessment is the process of collecting information for the purpose of making educational decisions. Administrators, government officials,
teachers, and other educators gather information about what happens in science
classrooms and what students learn as a result of their classroom experiences.
Administrators and government officials primarily use very general information
about student achievement to make judgements about the quality of teaching and
schools, to allocate resources, and to make educational policy. Teachers gather
more detailed and reliable information about student achievement and use it to
plan courses and lessons, to monitor student progress, to inform students about
the quality of their work, and to inform parents about their children's work. The
information collected by teachers has the greatest potential to increase the students' opportunity to learn science and, in turn, to improve student achievement.

Information collected outside the classroom has profound implications for the
lives of science teachers and their students. This information is collected in ways
different from those employed by teachers within the classroom and is used in
ways that vary from classroom purposes. Certain information is collected in controlled, rigorous ways and provides data that the district uses to communicate its
educational status to federal and state officials and to the members of its community. Student performance on standardized science tests and per capita expendi-
tures on science equipment are examples of data gathered by districts that are used
for these accounting purposes. Federal and state agencies use information of this
kind to allocate resources to districts. States use these kinds of data to sanction districts that are not performing up to standards. Community members use this information when voting for school tax bonds and school board members.

Because information collected outside the classroom has both direct and indi-
rect influences on the school science program and the classroom practice of science education, science teachers must monitor and be involved in the planning
and execution of these forms of assessment. But the focus of this chapter is on
the information district personnel and teachers collect and use to make district-
wide and classroom decisions; especially those related to students' understanding

of scientific inquiry, their ability to inquire in scientific ways, and the opportunity schools afford students to develop these understandings and abilities.

DEFINING INQUIRY

What evidence will teachers, administrators, and taxpayers accept that the students have learned to inquire and how will that evidence be collected? Defining expectations for performance and the conditions under which it is observed are parts of the assessment process. The nature of the evidence, in turn, depends on the district's expectations for the performance of students who have met successfully the district's standards. The expectations will reflect the district's definition of inquiry.

The *Benchmarks* and *Standards* provide a starting point for definition.[4] What is implied in both, but not made explicit, is that the practices of scientists and the scientific community serve as standards for the levels of inquiry expected of adults literate in science[5] and of students at various stages in the development toward adult science literacy.

The *Standards*, for instance, define as a component of science literacy the ability to ask questions. The rationale for this standard is that scientists pose questions and design inquiries, experiments, and investigations to answer them. While the practice of scientists is the implied standard, neither the attributes of the scientists' questions that makes them appropriate to scientific inquiry nor the scientific community's standards for question quality are defined explicitly in the *Standards*. The expectations for the levels of inquiry abilities and associated habits of mind to be attained by elementary, middle, and high school students identified in the *Benchmarks* and the *Standards* are not detailed enough to serve as the basis for the development of an inquiry assessment plan. Consequently, teachers must develop their own expectations building from these documents.

The *Benchmarks* and *Standards* employ the terms "scientific inquiry," "inquiry," "investigation," and "experimentation" with little indication of differences, if any, among them or indication of how they are related. More precise definition of these terms is a necessary condition for the development of strategies to assess them. In the absence of detailed definitions in the *Benchmarks* and *Standards*, we propose definitions on which we will base our discussion of strategies. Teachers may not agree with the distinctions we make among the different forms. That is all well and good. The essential principle is that the assessment strategy be congruent with the teachers' view of the nature of inquiry that they aim to develop in their students.

Forms of Inquiry

Table 1 presents a framework for inquiry with summary descriptions of two major forms of inquiry. The feature that distinguishes these two forms is the practitioners. Scientific inquiry is practiced by natural scientists. Science-related inquiry is practiced by adults and students who are science literate. Table 2 differentiates three forms of science-related inquiry distinguished primarily by the purposes the form of inquiry serves. In turn, the purpose determines the design of the inquiry and the kinds of data collected.

TABLE 1. TWO MAJOR FORMS OF INQUIRY

FORM	PURPOSES	PRACTITIONERS
SCIENTIFIC INQUIRY Inquiry as Practiced by Natural Scientists	Understand the natural world; Formulate, test, and apply scientific theories (as in making designer drugs)	Research Scientists
SCIENCE-RELATED INQUIRY Inquiry as Practiced by Science Literate Adults and Students	Obtain scientific information necessary to make reasoned decisions; Understand the natural world	Science Literate Adults K-12 Science Students

TABLE 2. THREE FORMS OF SCIENCE-RELATED INQUIRY

FORM	PURPOSES	PRACTITIONERS
Archival Investigation	Collect information from print, electronic and other archival sources on which to base personal, social, civic decisions Evaluate claims	Adults literate in science K-12 science students
Experimentation	Testing of theory	Secondary science students
Laboratory Investigation	Develop understanding of the natural world Develop understanding of inquiry Develop abilities of inquiry Collect empirical data on which to base personal, social, civic decisions Evaluate claims	K-12 science students

Scientific Inquiry

Scientific inquiry has as its primary purposes understanding the natural world by formulation and testing of theory and applying that understanding to practical problems such as the synthesis and testing of new drugs. The standards of practice of scientific inquiry are defined by the community of scientists, which specifies, among other things, the characteristics of empirical studies, methods of data collection, rules of evidence, and proper forms of reporting the results of experiments. Practices in the community of scientists are highly formalized and define a culture of science that includes a system of values, modes of behavior appropriate to scientific discussions, and characteristic vocabulary and forms of expression.

Science-Related Inquiry

The three forms of science-related inquiry are archival investigation, experimentation, and laboratory investigation. Archival investigation is practiced by adults who are literate in science and science students. Science literate adults inquire by gathering scientific information from paper, electronic, and other archival sources to make day-to-day personal, social, and economic decisions. Students also engage in archival investigation both in their personal lives and in their science programs.

In science class, students engage in science-related inquiry using empirical methods—employing observation, data collection, and the use of scientific apparatus and measuring instruments. Experimentation and laboratory-based investigations are two types of science-related inquiry that use empirical methods.

Experimentation is an inquiry for the purpose of testing a hypothesis that derives from a scientific theory.[6] An example is the test of a hypothesis that a mixture of carbon dioxide and natural gas behaves as an ideal gas. The hypothesis in the example comes from a theory of the behavior of gases. The experiment will test the behavior of the mixture of gases under certain conditions of temperature and pressure. If under the conditions tested the mixture behaves as an ideal gas, the hypothesis is supported. If, however, the gas mixture does not behave as an ideal gas, then the hypothesis is not supported. In neither case does the experiment prove or disprove the theory. It only demonstrates that under the conditions tested the results of the experiment support or do not support the theory. An instance of an investigation, also conducted in a laboratory, is a test to determine which of two brands of paper towels is more absorbent for the purpose of deciding which is the better buy. The question it is designed to answer does not relate to a theory of absorption or capillary action. The answer does not provide any insights into the nature of the natural world. Investigations require only that the test be fair and unbiased, while experiments must meet the rigid methodological requirements of the scientific community.

Before teachers can plan how to assess inquiry, they must decide which form of inquiry they expect their students to learn. Is the goal to develop graduates who will use the methods of scientific inquiry to develop new scientific knowledge or graduates who will have the inquiry abilities related to science that contribute to a satisfying and productive life? The answer to this question defines the form of inquiry a science program seeks to engender and consequently the nature of the evidence teachers will collect as proof that the goal has been met. The choice will also influence forms of inquiry that will be emphasized in the program's assessment plan.

If the goal is to develop a set of inquiry abilities, an assessment task for the members of a science class might be, as part of making a decision about selection of vitamin supplements, to plan an archival investigation to learn what vitamin supplements scientists and physicians recommend. If the goal is to cultivate the ability to conduct laboratory investigations, the task for a middle school student might be to plan and conduct an investigation to test the absorbency claim of a paper towel advertisement and, on the basis of the results of the test, decide which brand of paper towels to purchase. If developing new knowledge is the goal, the task for a high school student might be to design and conduct an experiment to determine the degree to which a mixture of gases behaves as an ideal gas. All are examples of inquiry but, if we are assessing the ability of a student to inquire by use of these examples, each will have its own performance standards. These can be effectively generated through an examination of each phase of any inquiry process.

PHASES OF INQUIRY

All forms of inquiry proceed in phases that we define as precursor, planning, implementation, and closure or extension. Within each of these phases, the student engages in two distinct aspects of inquiry: generation and evaluation. The student generates questions or hypotheses, plans for an investigation or experiment, and reports on the investigation or experiment. The student also evaluates the questions or hypotheses, inquiry plans, and reports of other students.

General Description of Inquiry Phases

In the precursor phase, experiences in the natural world, in reading, in interactions with others, or in response to personal needs make students aware of something related to science about which they are motivated to learn more. Through interactions with others, the student refines the question or hypothesis and enters the planning phase. There the student collects information, refines the hypothesis or question, and selects an appropriate method for the experiment or investigation. The student also presents the plan to others, criticizes the plans of other students, and makes refinements. Some preliminary or pilot laboratory work may be done as a part of the planning process. When a satisfactory plan is in place, the inquiry is conducted; information and data are collected and analyzed. It is possible during this implementation phase that problems with the plan may arise that require modifications. For instance, the need for additional data may become evident so that new data are collected and analyzed. During the final phase of the inquiry, the conclusions are evaluated to determine whether the inquiry has reached a satisfactory conclusion or further inquiry is required.

Successful engagement in inquiry requires knowledge of the nature of scientific inquiry, a variety of abilities including physical skills, and certain habits of mind. The inquirer must be able to pose questions and hypotheses amenable to scientific investigation and experimentation, to criticize plans for scientific investigations and experiments, to write a report about an inquiry, and to identify reliable sources of scientific information. These abilities, in turn, require information and the mental capabilities to process that information: that is, to reason scientifically.

Assessment Investigations and Phases

Our discussion of assessment strategies within the four phases of inquiry uses as an example an investigation conducted primarily in the laboratory but gathering some information outside it. The purpose is to gain information on which to base a decision. Investigation based in the laboratory is a form of inquiry that many districts have identified as the inquiry goal for their science programs. Investigations within the laboratory of social or other problems go a long way toward meeting the Inquiry Content Standard of the *National Science Education Standards* and the inquiry goal in the *Benchmarks*, which states:

> Before graduating from high school, students working individually or
> in teams should design and carry out at least one major investigation.
> They should frame the question, design the approach, estimate the
> time and costs involved, calibrate the instruments, conduct trial runs,
> write a report and finally, respond to criticism. (AAAS, 1993, p. 9)

While much of the knowledge and many of the abilities required for the investigations founded in the laboratory are the same as for experimentation, the two forms of inquiry differ in purposes (as shown previously in Table 2), in methodology (for instance, in the control of variables), and in the language for reporting results. Consequently, the discussion about investigation carried out in the laboratory cannot be applied to experiments without modification.

To assess the ability to inquire, teachers might ask students to design an investigation testing the claims made in breakfast cereal commercials about the vitamin and mineral content of the products. The results of the investigation could be used to decide which cereal to purchase. Do the products accurately report the USDA minimum daily requirements for each of the nutrients? This is a question that requires information gathering. Do the products actually contain the amount of each of the nutrients reported on the label? This might be answered empirically. To do so, the student must learn about how the nutrients are assayed and determine whether the chemicals and equipment necessary for the assays are available. What for each of the products is the ratio of nutrient to serving? Is the serving determined by volume or by weight? What is the ratio of serving to cost? What about taste appeal? Will the decision about which product to buy be based only on scientific information and cost? Not likely.

Assessment of a complete investigation of this type is intensive in resources. But if the ability to inquire is a central goal of a school science program, then students should have the opportunity to learn to conduct such investigations and be assessed on their ability to do so. Giving students the opportunity requires considerable time and equipment in laboratories as well as computers. If inquiry is to be a goal of a science program, these are considerations that should be taken into account. Once the decision is made, then districts must be ready to provide students with both the time and the equipment necessary to develop the ability, along with the opportunity to demonstrate their attainment of the ability. An alternative inquiry goal requires only that students know about scientific inquiry. That goal is less intensive in resources and the assessment strategies demanding only techniques for gathering information by paper and pencil.

If a district program is in compliance with the recommendations contained in the *Benchmarks* and *Standards*, much of the twelfth-grade science course for all students will be doing an extended investigation. Data collected as students engage in the investigation may be used to grade the students or to evaluate the effectiveness of the K-12 science program. Or as suggested in the *Benchmarks*, the purpose of engaging students in a complete investigation is to provide the students with the experience of putting together into a coherent whole the various elements of inquiry to which they have been exposed throughout the K-12 program. Either way, doing a complete investigation should be viewed as an occasion for learning. As some assessment experts have observed, good assessment practices cannot be distinguished from good learning activities. The process of assessing a complete investigation involves by definition cooperative effort among peers. A component of the peer interaction is giving and receiving criticism of one another's work.

TABLE 3. PRODUCTS, ABILITIES, AND INFORMATION ASSESSED IN THE PRECURSOR PHASE OF LABORATORY-BASED INVESTIGATIONS

PRODUCTS	ABILITIES	INFORMATION
Question that will guide the investigation	Obtain and analyze information for the purpose of formulating a question to guide an investigation	About conducting searches for information About how to assess the quality of information sources
Rationale for the question	Formulate questions appropriate to scientific investigation	About scientific inquiry:
Critiques of peers' questions and rationales	Construct a coherent argument in support of a question	◗ Distinguish among terms related to scientific inquiry (for example, variable, control, experiment, hypothesis) ◗ Characteristics of well-formulated questions appropriate for scientific inquiry
Reflective report documenting the evolution of the question	Communicate the questions and rationale with peers	◗ Practical characteristics
	Respond reasonably to criticism by peers	About the natural world: ◗ Scientific facts and principles
	Judge the scientific quality of a question	About the characteristics of well-structured scientific arguments
	Criticize questions and rationale of peers	About social, technological, civic aspects of the personal context of the question guiding the inquiry
	Keep written records documenting the evolution of the question and the rationale.	

Tables 3, 4, 5, and 6 summarize the products, abilities, and information assessed in each of the four phases of an investigation based in a laboratory. In each phase, the student produces questions, plans, and reports and is responsible for criticizing the products of others.[7]

Tables 3–6 provide a framework for assessing investigations but do not include the performance standards for the abilities and information. After examining the abilities and information called upon in the successful performance of investigations, however, we describe in the next section of this chapter how teachers and curriculum specialists need to work together to develop performance standards for students in grades K-12.

Precursor Phase. As shown in Table 3, during the precursor phase students formulate a preliminary question and a rationale for it: a discussion that relates the question to some experience the student has had and an explanation for why the question is appropriate for scientific investigation. Students present to their peers the draft question and the rationale. The peers have the responsibility for commenting on the question and the rationale. On the basis of peer review, students revise questions and rationale.

Work produced by the student during the precursor phase includes the question the student has formulated to guide the investigation. It contains also a rationale for that question: documentation of its origins, a description of its context, an account of the scientific information relevant to it. A reflective record documenting the evolution of the question is called for; so is the student's criticisms of the questions of peers and their rationales.

Among abilities involved in the precursor phase is that of obtaining and analyzing information for the purpose of formulating a question appropriate to scientific investigation. The student must be able to present a coherent argument in support of a question, communicate the question and rationale to peers, respond to criticism, comment on the questions and rationales of peers, and keep written records documenting the evolution of the question.

These abilities, in turn, require the students to have a large store of information and the mental capacity to process it in scientifically acceptable ways. The information necessary for successful engagement in laboratory investigations is considerable. The student must have information about conducting searches, knowledge about the characteristics of good sources of information, and knowledge of scientific inquiry and of characteristics of well-structured scientific arguments. Knowledge of scientific facts and principles must be supplemented by information about the social, technological, civic, and personal context of the decision that the student seeks to make. The student must have not only a vast store of information but also the cognitive capability to process it in scientifically acceptable ways.

TABLE 4. PRODUCTS, ABILITIES, AND INFORMATION ASSESSED IN THE PLANNING PHASE OF LABORATORY-BASED INVESTIGATIONS

PRODUCTS	ABILITIES	INFORMATION
Plan	Collect information from reliable sources to inform the development of a plan	About scientific inquiry
Rationale for the plan		About characteristics of investigations
Comments on the plans of peers	Develop a plan for an investigation (including appropriate scientific apparatus)	About characteristics of investigations whose methods are well matched to the question under investigation
Reflective report documenting the evolution of the plan	Explain how the plan meets the quality standards related to science inquiry	About the characteristics of investigations that meet scientific standards
	Communicate about the plan with peers in written and oral form, and in sufficient detail that another student could do the investigation	About scientific equipment and instruments
	Develop a well-structured argument	About the appropriate use of equipment and instruments
	Revise the plan on the basis of peer critique	About safety precautions
	Criticize the plans of peers	

Planning Phase. Table 4 shows that during the planning phase students in consultation with their peers develop plans for the laboratory investigation and comment on the plans of their peers. The products produced by students during this phase include the plan, a rationale for it that indicates how it meets the standards of an inquiry related to science, criticisms of their peers' plans, and a reflective report documenting the evolution of the plan.

The abilities assessed in this phase are numerous and complex and much like those assessed in the precursor phase. The major difference is that the abilities require knowledge about the characteristics of well-designed laboratory investigations and scientific apparatus. The application of this knowledge to the plan requires considerable cognitive capability. The ability to develop reasoned arguments, to communicate with peers, to comment on experiments, and to modify work in response to criticisms offered by others are assessed in the planning phase.

Implementation Phase. As shown in Table 5, during the implementation phase students set up the laboratory investigation, collect and analyze data, and develop conclusions from their analysis. As in the previous phases, students consult with their peers to learn how well they have made the logical connections between their data and their conclusions. The products of the implementation phase include the students' data, the data analysis, their conclusions drawn from the data, the argument leading from the data to the conclusions, critiques of the work of peers, and a reflective record of the events and conclusions from the implementation phase. The abilities, skills, and knowledge applied during this phase are different from that applied in the earlier phases in being related primarily to the nature, collection, and analysis of data.

Closure and Extension Phase. The phase sketched in Table 6 has students apply the information they collected in the laboratory investigation to make the practical decision. At this time, information and factors that are related to the context but not to science are brought into the investigation process. Often the decision is not clear cut. In the example of breakfast cereal, economics and taste preference may come into play and even override the scientific information collected by the student. The products of this phase are the students' reports of the investigation and commentaries on other students' reports. The abilities used during this phase are related to the forms of acceptable scientific communication, as well as the reasoning structure of convincing arguments.

As students engage in the investigation, the teacher collects a variety of information. The written products produced by the students, supplemented by observations the teacher makes of students as they interact with their peers and

perform laboratory work, provide evidence about the extent to which students
have met the inquiry goal.

**TABLE 5. PRODUCTS, ABILITIES, AND INFORMATION ASSESSED
IN THE IMPLEMENTATION PHASE OF LABORATORY-BASED
INVESTIGATIONS**

PRODUCTS	ABILITIES	INFORMATION
Data	Use laboratory equipment and measurement devices	About laboratory techniques
Data analysis		About proper use of laboratory equipment
Conclusions drawn from the data	Make accurate measurements	
	Choose appropriate representation	About acceptable forms for data tables and graphs
Argument leading from the data to the conclusions	of data	
	Interpret data appropriately	About significant figures
	Report data in properly formatted tables and graphs	About measurement units
Comments on the work of peers		About measures of central tendency—their calculation and appropriate use
	Construct an argument developing reasonable conclusions based on data	
Reflective record of the events and conclusions from implementation phase		About multiple repetition of empirical investigations
	Criticize work of peers.	

**TABLE 6. PRODUCTS, ABILITIES, AND INFORMATION ASSESSED
IN THE CLOSURE AND EXTENSION PHASE OF LABORATORY-
BASED INVESTIGATION**

PRODUCTS	ABILITIES	INFORMATION
Report	Weigh scientific information against other kinds of information in making a personal, social, civic or technological decision	About acceptable forms of reports and scientifically acceptable arguments.
Criticisms of reports by other students		
	Write a well-reasoned report	
	Develop a complete description of an investigation	
	Develop an argument leading from the original question through the data collection and analysis to the decision	
	Devise a well-structured argument in the application of conclusions from the laboratory investigation to the civic, social, or personal context	
	Consider alternative ways in which the conclusions from the laboratory investigation might be applied in a particular context	

SETTING PERFORMANCE STANDARDS

Setting performance standards for inquiry is the most challenging task faced by science teachers. It involves consensus between teachers and other district personnel responsible for the science curriculum. The framework for investigation in Tables 3–6 guides the process of setting performance standards for investigations, whether the laboratory or other sources of information are at issue. The performance standards must be set for students at each grade level. The preliminary step in a process for setting performance standards that we have found successful brings together groups of teachers who develop descriptions of performance that would convince them that students at their grade level are progressing toward meeting district standards. The teachers then design a task that will elicit the performance from their students. Teachers collect samples of resulting student work and match these against their expectations. In collaboration with their colleagues, they reach final consensus on reasonable performance standards for students at the grade level they teach.

For instance, teachers at the sixth grade would need to set performance standards for investigations by students at that level and the quality of students' explanations for why their plans are scientifically sound. Working in grade level committees, teachers would develop descriptions of the characteristics they expect for their students' investigations and explanations and then decide how they will assess their students' abilities to meet their expectations. After students have had ample opportunity to learn to design investigations, teachers would have students design plans. All sixth-grade teachers in a school or district would bring their students' plans and explanations to a meeting, where they discuss how well the work matches the expectations. Through a process involving the refining of the expectations in light of student work, performance standards for plans and explanations for the sixth grade would be developed. To achieve articulation for this ability and knowledge across grade levels, teachers from all grades need to work together to compare expectations as a means of identifying consistencies, redundancies, and inconsistencies in characteristics they have identified.

The process we have described is time consuming, but we believe that teachers who engage in it will come to a deeper understanding of what it means to inquire and the challenges of teaching students to inquire.

CLASSROOM-BASED ASSESSMENT

To plan and evaluate their classroom practices, teachers use the information they collect in their classrooms as summarized in Table 7. Teachers gather it week by week, day to day, minute by minute. They use this information to make judgements about the effectiveness of teaching methods and decisions about which methods will be used. The most recognized use of the information collected in class is to monitor and report student progress, especially grading students and reporting their achievement to parents.

TABLE 7. CLASSROOM-BASED ASSESSMENT

PURPOSE	DECISIONS	ASSESSMENT STRATEGY	RESPONSIBLE INDIVIDUALS
Inform Classroom Practice			Committees of teachers and students responsible for the course
			Individual teachers
Daily planning	To make on-the-spot modifications in teaching plans To provide individual students special instruction	Monitoring Student Learning Teacher gathers information on a minute-to-minute basis about students' understanding and abilities ▶ Questions students during lessons ▶ Cursory reviews of student work ▶ Short-term observations of student performance Monitor Teaching strategies	
Weekly and longer-term planning	To make extensive modifications in teaching plans	Teacher gathers information on daily or weekly basis about students' understanding and abilities ▶ Gives short quizzes ▶ Reviews student work ▶ Makes long-term systematic observations of student performance	

Testing the effectiveness of teaching methods	To select effective teaching methods	Measures of student achievement Measures of opportunity to learn
Reporting Student Progress Grading Reports to parents	What grade to assign the student Whether the student will pass or fail the grade	Measures of student achievement

Some decisions made by teachers using the information they collect have profound and long-term effects on the lives of students and teachers. Such decisions must be based on information in which teachers have great confidence. For instance, decisions to promote have a more profound effect on the lives of students than decisions to modify a student's daily assignment. Decisions about assignment changes can be easily rectified in light of new and better information without serious consequences. Such is not the case for promotion decisions.

The quality of educational information is determined in large measure by its accuracy and congruence with the decisions that will be made using it. If inquiry is a central objective of a course, grades in the course should be based on information about students' ability to inquire. Grades based on information gathered using multiple-choice tests are suspect because the information is not congruent with the course objectives.

The role of teachers in the assessment of inquiry cannot be confined to the grade level they teach or to assessment made in the classroom. They must also be involved in the design and implementation of the district's overall plan for inquiry in the science curriculum. Assessment plays an integral part in the planning process at the district level.

ASSESSMENT IN PROGRAM AND COURSE PLANNING

This chapter is predicated on the assumption that the development of effective assessment strategies for inquiry is a collaborative process. At a minimum, science curriculum specialists and science teachers must work together to develop assessment plans that include methods of data collection and analysis. Expectations for proficiency in inquiry that all students should attain, as well as strategies for the assessment of inquiry, are too poorly defined and their development too challenging for individual teachers working in isolation to accomplish.

Developing such strategies is an integral part of K-12 planning for science programs and courses. Program and course planning is facilitated by the development of assessment strategies at the beginning of the planning process rather than treating assessment like the little red caboose that always came last. Definition of a science program's goals and objectives in measurable terms—an essential step in the assessment development process—facilitates communication among program planners about the goals and objectives and enhances the reaching of consensus. Such consensus is the essential precursor to further steps in the program planning process: development of scope and sequence, selection of pedagogical practices, and textbook choice. Definition of the goals of a program or course and objectives in measurable form not only facilitates further program planning but also is the beginning to the design of the assessment plan for the program.

If a program goal is to teach its students to inquire, members of the program planning team are challenged to reach consensus on an essential question: what evidence will they accept that their students have met the goal? Answering this question requires that individual program planners agree on the nature of the evidence that will be collected and how that evidence will be interpreted. The purpose of the assessment plan is to decide whether the district's students have met the program goals. The assessment planning process requires consensus among the responsible individuals about how the data that will be used as evidence will be collected.

The assessment plan should question not only the achievement of students but the extent to which the school science program provides the students with the opportunity to develop the knowledge and abilities required for proficient inquiry.[8] The opportunity to learn question is much the same in form as that of student attainment: what evidence will we accept that we have provided our students adequate opportunity to develop the abilities of science inquiry? Designing the strategies that will be used to collect the evidence is integral to the process of curriculum planning. Reaching consensus on the strategies for assessment of opportunity to learn ensures that the responsible individuals are in agreement on resources that need to be in place if the student attainment goal is to be met.

Table 8 summarizes purposes, decisions, assessment information, and responsible individuals for program and course planning and evaluation.

TABLE 8. ASSESSMENT IN K-12 PROGRAM AND COURSE PLANNING AND EVALUATION

PURPOSE	DECISIONS	ASSESSMENT INFORMATION	RESPONSIBLE INDIVIDUALS
K–12 Program and Course Planning			
Reach consensus on program or course objectives Reach consensus on the conditions under which students can meet the objectives Identify resources that will be necessary to meet the conditions	Which of the possible program or course objectives will the K-12 program select? Which teaching materials and strategies to select? Does the district have the resources to meet the conditions?	Definition of program or course objectives in measurable form: design of prompts and performance expectations Definition of the conditions in measurable form	Program Planning: Committee consisting of teachers, administrators, curriculum specialists, parents, community representatives, and students Course Planning: Committee of teachers (and students) responsible for the course
K–12 Program and Course Evaluation			
Determination of how well the existing program or course is meeting the district's expectations	Whether the program or course will be continued or requires modification	Measure students' attainment of the program or course objectives: use the prompts and performance expectations developed in the program or course planning phase to measure and evaluate the effectiveness of the program or course Measure the extent to which the conditions necessary for students to meet the program or course objectives were met	Program Planning: Committee consisting of teachers, administrators, curriculum specialists, parents, community representatives, and students Course Planning: Committee of teachers (and students) responsible for the course

Program and Course Planning

As indicated in Table 8, consensus on program goals and course objectives and the classroom conditions is essential to planning programs and courses. Assessment plays a central role in the process of reaching consensus on these matters. Definition of the objectives in measurable form, that is, specification of the prompts and performance expectations for the objectives, is an effective way of improving communication among the individuals responsible for making the decisions.[9] The planning process also requires consideration of the resources necessary to provide students with adequate opportunity to achieve the objectives. Assessment, in this case, defining in measurable terms the resources required to provide that opportunity, is an effective mechanism for clarifying the necessary resources. Once the requisite resources have been identified, the responsible individuals have the information necessary to decide whether the district can provide the resources.

Program and Course Evaluation

Program and course evaluation is a generally recognized purpose for assessment, and is therefore included in Table 8. But far too often evaluation planning comes after the program or course is developed and implemented. When objectives and resource requirements are defined in measurable terms in the program and course planning processes, designers know how the program or course will be evaluated. Such information guides the development process in ways that increase the likelihood of a favorable evaluation.

CONCLUSION

This chapter has shown assessment of inquiry to be a complex process, intensive in the need of resources. The principles of good assessment are more easily defined than implemented. The *Benchmarks* and *Standards* can provide guidance in defining goals. They do not, however, give details on how a school district, classroom teacher or student of inquiry will actually meet those goals. If inquiry is a district goal, the district must be ready to devote considerable resources to providing students with adequate opportunity to meet that goal, to supplying teachers with adequate occasion to develop strategies to teach inquiry abilities, and to assessing attainment of the goal of inquiry.

ENDNOTES

1. Both documents call for students to become knowledgeable about inquiry as it is practiced by scientists and for engaging students in inquiry to develop their understanding of the natural world. The *Benchmarks for Science Literacy* (AAAS, 1993) place greater emphasis on students knowing about inquiry than on their acquiring the ability to conduct inquiries. The *National Science Education Standards* (NRC, 1996) place greater emphasis than the *Benchmarks* on students' acquisition of the ability to conduct scientific inquiries (Kouba & Champagne, 1998).

2. For psychologists the terms "learning" and "development" have different meanings. The terms as they are used in this chapter emphasize the complexity of the knowledge, abilities, and inclinations possessed by the competent adult inquirer and acknowledge that the requisite knowledge and abilities are acquired over a long period. Consequently, as we speak of inquiry in the context of the K-12 program, we used the term "development." When referring to the individual components of information and the many individual abilities that may be acquired in relative short periods, we speak of "learning."

3. In education the terms "program" and curriculum" are used interchangeably. We have chosen to use "program," which we define to include goals, content, preferred teaching methods, and assessment strategies; as "objectives" contain detailed and explicit statements about what students are expected to learn and the order in which students are introduced to the content. Typically a curriculum contains, in addition to the elements of a program, statements about its philosophical, psychological, social, political, and disciplinary foundations.

4. For a detailed discussion of how assessment can be used by curriculum developers to match curricula with standards, see Champagne (1996).

5. The *Benchmarks* (AAAS, 1993) and *Standards* (NRC, 1996) call for schools to produce students who are literate in science as well as for the development of the abilities of science inquiry. But it is not clear whether the ability to inquire is equivalent to being literate in science (Kouba & Champagne, 1998).

6. Our placement of experimentation in high school is consistent with its placement in the *Standards* (NRC, 1996).

7. As outcomes for science education, the *Benchmarks* (AAAS, 1993) and *Standards* (NRC, 1996) include cognitive processes such as scientific reasoning, logical reasoning, critical thinking, analysis, and evaluation. Cognitive processes cannot be observed directly. Teachers must base their inferences on samples of a student's work or observation of performance. Our analyses draw on particular models of cognition, such as those proposed by theorists of information processing (Simon, 1962; Bloom et al., 1956; Inhelder & Piaget, 1958; or Gagne, 1970). Assessing cognitive process is challenging because it is so inferential and dependent on theory. Consequently we have not included in our discussion of the assessment of laboratory inquiry descriptions of the cognitive process that are the foundations of the abilities of the proficient inquirer.

8. The *Benchmarks* (AAAS, 1993) and *Standards* (NRC, 1996) advocate inquiry as a pedagogical method to develop understanding of the natural world and the character of scientific inquiry. Because understanding both contributes to the ability to inquire, assessing the degree to which students are exposed to inquiry speaks not only to the abilities of inquiry but also to the knowledge required to be a proficient inquirer.

9. In defining the concept of inquiry, we follow the suggestion that Alfred North Whitehead made in 1925 that scientific concepts must be defined in terms of their measurement.

REFERENCES

American Association for the Advancement of Science. 1993. *Benchmarks for science literacy*. New York: Oxford University Press.

Bloom, B. S., M. B. Engelhart, E. J. Furst, W. H. Hill, and D. R. Krathwohl. 1956. *Taxonomy of educational objectives. Handbook I: Cognitive domain*. New York: David McKay.

Champagne, A. B. 1996. Assessment and science curriculum design. In *National standards and the science curriculum: Challenges, opportunities, and recommendations*, edited by R.W. Bybee, 75-82. Colorado Springs, CO: BSCS.

Kouba, V.L., and A.B. Champagne. 1998. *Literacy components in natural science and mathematics standards documents: Communication and reasoning*. Albany, NY: National Research Center on English Learning & Achievement.

Gagne, R.M. 1970. *The conditions of learning*. New York: Holt.

Inhelder, B., and J. Piaget. 1958. *The growth of logical thinking from childhood to adolescence*. New York: Basic Books.

National Research Council. 1996. *National science education standards*. Washington, DC: National Academy Press.

Simon, H. A. 1962. *An information processing theory of intellectual development*. Monographs of the Society for Research in Child Development 27 (2, Serial No. 82).

Whitehead, A.N. 1925. *Science and the modern world*. New York: New American Library, The MacMillan Company.

Implications for Teaching and Learning Inquiry: A Summary

Jim Minstrell

INTRODUCTION

When Emily van Zee and I were invited to edit this book on inquiry, I accepted because I thought the process and product would help me better understand the nature of scientific inquiry and how it might be taught and learned. We got to choose the authors. Although there were others whose work and ideas we would like to have included, the set here makes up a very good team in having both diverse and overlapping ideas. There are some minor differences of opinion, but on the whole, the chapters help to create a coherent vision of many attributes of inquiry.

The first section of this chapter presents my brief summary of some aspects of inquiry I see exhibited in the chapters in this book. The second presents through a vignette a vision of what inquiry might look like in my classroom. In my short space, there is no way I can do justice to all the efforts and ideas expressed by the other authors. I strongly suggest that you convene a group of educators to have discussions around each of the chapters and come up with your own summary, and then work together on implementing your vision in the learning environments you create.

BRIEF SUMMARY OF LESSONS ON INQUIRY

Inquiry Seems to Mean Different Things to Different People

At the present, we may agree more easily on what inquiry isn't than on what it is. We seem to know when what we are seeing is not inquiry. Perhaps we can know inquiry by knowing what intentions it serves. Is the central purpose of your inquiry to prepare students to develop new knowledge or to contribute to a satisfying and productive life? What goal we choose determines how we instruct and how we assess the achievement of those learning targets.

What does it mean to do inquiry? One definition is that it involves fostering inquisitiveness; inquiry is a habit of mind. Another holds it to be a teaching strategy for motivating learning. But what does that mean? Some say inquiry means "hands-on." Others say that is too simple: that hands-on is not necessarily minds-on, and inquiry requires mental reflection on experiences. Some educators add that inquiry includes manipulating materials to become acquainted with phenomena and to stimulate questions, as well as using the materials to answer the questions. Scientific inquiry is a complex process. There is no single, magic bullet. Some instructors, in the interest of a focus on inquiry, address only one aspect of the complex process. That contributes to the impression that inquiry means different things to different people. So we need to identify the various aspects of the process and see them as a whole.

We need to encourage and support personal curiosity when it occurs spontaneously and stimulate it when doesn't occur naturally. Then we need to challenge students to do deep thinking and find answers for themselves. When learners are involved in the struggle to understand something for themselves, they are more likely to feel the sense of pleasure when the understanding does come. I'm reminded here of the intellectual pleasure and sense of personal accomplishment I felt when I brought myself to deeper understanding with only the assistance of questions from my mentor to help guide my personal investigation. People like to confront the unknown. They are motivated to understand what they presently do not. In general, we have not taken the time to understand students' preconceptions; we give their ideas little credit or status. But for students to engage in inquiry, the questions and ideas being tested best come from them, or we run the risk of continuing to lose their intellectual interest.

Scientific inquiry is not new. For some time science educators have been trying to define and emulate the examples of inquiry conducted by early scientists. My favorite is Galileo's "Dialogues Concerning Two Chief World Systems," which gives a glimpse of scientific inquiry in the interaction among three discussants. Clearly one lesson is that scientific argument is about making sense of observations, not standing on the opinions of authority. When we have finished the inquiry we should know something we did not know before we started. We should be able to make more accurate predictions about events in nature. Even when our investigation fails to find the answer, at least the inquiry should have yielded a greater understanding of factors that are not involved in the solution.

What are the assumptions behind learning and teaching inquiry? Here is my summary of assumptions that follow from the other chapters of this book.

Almost Anyone Can Learn by Inquiry

Practicing scientists and young children, nearly everyone can learn from inquiry. A common interpretation of the work of Piaget is that young children cannot do inquiry, for they are not capable of formal, hypothetical, and deductive reasoning. But several of our authors are quick to point out that Piaget's work describes uninstructed children's capacity for engaging in the reasoning involved in scientific inquiry. When children have an opportunity to learn in a scaffolded but challenging reasoning environment offering evidence and built from personal observation and experiences, they show that they can reason at much higher levels than even the national standards documents suggest. Several chapters have described special populations, such as inner city, economically disadvantaged children or children with various physical or learning disabilities (with the possible exception of mentally retarded) learning by inquiry, as well as they learn by more traditional methods or better. But that requires specific teaching practices and supportive curriculum.

Inquiry Should Be Incorporated into Planned Learning Goals and Teaching Methods

We need to improve learners' knowledge about inquiry and their ability to do it. Past approaches to learning will not be good enough for the future, for this new century. Scientific information and procedures can be helpful in making day-to-day personal, social, and economic decisions. In general, people will need to be able to ask questions and seek and evaluate information that may help them answer those questions. Of course developing scientists will need to be able to conduct inquiry. But the general population can also profit from scientific inquiry that will satisfy intellectual curiosity or enhance the quality of life as well as solve practical problems. Some say inquiry when properly done can motivate life-long learning.

Schools will likely need to make some changes to meet this future. Many of today's students are disengaged with school activities. As several of the authors in this volume suggest, colleges and universities need to reform their practices as well. Teachers and others can benefit from courses that are challenging. Even recent science Ph.D. graduates drafted to become pre-college science teachers will need to learn about the nature of learning and understand the relation between learners' present knowledge and the sort of experiences that can help foster growth to a deeper, more principled scientific understanding.

National and many state standards documents recommend these kinds of learning connected with inquiry: learning about inquiry, learning to do inquiry, and learning by inquiry. While learning about inquiry can be through studying the work of others, the last two involve active participation. But even if learning subject content is by active involvement in inquiry processes, students may not learn to use the processes of inquiry unless teachers explicitly target and address the learning of specific aspects of inquiry. More than one of the authors point out that teaching both subject matter and inquiry processes at the same time most frequently brings students to learn one or the other but not both. Each goal of learning needs specifically to be identified, taught, and assessed.

The Inquiry Process Is Motivated and Sustained Through Dialogic Questioning and Reflection

Although the authors differ on whose question initiates the inquiry, all agree that the learner must own the questions around which the inquiry takes place: that the questions are understandable and intellectually interesting to the learner. Some of the teacher authors give examples of how they generate interest in the questions for the learners. At the elementary level, sometimes the interest can come from stories in which the children have already shown curiosity. Other times we have to work a bit harder and relate our questions to aspects of their daily experience.

As a teacher I worked hard to formulate questions that both addressed content important to physical science and yet were interesting to the general high school population. For example, to get kids interested in questions of measurement, I designed a scenario involving multiple measurements of blood alcohol for a particular driver suspected of driving under the influence. "What should be reported as the blood alcohol count for the driver?" "Why would there be different readings from the same sample of blood?" "So should the driver be presumed to be under the influence or not? How did you decide?"

These questions, while I initiated them, got nearly all my high school juniors and seniors mentally actively involved in an inquiry about the nature of measurement and related issues. Whenever during the rest of the year we again encountered issues of measurement, students referred to what they had learned from this inquiry.

In a community of inquiry, whether in the classroom or in the informal environment of a science museum, the learning of each member is affected by other learners and guides as well as by the physical environment. Carrying on a dialogue with one's self and with others stimulates deeper reflection. As we share our observations, questions, and ideas with others, we gain clarity of meaning for ourselves. Many of the authors focus on creating cooperative or collegial arrangements for learning with and from fellow learners. But these environments are not easy to manage. Learners need to be guided to probe their own thinking and the thinking of others.

The verbal interaction between teacher and students and among students needs to embody respect for one another at the same time it is critical of the ideas being expressed. When learners share their questions and their ideas for making sense, listeners foster growth by helping the speaker to express and represent his or her ideas clearly. Appropriate scaffolding includes respectfully helping the speaker identify possible reasoning errors or a need for more experimentation to clear up ambiguities in the data and to support an argument. "Metacognitive reflection" on the part of the individual learner is enhanced by support from the learning community.

Inquiry Includes Reasoning From Experience and From Ideas That Describe and Summarize Experience

Virtually every chapter talks about mental activity on the part of the learner involved in scientific inquiry. Reasoning doesn't happen in a vacuum. Experience with materials, phenomena, and ideas provide the grist for reasoning in science. In the process of generating new ideas, they are induced from experiences with phenomena. Predictions are reasoned from phenomena or from previously generalized experiences. Conclusions are generalized from results. When learning about scientific inquiry, students need opportunities (and scaffolding) to separate their observations from their inferences. Several of the authors point out the common difficulty of separating conclusions (which require personal reflection) from results (which are organized observations of the experience). To stimulate reasoning from experience, critical questions to ask learners include "What happened, what did you see or hear?" (What did you observe?) and "Why do you think that happened, what sense do you make from these observations?" (What do you infer? What is nature trying to tell us?)

Although many of the chapters suggest hands-on experiences, most authors are quick to point out that hands-on should not be equated with inquiry. Reasoning about experiences that have become personally viable to the learner is what matters. The experiences might have been initiated by hands-on or by a shared, reliable (and probably repeatable) experience by others. First, learners need to "know" the phenomena of the experience. Doing it or seeing it for themselves can help. Then, they can be stimulated to reason to make sense of the experience.

In science it is important always to couple reasoning with experiential evidence. Observations and related inferences of meaning are primary intellectual activities. Opinions are not a necessary part of science. Personal opinions may be formed as a result of scientific inquiry, but opinions can also arise out of personal whim, what a learner wants to happen or dreads might happen. A scientific inquiry goes on to ask for the experiences that suggest that that phenomenon might happen and the reasoning that leads from those experiences to a conclusion. Or opinions may be formed from having done the science. The scientific inquiry may have provided the reasoning behind the formation of an opinion.

The Materials Used in an Inquiry Environment Need to Support the Inquiry Learning Expected

The worksheets, the laboratory or demonstration equipment, the software, the text resources, and assessment activities all should support the inquiry goals of the program. For example, the printed materials can invite and guide thinking in what to think about, but should not short-cut thinking by directing students what to think. Materials can be made available to learners after questions have been initiated by learners, or materials and minimal instructions for what to do with the materials can be provided to stimulate more questions and thinking. Software can assist students in recording their results and can scaffold students in the development of conclusions from results without directing students in what they should think. Text and technological resources of information can be useful to learners, when they have the questions for which they want answers. These resources can assist learners in making more efficient the search for specific information. Assessment tasks, whether on paper or served through technology, can help learners and teachers identify a needed relevant experience or a weakness in the learner's reasoning from prior experiences. Assessment of inquiry skills, aligned with the specific inquiry goals of the educational program, must be as explicit as assessment of understanding of ideas.

Meanwhile, the tone of the environment by teachers, software, or demonstration medium needs to be supportive of thinking itself rather than set by extrinsic rewards for getting the right or punishment for arriving at the wrong answer. If learners obtain results that are not consistent with what others are getting, that needs to be identified. Then, help learners identify why their results may be different and encourage replication. If conclusions are not deemed as correct, help the learners identify inconsistencies between their thinking and prior experiences. Frequently teachers may find that it is the curriculum materials that are lacking rather than the students' thinking.

The point is to create an atmosphere of identifying and supporting learning needs rather than dictating what learners are to do and think. Setting a proper set of experiences before the learners will enable them to make their own good sense of the experience.

Learning to Do Inquiry and to Teach Inquiry Are Best Done Through Experience in Learning by Inquiry

Learning itself is an active process, and inquiry is best learned through actively participating in the process. Children should have opportunities to learn to do inquiry by participating in the inquiry process in class, or in the informal setting, such as the museum or group or family experiences in natural environments. The same is true of teachers, whether in elementary or in secondary schools, and the principle applies to young scientists switching careers to teaching. Since teachers typically teach as they have been taught, learning by inquiry will increase the probability that they will teach by inquiry.

Teachers also need opportunities to engage in their own professional inquiry into the teaching and learning of their students: to talk with others about what sorts of conceptions and procedures students are likely to develop, what kinds of instructional activities will dissuade students from inquisitive ideas and skills, and what sorts of assessment activities will allow the teacher and students to monitor development. Teachers need opportunities to conduct teaching experiments to understand better their students' learning.

PUTTING IT ALL TOGETHER

Many of the chapters give examples of the authors' views of inquiry in the classroom. I can't resist the opportunity to share an example from my experience. While I can't possibly exhibit all the ideas presented in this book, I would like to address some of the critical assumptions and issues expressed in the prior section, particularly the concern about trying to do many things without cheating on any of them. How is it that a teacher might concentrate on teaching by inquiry both specific content and specific aspects of inquiry? How might a teacher create and maintain an environment for inquiry while centering on development of specific concepts and skills? In this vignette, I will give a flavor for how I might teach about and by inquiry and have students develop their conceptual understanding as well.

The Context for This Vignette

We are about to begin a unit on forces and explanation of motion, after which I expect that the students will explain various motions, using ideas more consistent with Newton's Laws. But I want them also to gain a better understanding of scientific inquiry and skills for doing it.

Raising Questions

At the beginning of a unit I find it helpful to pose questions that survey students' present understanding of ideas or situations that will be the source of our class inquiry for the next several days or weeks. As the students construct their answers individually, they start to generate their own questions about the issues of the unit, and they become motivated to answer questions they and others have initiated.

From research on students' understanding, I know that one of the alternative explanations for motion that makes sense to them is something like: "to explain motion, one needs a net force, acting on the object. The size (magnitude) of the net force should be proportional to the speed of the object" from their point of view. I call such constructions of ideas "facets of students' thinking" (Minstrell, 1992). Opportunities for students to express these facets of their thinking need to be built into my initial elicitation questions.

A first question might involve pushing a book across the table top and asking students to use arrows to represent each of the forces acting on the book while it is moving with a constant velocity, and it includes asking students to clarify what is exerting each force. A follow-up question asks students for diagrams that represent the forces on the book as my hand pushes and the book accelerates across the table. For each case, I ask for five diagrams at equal time intervals. Students want to talk about getting the object going and stopping it at the end. The three images I require for the middle of each motion story gets them to think instead about the constant velocity or the constant acceleration (Minstrell, 1984).

Determining the Hypotheses

Following the pre-instruction questions, I conduct a full-class discussion. As students share their answers and ideas, I write the ideas on the front board for all to see. In this case the two events we are trying to explain consist of constant velocity of an object and its constant acceleration. Later we will worry about more exotic or unusual motions. The answers and ideas evolve into initial ideas (hypotheses) to be tested through classroom laboratory experiences.

In this case for the constant velocity in a straight line, typically the class suggests these two ideas: A constant extra (unbalanced) force is necessary to keep an object moving with a constant speed (stated in "if.., then..." form, it would be "An object moves with a constant velocity if, and only if, there is a constant net force acting on the object"); and Once the object is moving, no (zero) extra force is required to keep it moving with a constant speed. For the constant acceleration cases the two typical hypotheses run something like these: An extra force that increases at a constant rate is necessary to keep the object accelerating at a constant rate; and A constant extra force will be required for the object to accelerate at a constant rate. In my classes, typically in excess of ninety percent of the students believe the first hypothesis for explaining constant velocity and the first hypothesis for explaining constant acceleration. About five percent suggest the second hypothesis for explaining constant velocity and the second hypothesis for explaining constant acceleration. Occasionally another hypothesis is listed.

"Are the ideas we've written on the board true?" I ask the students. While some initially say yes, others are quick to point out that they can't all be true, because they conflict with each other. After some discussion among students, they suggest that each statement is an idea that somebody in the class thought was true. After students have articulated a recognition that the suggested explanatory ideas are tentative, I observe that in science these testable ideas we think might be true are called "hypotheses." Thus, I write "Class Hypotheses" above the set of ideas we are about to test. But before I used the more technical term "hypothesis," I had the students articulate the notion of a tentative (and testable) idea.

I usually run the survey and the preceding discussion in about twenty minutes at the end of one class period. I may ask students to consider (and record in their journals) various situations in their lives that seem to support one or another of these hypotheses. The next day I have the initial class hypotheses written on the board and the discussion continues.

Identifying Relevant Variables

"So if we are going to do an experiment to determine what idea best fits careful experimentation, what variables are we going to need to measure?" With some discussion the students decide that the relevant variables are speed and force. Some students are considering force as simply the pulling or pushing force by whatever is making the object move forward. Others are concerned that we need to consider any retarding forces like friction as well. Finally, it is decided that the extra force (net force) is the important variable, and that will involve a resolution of any backward forces as well as the forward force on the object.

Simplifying the Experiment

In subsequent discussion, some students voice concern that the backward force of friction might be difficult to measure. Preferring to avoid the topic of friction at this point in our class inquiry, I guide the students in setting aside this concern. Several students readily suggest we design the experiment so we have no (or negligible) friction. They choose to do the experiment using a low friction cart as the object that will move rather than a book sliding across the tabletop.

Measuring the Relevant Variables

"What measurements will we need to make?" The students suggest that if the cart wheels are exerting very low friction, we only need to measure the force of the string pulling the cart. That will be the extra force causing the motion. "Any other measurements we will need?" Some students suggest we will need to measure the motion. "How might we do that? What do you mean by motion here?" Students suggest we will need to measure the speed of the cart. "What speed? The average speed or the instantaneous speed?" Some students suggest we just need to get the average speed because we are looking for how big the speed is in relation to how big the pulling force is. "Is that right, do we know the speed will be constant? For sure?" Other students are quick to point out that that is the purpose of our experiment, to see what sort of motion is related to the extra force. They suggest we do what we had done during earlier studies of kinematics, the description of motion: measure several positions and the corresponding clock-readings (times), during the motion of the cart across the table. From those data, they suggest, we will be able to construct a graph of speed against time and describe the motion. Measurement and kinematics were among our earlier units, so they are familiar with issues of best value and uncertainty in measurement of quantities including position and time. In the preceding sub-unit on explaining the condition of the object at rest, we have used spring scales for measuring forces (Minstrell, 1982).

In this case I have used what Emily van Zee and I (1997) have called a "reflective toss," taking one student's answer and reflecting it back to that student or to the class as a whole. Rather than having students key on the teacher for knowing what to do, I want the students to be responsible for building their own understanding. As their teacher, I am there to juxtapose related experiences and to use questions to guide their thinking about the experiences and ideas. They need to decide what to do and why and what the results will mean.

The Experiment

In the back of the classroom I have anticipated and already set out the sort of apparatus the students will need. They have carts with spring scales taped to them. A string is attached to the spring scale and leads down the table around a pulley, to the top of a ringstand, and around another pulley and is attached to a 100 gm mass that hangs out over the end of the lab table. This is a variation of an experiment from *Project Physics* (Rutherford, Holton, & Watson, 1972). An even more primitive version was in *PSSC* (Haber-Shaim et al., 1959/1971). Newer versions are available using micro-computer based force probes and motion sensors, but I prefer that the initial experiment be done with the spring scale, meter stick, and stopwatch, the measurement tools that are more transparent than the computerized probes and are more accessible to students out of school. I want to encourage students to put together experiments at home. Because the movement across the table only takes a couple of seconds, I have had to use a tickertape timer to get a representation of the positions across uniform "dots" of time. Students have used this same apparatus in earlier units.

Obtaining the Results

With the suggested apparatus, when students let go of the 100 gm mass hanging from the pulley, the cart accelerates across the table leaving a set of dots at regular time intervals on the tickertape. While the cart is moving across the table, students read the spring scale. When they graph the motion it turns out to be a straight diagonal line for speed vs. time. That represents a constant acceleration. While the spring scale reading drops slightly from its initial value, it then stays constant (except for vibration) all the way until the cart is caught. This suggests the constant extra pulling force resulted in a constant acceleration. The students then repeat the experiment with a 200 gm mass and then a 300 gm mass hanging over the end of the table. For these the acceleration is roughly proportionately greater which is represented in a steepening of the slopes of the speed-time graphs, and the spring scale readings are also proportionately greater.

Inclination to Deny the Results

Although some students just did the experiment and got their results without thinking, others asked whether they could have another spring scale. They suggested that their results had not come out right. They observed that while they were obviously getting accelerated motion—dots were getting progressively farther apart—the spring scale seemed to stick on one number. This, of course, was what I had expected to happen, but I gave them a different spring scale without suggesting that their results were OK. Others came and said they could never do science and get the right answers. Wanting to boost their confidence in their capabilities, but also wanting them to discover that their results were all right, I proposed that they check their results against the results obtained by other groups. An occasional group would excitedly come to share with me their recognition that the dominant class hypothesis was not right. Several recognized their original intuition had not been supported. Students finished their analysis of the data for the three different masses and got ready for a large group discussion of conclusions from these experiments.

Post-lab Discussion and Conclusions

We assembled for a whole class discussion. Still on the board was written:

Class hypotheses

To explain movement with a constant speed in a straight line

1. A constant extra (unbalanced) force is necessary to keep an object moving with a constant speed.
2. Once the object is moving, no (zero) extra force is required to keep it moving with a constant speed.

To explain movement with a constant acceleration

1. An extra force that increases at a constant rate is necessary to keep the object accelerating at a constant rate.
2. A constant extra force will be required to keep an object accelerating at a constant rate.

Which hypothesis for each motion seems to work?

Here is a paraphrase of typical past discussions:

Teacher: So, what do your results tell you about the relation between force and motion? What can you conclude? (I am trying to get the students to move from results to conclusions, a difficult differentiation for many students, as noted by a few authors in this volume.)
Student: We got constant acceleration.

Teacher: For how many of you did the acceleration turn out to be constant? (Most raise a hand or otherwise acknowledge that result.)
And how did you know the acceleration was constant? (Several students observe that the graphs of speed against time had constant slopes or that the spaces on the ticker tape were getting larger in a regular way. My question provided an opportunity for students to visit, and practice, analyses they had done in a previous unit.)

Teacher: OK, but was that the purpose of the experiment—to see whether the motion was a constant acceleration?
Student: We were also trying to see what the force would be like.

Student: We were trying to see whether the force changed or stayed constant.
Student: The force changed. It went down.
Student: Well, not really. Actually, it only went down right at the start. Then it stayed constant.

Teacher: Who observed that the scale reading dropped off? (Nearly everyone agrees.)

Teacher: Did the force scale reading gradually drop off, all the way across the table, or did it just drop down a bit at the beginning and then stay more or less constant the rest of the way before the cart was caught? (Students confer with one another and, after some argument among themselves, agree that it dropped right at the beginning but then stayed constant. One of the groups that wasn't sure goes back to the tables with the apparatus set up and quickly retries a run and announces that it dropped at first and then stayed constant. Another of my goals for interaction in my classroom is that students question one another and argue on the basis of evidence even if it means they need to repeat the experiment to get clear on the results.)

Initially I was trying to get the students to focus spontaneously on formulating a conclusion about the relation between force and motion. They concentrated instead on citing specific observations, so I changed my immediate focus to get them to clarify their observations. But now I do want students to move from results to conclusions, in this case a conclusion about the relation between force and motion.

Teacher: Suppose we look back at these initial hypotheses (gesturing to the hypotheses on the board) that represented our class ideas about the possible relations between force and motion. We called these our "class hypotheses." Did these experiments relate more directly to explaining the constant velocity or the constant acceleration?
Student: The cart didn't move with a constant speed.
Student: The cart was accelerating.
Student: The constant acceleration, 'cuz the cart accelerated at a constant rate.
Student: The bottom two hypotheses about constant acceleration.
Student: But both of those can't be true, because one says an increas-

ing extra force and the other says a constant extra force, and it can't be constant and increasing at the same time.

Student: Number 1 can't be true because it says an increasing extra force, but ours decreased and then stayed constant.

Student: So number 2 must be the one that is true. We got a constant extra force reading and we got constant acceleration.

Student: But I thought we decided that the force should be increasing constantly to get a constant acceleration?

Student: Well, that was what we thought to begin with, but that is not what happened in the experiment. So we must have been wrong with our first ideas.

Student: Our first hypothesis was wrong. The experiment comes out more like hypothesis number 2.

Teacher: OK, does that make sense? When you did your experiments, your results involved getting constant acceleration of some number and you got a constant spring scale reading, while it was moving across the table. Each of you got slightly different numbers for the force and for the acceleration, but in every case the results involved a constant acceleration and a constant extra force. We can now write a general conclusion: "To get an object to accelerate at a constant rate requires a constant extra force." From our experiments, this seems like a reasonable conclusion. Our only other hypothesis got eliminated. Right? Does that make sense? (Most students chime in with the affirmative.) Notice that is different from what most of you thought to begin with. That is OK. Experiments often can tell us that our present thinking is NOT correct. That is valuable to know.

Student: I see that. It makes sense, but then, how do we explain the constant velocity? Both motions can't be explained the same way.

Teacher: Good next question. But first, are there any other questions about the constant acceleration cases? (After asking a question that requires deep thinking or for which students will need time to formulate their ideas or concerns, it is advisable to wait for something over three seconds [Rowe, 1974].)

Teacher: OK, so now let's consider an object moving along at a constant speed in a straight line. How might we explain that? Do these past experiments tell us anything about that? (More wait-time.)

Student: (Looking at the board) Well it can't be a constant extra force, because we just said that explains constant acceleration.

Student: But didn't we agree yesterday, even before we did the experiment, that constant force makes for constant speed?

Student: Yeah, that's what we all, well most of us, thought before we did the experiment. But the experiments proved us wrong.

Student: It can't be constant extra force for constant speed in a line, because we just got a constant force and a constant acceleration. Constant force can't do both.

Student: Well, it wouldn't make sense to say no force is needed to keep it going with a constant speed. (Actually, eventually I am hoping that the students will see that this is the only valid conclusion given the results.)

Student: What else could it be?

Student: We eliminated constant force, so it has to be either an increasing force, a decreasing force, or no force. Those are the only other possibilities.

Teacher: When you say "force," we are talking about the extra or "net" leftover force, right?

Student: Yeah. Those are the only possibilities for constant velocity. Either the net force is constant, which we eliminated by doing the experiment, or the net, extra force is decreasing, increasing, or zero.

Student: Well we know it can't be increasing, 'cuz that just wouldn't make sense that a constant force gives constant acceleration and an increasing force would give constant velocity.

Teacher: Is that right, can we eliminate increasing net force then? (Most students agree, even though this was the idea that most believed true when we started.)

Student: So it's gotta be a decreasing extra force. (Other students seem to be agreeing, and I want to be sure they don't just accept this idea without its being challenged.)

Teacher: Does that make sense that it could be a decreasing net force? (This seems to provide an opening for other students to challenge that hypothesis.)

Student: No, that won't work. (Pause) Suppose at first we have a big net force. The cart would have a big acceleration, right? Then, if net force decreases, we would have a smaller acceleration. Then if net force decreases even more, we have an even smaller acceleration. But we always have an acceleration. It is just getting smaller and smaller, but it is still acceleration.

Student: Oh, yeah. In our experiments when there was a big pulling force, we got a big constant acceleration, and with a small pulling force we got a small acceleration.

Student: So it has to be zero net force to explain constant speed in a line, logically.

Teacher: And why do you say "logically"? (I want all the students to follow and reflect on the argument.)

Student: The only possibilities are increasing, decreasing, staying the same or zero. If we eliminated the first three, zero net force is all that is left.

Student: But if it was zero net force, how would the cart even get going?

Student: In the beginning, the net force would not be zero: that would not accelerate the cart. But once it got going, no net force would be needed. You might need a little bit of force to match any friction, but there wouldn't need to be any extra, leftover force bigger than friction.

Student: Wow. That would be like in space. You would just push a little to get the thing going, and then it would go on its own out through space with no force acting on it.

Teacher: Provided you didn't need to worry about any gravitational forces.

OK, summarize the results from your experiments. What are your conclusions about the relation between force and motion of various kinds. Also, go back to what you wrote in your journals about force and motion. Are there any situations that you cannot explain with the ideas we developed in class in the last couple of days?

It turns out that these ideas we came up with are consistent with those expressed by Sir Isaac Newton. He wrestled with some of the same concerns you all have expressed before he came to formulate Newton's Laws of Motion. Good job. Good thinking. Tomorrow I want also to summarize the thinking processes by which we came to our conclusions.

The next day I will get the students to summarize the conclusions and to share how the ideas apply to the various situations they had written. They need to see that the new ideas are useful in everyday situations as well as in the class laboratory.

For the last few days the agenda has primarily concerned coming to understand concepts of force and motion. We could not do this without also addressing some aspects of scientific reasoning processes and skills. In this next discussion I want the ideas about the nature and processes of science to be the main question. I want students to be able to answer questions relating to how we come to know in science. Prior units of work—measurement and kinematics—were primarily descriptive. This unit on dynamics and several subsequent units, all of them involving making sensible explanations for phenomena, should offer a promising start to understanding the processes and skills of scientific inquiry.

Coming to Know and Understand Through Experiment

Teacher: We have reviewed the explanations for constant accelerated motion and for constant speed in a straight line. We also have seen how those explanations can be applied to make sense of everyday situations. Now I want us to review how we came to know and believe these explanations for constant acceleration and constant speed in a straight line. What were our thinking processes in coming to those conclusions that might tell us something about doing science?

We now know that constant unbalanced force is needed for constant acceleration and no net force is needed to keep an object going at a constant speed in a straight line, once the object is moving. How did we come to know these ideas? Did you need some authority like the teacher or textbook to tell you the ideas were true? What did we do in class that brought you to believe these ideas?

Student: We did experiments.

Teacher: But how did we decide what experiments to do?
Student: We had some ideas to experiment.
Student: The experiments were to test the ideas.
Student: Test the hypotheses.

Teacher: But where did the ideas or hypotheses come from?
Student: You gave us some questions.
Student: Those questions were just to get us thinking.

Student: Well, but the questions also made us know what our ideas were.
Student: We put up the possible ideas we believed at the time, and then the experiment was to test the ideas we thought were true.

Teacher: OK, I hear you saying the original questions were important because they helped you identify what your ideas or hypotheses were for explaining the situations. Were your original ideas right? Did the hypotheses have to be right before you could do the experiment?
Student: We thought they were right. But they turned out to be wrong.
Student: The hypotheses were just our ideas we thought were true before we did the experiment.

Teacher: Scientists wonder about things. They might come up with their own questions or they start thinking about questions somebody else has asked. They think they have some ideas to explain what happens. The scientists treat these hypotheses as tentative: maybe they are true, maybe not. The ideas have to be tested. Does this sound like what happened in our class? (Note this was a mini-lecture, which in this case ended with a rhetorical question.) Take a couple of minutes to summarize in your learning journal:
    1. What was the role of questions in this investigation?
    2. What was the role of the hypotheses?
    3. What were the variables involved in our hypotheses?

(After about two minutes the discussion continues.)

Student: I think I'm OK with the first two questions, but what were the variables involved?
Student: The motion of the cart and the extra force.
Student: Oh, yeah. The spring scale reading for force and the dots for motion.

Teacher: OK? Any other questions there? (Wait-time.)
In addition to questions and hypotheses, we were talking about experiments. What came out of the experiments? What was the value of doing them?
Student: It let us test our hypotheses, to find out whether they were right.

Teacher: Say more. How did the experiment do that, specifically with respect to what you did in the experiment?

Student: Well, we ran the carts across the table and we graphed the motion.

Student: And we read the spring scale while the cart was moving.

Teacher: And what did you see? What did you observe? (Note that here the questioning is more of a quick give and take, because we are summarizing the argument, our observations, and our inferences.)

Student: That the dots got farther and farther apart.

Teacher: Good, so what did that tell us?

Student: That the motion of the cart was a constant acceleration.

Teacher: That the motion of <u>this</u> cart was constant acceleration. And what did we learn from observing the spring scale?

Student: It stayed constant, after it dropped. Why did it drop in the beginning? I'm still not sure about that.

Teacher: That is a good question, and we should be able to answer that, but let's stay with this argument for now, OK? (I want to encourage students to ask their own questions, but I want them not to lose sight of the path of this discussion.) OK, so the pulling force stayed constant while this cart was moving across the table.

The constant acceleration and the constant force scale reading were our results from this particular run of the cart going across the table. These were our results of the particular experiment. What happened in the rest of the experiments done by people at other tables and using different masses hanging on the spring?

Student: We all got constant acceleration and constant spring force.

Teacher: Is that true? (Most students chime in, in agreement.)

Teacher: Good. Now notice that you each did your experiment, and you specifically may have gotten a pulling force of 0.8 newtons and you may have gotten 0.7 m/s/s for the acceleration. These are the specific <u>results</u> of the one experiment. But others did similar experiments and you did similar experiments with other specific results, but when we look across all the experiments, we are able to make some tentative conclusions about force and motion. Our <u>conclusions</u> are that when we have any object undergoing a constant acceleration, a constant unbalanced force will be needed to explain the motion. Results are our specific, observable outcomes (or nearly observable, if we have to do a little bit of work to get the acceleration numbers) from the experiment. Conclusions are our inferences we make to try to summarize the meaning from all those results. Once we have the general conclusions, we try to apply them to lots of specific situations. If the conclusions seem to work, we think we have learned something about the world.

Teacher: Now take a couple of minutes to summarize in your learning journals:
4. What is the role of experiments in doing science?
5. What is the difference between results of experiments and conclusions from experiments.

After about three minutes I asked the students to summarize the various answers they had written for each question 1 through 5, since these were the ideas and processes that were the focus of this discussion. Next we needed a short discussion to consider the logical argument used in developing an understanding of forces on the object moving in a straight line with constant speed.

Coming to Know and Understand Through Logical Reasoning

Teacher: Let's think a bit about how we came to believe that straight line motion with a constant speed could be explained by a zero net force. We couldn't do a direct experiment here, but Chris suggested a "logical" argument. Can somebody summarize that for us?
Student: Well there were only certain possibilities. Either a constant net force, a decreasing net force, an increasing net force, or no net force was possible.
Student: No other outcomes were possible.

Teacher: Then what?
Student: We eliminated each of the other possibilities, so the only one left was zero net force.

Teacher: So only "zero net force to explain constant velocity" was left. Did we verify this by experiment?
Student: Not exactly. All our experiments showed if we didn't have zero net force, we got acceleration.

Teacher: OK, good. Notice then that sometimes we need to use logical reasoning to determine all the possible explanations and then eliminate as many as we can on the basis of experiment. So, some-times experiments tell us more about what the explanation CAN'T be than about what is the right explanation.

Now we have some ways of explaining constant accelerated motion and constant speed in a straight line. We believe these ideas because our experiments together with logical argument seem to support them. Be sure you have these ideas summarized in your learning journal. We will have lots more opportunities to practice these and other processes of scientific inquiry.

SUMMARY OF LESSONS APPLIED TO THE VIGNETTE

Setting Learning Goals

In the previous dialogue, my goals included both development of conceptual understanding and development of skills for conducting inquiry. To do that I put the conceptual understanding in the foreground of the activities initially and the inquiry skills, while important, were in the background. In the later discussion, I put in the foreground the development of inquiry skills and the conceptual understanding, while still important, was in the background. This is my way of addressing one of the concerns brought out in other chapters of this book. I cannot teach content without reasoning processes, and I cannot teach reasoning processes without content. But I need to bring explicitly into the foreground whichever I want students to attend to at a particular time. I cannot just teach for development of subject matter using processes and expect students to grasp the skills without setting class time when I ask students explicitly to pay attention to them. If I had just taught for conceptual development, students would not likely have learned inquiry skills, even though I believed I was teaching by inquiry. Learners need to have their attention concentrated on the goals of the learning.

These goals for content and process are included in the national learning goals and in my state guidelines, which suggests that they are for all students. I have taught these ideas to high school physics students and to special education classes of younger students in a physical science course. The latter group requires more time and more carefully centered attention, but nearly all students at upper middle and high school are capable of learning these important ideas and skills.

The laboratory experiences and discussion would not have made sense without some prerequisite knowledge. We had studied the topics of measurement that allowed us to have some idea of how certain we could be with respect to the measurements we made. We developed skills in the use of measurement apparatus. Our prior study of the description of motion allowed us quickly to analyze the motion and recognize constant acceleration when it occurred. Immediately prior to this investigation of forces in moving situations, we studied forces on objects in the static situations. That sub-unit included using arrows to represent forces and some beginning ideas about the nature of force as an idea used to explain situations. These and other earlier learning experiences helped prepare the way to develop understanding of these difficult ideas about force and motion.

This was not a full, open-ended investigation. I elected to guide it. "Guided inquiry" is probably an accurate description of the process. The ideas to be developed are subtle and not likely to be invented and justified without guidance. These are ideas, moreover, on which much of the rest of our development of content will rest. Yet the investigation was still centered in the students in the sense that their ideas were being tested and the final explanations were constructed by the students from their observations and inferences. This sort of guided inquiry has occupied much of the activity of my classes.

Sometimes I do have the students do open inquiry. But it is typically about ideas that are not going to be so important to our learning story line. For example, I frequently set an open-ended investigation around factors that affect friction: "Identify factors you believe might affect frictional force. Choose two of those factors and design experiments to see whether each factor affects friction, and if so, how each factor is related (mathematically) to friction." Here the primary concern is for inquiry skills. Conceptual understanding of friction is secondary. If I let students do the full investigation with no guidance, they are not likely to come to the idealized conclusion of the science text.

Dialogues

While questions of fact, such as asking for an observation, can be asked and answered rather quickly, questions for which thinking is required to develop an answer take time and need environments free of distractions. In this vignette, questions from the teacher were used to stimulate thinking on the part of the participants. Many of the important questions to foster development are higher level questions, requiring substantial inferencing.

Dialogues between teacher and students, between student and student, and even within the student are all important. In the preceding discussion, the teacher modeled appropriate interaction between learners by listening to students' ideas and responding by respectfully asking questions and gently challenging ideas and reasoning that didn't yet make sense. The intention was to stimulate reflection by the students. Students were encouraged to ask and attempt to answer one another's concerns. Wait time and other appropriate listening techniques were used to foster personal reflection by students. It was hoped that eventually students would internalize the dialogue and probe and stimulate their own thinking.

Notice that there were virtually no evaluative comments by the teacher. Typical verbal interaction in a classroom involves the teacher's asking a question, the student responding, and the teacher's evaluating the response. What happens in my class is more like a conversation among the several participants. I attempt to build an atmosphere of collaboration among participants who are collectively constructing meaning through collective inquiry. This classroom climate requires mutual respect among the participants. Students need to support one another as they share their ideas and gently probe for understanding. Students need to trust the guidance of the teacher: to believe that the teacher is putting experiences before them that will allow them to make sense. Teachers need to respect learners in their attempts to make sense, respecting that they are capable of the skills and reasoning that will allow them to develop desired understanding.

From Experiences to Ideas

In this vignette, students were not told Newton's Laws at the outset. They began with typical experiences of books sliding across tables. They suggested hypotheses about motion that made sense to them at the outset. After experiences with the experiments they reconsidered their initial ideas and suggested new understandings about force and motion. Then I confirmed that the conclusions they had formed were consistent with Newton's Laws. The experiences came first, then the idea, and finally the confirmation from authority. Following this line of development, students suggest, makes them feel empowered, capable of coming to understanding through their own resources and capabilities. They don't need some authority to tell them what is true about the natural world. I did conduct short lectures, but they were after students had sufficient experiences with hand and mind that what I had to say would make sense.

Materials to Support Inquiry

Equipment and materials to support inquiry do not need to be extensive. The activities in the vignette consisted of a handout of elicitation questions to which students were to respond and the simple apparatus involving cart, spring scale, string and pulleys. The tickertape timer was more sophisticated, but students had enjoyed an opportunity earlier to learn how to use that to create a motion story. How the equipment worked was relatively transparent. More important is that the equipment was just enough to address the goals for learning, to measure the relevant variables from which students could test their initial ideas.

After students had developed the ideas consistent with Newton's Laws, text materials were made available to students. My course was not directed by a textbook. At its beginning, students had three texts from which to choose for background reading. That not all students were checking out the same text helped to emphasize the freedom of the course from the constraints of a textbook. I did use technology that supported the inquiry students had experienced. After guiding their inquiry on the development of the ideas for explaining different sorts of motion, I asked them to interact with a computerized Diagnoser to check whether their knowledge and reasoning were consistent with the results of our inquiry. Diagnoser serves questions in pairs. Following a phenomenological question essentially asking "What would happen if...?" is a question asking "What reasoning best supports the answer you chose?" Expected responses were associated with one or another's attempt at understanding that falls short of the learning target. Immediate feedback for such responses suggested what students need to think about, how what they said or did was inconsistent with experiences from classroom activities or from daily life (Hunt & Minstrell, 1994, and see www.talariainc.com/facet). To support the inquiry, the computerized Diagnoser works to build ideas from experiences and in a manner consistent with dialogues.

Final Comment

Inquiry teaching is hard work. Students may ask questions for which I don't know the answer. For me this has been stimulating and a great source of learning. It would be much simpler to lecture, but I wouldn't learn as much myself. I have to learn the various directions students might go. I need to make decisions about which of those directions I am willing to invest the intellectual energy of the class in investigating.

Inquiry teaching takes time. It would be faster to tell the students the "correct" scientific ideas. But to learn to do inquiry, students need opportunities to participate in learning by inquiry. The fruits of the additional time investment include deeper and lasting understanding. It also fosters increased opportunities for life-long learning for both students and teacher.

REFERENCES

Haber-Shaim, U., J. Cross, J. Dodge, and J. Walter. 1971. *PSSC physics*. Lexington MA: D.C. Heath.

Hunt, E., and J. Minstrell. 1994. A cognitive approach to the teaching of physics. In *Classroom lessons: Integrating cognitive theory and classroom practice*, edited by K. McGilly, 51-74. Cambridge, MA: MIT Press.

Minstrell, J. 1982. Explaining the "at rest" condition of an object. *The Physics Teacher* 20: 10-14.

Minstrell, J. 1984. Teaching for the development of understanding of ideas: Forces on moving objects. In *AETS Yearbook: Observing Science Classrooms*, edited by C. Anderson, 53-72. Columbus OH: Ohio State University. Association of Educators of Teachers of Science.

Minstrell, J. 1992. Facets of students' knowledge and relevant instruction. In *Proceedings of the international workshop: Research in physics learning—theoretical issues and empirical studies*, edited by R. Duit, F. Goldberg, and H. Niedderer, 110-128. Kiel, Germany: The Institute for Science Education at the University of Kiel (IPN).

Rowe, M.B. 1974. Wait-time and rewards as instructional variables, their influence on language, logic, and fate control: Part one—wait time. *Journal of Research in Science Teaching* 11:81-94.

Rutherford, J., H. Holton, and F. Watson. 1972. *Project physics*. New York: Holt, Rinehart and Winston, Inc.

van Zee, E.H. and J. Minstrell 1997. Using questioning to guide student thinking. *The Journal of the Learning Sciences* 6: 227-269.